MOLECULAR GENETICS OF PLANT-MICROBE INTERACTIONS

CURRENT PLANT SCIENCE AND BIOTECHNOLOGY IN AGRICULTURE

Aims and Scope of the Bookseries

The bookseries is intended for readers ranging from advanced students to senior research scientists and corporate directors interested in acquiring in-depth, state-of-the-art knowledge about research findings and techniques related to plant science and biotechnology. While the subject matter will relate more particularly to agricultural applications, timely topics in basic plant science and biotechnology will be explored as well. Some volumes will report progress in rapidly advancing disciplines through proceedings of symposia and workshops while others will detail fundamental information of an enduring nature that will be referenced repeatedly.

Evans, H.J., Bottomley, P.J. and Newton, W.E. (eds): Nitrogen fixation research progress. 1985. ISBN 90-247-3255-7

Zimmerman, R.H., Griesbach, R.J., Hammerschlag, F.A. and Lawson, R.H. (eds): Tissue culture as a plant production system for horticultural crops. 1986. ISBN 90-247-3378-2

Verma, D.P.S. and Brisson, N. (eds): Molecular Genetics of Plant-Microbe Interactions. 1987. ISBN 90-247-3426-6

Molecular genetics of plant-microbe interactions

Proceedings of the Third International Symposium on the Molecular Genetics of Plant-Microbe Associations, Montréal, Québec, Canada, July 27–31, 1986

edited by

DESH PAL S. VERMA
McGill University, Montréal
Québec, Canada

NORMAND BRISSON
Université de Montréal, Montréal
Québec, Canada

1987 **MARTINUS NIJHOFF PUBLISHERS**
a member of the KLUWER ACADEMIC PUBLISHERS GROUP
DORDRECHT / BOSTON / LANCASTER

Distributors

for the United States and Canada: Kluwer Academic Publishers, 101 Philip Drive, Assinippi Park, Norwell, MA 02061, USA
for the UK and Ireland: Kluwer Academic Publishers, MTP Press Limited, Falcon House, Queen Square, Lancaster LA1 1RN, UK
for all other countries: Kluwer Academic Publishers Group, Distribution Center, P.O. Box 322, 3300 AH Dordrecht, The Netherlands

Library of Congress Cataloging in Publication Data

International Symposium on the Molecular Genetics of Plant-Microbe Associations (3rd : 1986 : Montreal)
Molecular genetics of plant-microbe interactions.

(Current plant science and biotechnology in agriculture ; 3)
Includes index.
1. Microbial genetics--Congresses. 2. Molecular genetics--Congresses. 3. Micro-organisms, Phyto-pathogenic--Host plants--Congresses. 4. Plant molecular genetics--Congresses. 5. Symbiosis--Congresses. I. Verma, D. P. S. (Desh Pal S.), 1944- II. Brisson, Normand, 1955- . III. Title. IV. Series.
QH434.I58 1986 576'.139 86-23702

ISBN-13: 978-94-010-8496-3 e-ISBN-13: 978-94-009-4482-4
DOI: 10.1007/978-94-009-4482-4

This Symposium was made possible through the generous support of the following:

National Science Foundation (USA); The Rockefeller Foundation; International Society for Plant Molecular Biology; Agriculture Canada; Natural Research Council Canada; Natural Sciences and Engineering Research Council of Canada; United States Department of Agriculture; United States Department of Energy; McGill University Faculty of Graduate Studies and Research; Agracetus; Agrigenetics Corporation; Allelix Inc.; Amersham Corporation; Bio-Agral; BioTechnica International, Inc.; Boehringer Mannheim Canada; CIBA-GEIGY Corporation; Conviron Products Company; E.I. DuPont de Nemours Company; ICN Biomedicals; Lilly Research Laboratories; New England Nuclear Canada; Northrup King Co.; Pharmacia (Canada) Inc.

PREFACE

Increased interest in the basic biology of plants and microorganisms stems from the fact that crop productivity is directly affected by plant-microbe interactions. In spite of the fact that plants exist in the environment amongst diverse species of microorganisms, only a few ever establish a direct relationship. Emerging awareness concerning the indirect effect of microbial association on plant growth and the possibility of using one microbe against another for controlling pathogenic interactions is at the genesis of new fields of studies. The primary reason for a microbe to associate with photoautotrophic organisms (plants) is to tap its nutritional requirements, fixed carbon, as a source of energy.

By hook or by crook, a microbe must survive. Some have evolved mechanisms to exploit plants to develop a niche for their biotropic demands. When in contact with a living plant, microorganisms may live in a passive association using exudates from the plant, invade it pathogenically or coexist with it in symbiosis. The plant responds to the interloper, either reacting in a hypersensitive manner to contain the invasion of pathogens, or by inducing a set of genes that leads toward symbiosis, or by simply succumbing to the invader. Thus, prior to contact with the plant, microorganism is able to sense the presence of the host and activate accordingly a set of genes required for the forthcoming interaction, whether symbiotic or pathogenic.

Understanding the "language" of communication between the host plant and the microorganism may lead to eventual manipulation of these associations for the benefit of mankind. This Symposium marked the beginning of an era towards deciphering the "alphabets" of this language. It is apparent that it is a two-way communication. The microbe has developed means of avoiding the defense responses of the plant and has exploited plant products as environmental cues for controlling genes which are only required during this interaction. With the development of genetic tools in fungi, the doors have been opened for understanding and manipulating devastating diseases of crop plants.

Over 400 scientists from 27 countries took part in this Third International Symposium on the Molecular Genetics of Plant-Microbe Interactions. Eight sessions with 66 oral presentations and 3 poster sessions attracted the interest of participating scientists from universities, research institutes, and private sectors concentrating on basic research and its potential application.

The Montreal Symposium would not have been possible without the generous support by the various foundations, research organizations and private corporations as well as the assistance from the host institution, McGill University. We are indebted to the number of people who aided in making this Symposium a successful event.

DESH PAL S. VERMA

TABLE OF CONTENTS

Page

CONTRIBUTORS . . . .

Section I: MOLECULAR GENETICS OF AGROBACTERIUM AND PLANT TRANSFORMATION

"Ecology of Agrobacterium: plasmids and biovars"
Kathy Ophel and Allen Kerr. 3

"The Agrobacterium rhizogenes root-inducing system"
F. Richaud, C. Aubry, A. Beyou, F. Boulanger,
C. Estramareix, A.-M. Fleury-Guerout, C. Mignotte,
and O. Reyes. 6

"Effect of the presence of the plasmid pSA and of auxin on the attachment of Agrobacterium tumefaciens to plant host cells"
Ann G. Matthysse 11

"Dual regulation of virulence genes of Agrobacterium plasmid pTiC58"
P. Rogowsky, T.J. Close, and C.I. Kado. 14

"Overdrive, a T-DNA transmission enhancer on the A. tumefaciens tumor-inducing plasmid"
Ernest G. Peralta, Renate Hellmiss, Joon M. Ji,
Wendy H. Berger, and Walt Ream 20

"Physical structure and genetics of the T-DNA in plants transformed by Agrobacterium tumefaciens"
Albert Spielmann and Robert B. Simpson. 27

"Mammalian metallothionein functions in plants"
D.D. Lefebvre and J.-F. Laliberte 32

"Tumorigenesis and root nodulation by Agrobacterium tumefaciens carrying Rhizobium symplasmids"
Pauline A. Donaldson and V.N. Iyer. 35

Supplementary articles (see Section VI)

Section II: MOLECULAR GENETICS OF PHYTOPATHOGENIC BACTERIA AND FUNGI

"Cutinase and pectinase in host-pathogen and plant-bacterial interaction"
P.E. Kolattukudy, Joseph Sebastian, William F. Ettinger,
and Mark S. Crawford. 43

Page

"Siderophore biosynthesis, uptake and effect on potato growth of rhizosphere strains"
Peter Weisbeek, Joey Marugg, Gerard van der Hofstad, Peter Bakker and Bob Schippers. 51

"A gene cluster in Xanthomonas campestris PV Campestris required for pathogenicity controls the excretion of enzymes"
J.M. Dow and G. Scofield. 54

"Direct analysis of the invasiveness of Xanthomonas campestris mutants generated by Tn4431, a transposon containing a promoterless luciferase cassette for monitoring gene expression"
Joe J. Shaw and Clarence I. Kado. 57

"Analysis of the spontaneous mutation to avirulence by Pseudomonas solanacearum"
Mark A. Schell, Daniel P. Roberts, and Timothy P. Denny . 61

"Characterization of pathogenicity genes of Erwinia carotovora"
A.K. Handa, R.A. Bressan, L. Lee, D.J. Charles, R.K. Jayaswal, J. Chiu and J.L. Bennetzen. 67

"Characterization of a novel esterase produced by plant pathogenic Streptomyces"
D.A.R. McQueen and J.L. Schottel. 73

Supplementary articles (see Section VI)

Section III: MOLECULAR GENETICS OF THE HOST = (SYMBIOSIS/PATHOGENICITY)

"Induced symbiosis mutants of Pisum sativum"
B.E. Kneen, D. Vam Vikites and T.A. Larue 79

"Plant host genetics of nodulation initiation in soybean"
Peter M. Gresshoff, Jane E. Olsson, David A. Day, Kathryn A. Schuller, Anne Mathews, Angela C. Delves, Arno Krotzky, G. Dean Price and Bernard J. Carroll. . . . 85

"A mutant of pea (Pisum sativum) possibly disturbed in the production of a compound required for the induction of nitrogenase activity in bacteroids"
Jenne G. Postma, Evert Jacobsen, Ton Bisseling and Willem J. Feenstra. 91

"Non-nodulation mutants of soybean"
Anne Mathews, Bernard J. Carroll, and Peter M. Gresshoff. 94

"Early nodulins in root nodule development"
Jan-Peter Nap, Marja Moerman, Albert van Kammen, Francine Govers, Ton Gloudemans, Henk Franssen and Ton Bisseling 96

Page

"Peribacteroid membrane nodulins of soybean"
Marc G. Fortin and Desh Pal S. Verma. 102

"Isolation of nodule specific c-DNA clones from *Medicago sativa*"
György B. Kiss, Eva Vincze and Zoltan Vegh. 108

"Analysis of nodule-specific gene expression in ineffective alfalfa root nodules and callus cultures derived from ineffective root nodules"
Joanna F. Hanks, Lisa A. Macol, Jonathan Goldthwaite, and Ann M. Hirsch 112

"Nodule specific genes in *Phaseolus vulgaris*"
F. Campos, M. Vazquez, J. Padilla, C. Enriquez, and F. Sanchez 115

"Investigation of plant genes expressed during symbiotic nitrogen fixation"
S.G. Gottlob-McHugh and D.A. Johnson. 118

"*Rhizobium* induced plant proteins in target root epidermal cells of *Vigna unguiculata*"
Arvind Bhagwat and Joseph Thomas. 120

"Four soybean nodulin genes evolved from a common ancestor"
F. Jacobs, M. Zhang, M. Fortin, and Desh Pal S. Verma . . 123

"Coordinated expression of nodule-specific and root genes in yellow lupin"
M. Sikorski, U. Szybiak-Strozycka, P. Strozycki, B. Golinska, C.J. Madrzak, R. Kamp, B. Wittmann-Liebold, and A.B. Legocki. 127

"Plant gene expression during effect and ineffective nodule development of the tropical stem-nodulated legume *Sesbania rostrata*"
P. de Lajudie and T. Huguet 130

"Expression of two enzymes involved in ureide formation in soybean regulated by oxygen"
Knud Larsen and Bjarne Jochimsen. 133

"Probing cell wall structure in the soybean root nodule"
S.-S.T. Hua, K.L. Miller, V.J. Vreeland, and W.M. Laetsch. 138

"Monoclonal antibodies to components of *Rhizobium*-induced pea nodules"
Desmond J. Bradley, Elizabeth A. Wood, Geoffrey W. Butcher, Giovanni Galfre, and Nicholas J. Brewin. 141

Page

"Localization of the glutamine synthetase polypeptides in Phaseolus root nodules"
M. Lara, J.L. Ortega and B. Valderrama. 142

"Changes in protein and mRNA accumulation in potato tubers treated with an elicitor"
H. Giroux, C. Marineau, and N. Brisson. 145

Section IV: MOLECULAR GENETICS OF RHIZOBIUM

"Organization of the Rhizobium phaseoli genome"
Rafael Palacios, Margarita Flores, Susana Brom, Esperanza Martinez, Victor Gonzalez, Silvia Frenk, Carmen Quinto, Miguel Angel Cevallos, Lorenzo Segovia, David Romero, Alejandro Garciarrubio, Daniel Pinero, and Guillermo Davila 151

"Rifampin resistance and nodulating competitiveness in Rhizobium meliloti"
D. Mark Lewis, Eden S.P. Bromfield, Leslie R. Barran. . . 157

"A method for isolating competition defective mutants in Rhizobium"
Thomas J. McLoughlin, Ann Owens Merlo, and Eric Johansen 159

"Genetic determinants of nodulation in pRle 1001a: nodD"
A. Squartini, P.J.J. Hooykaas, and M.P. Nuti. 162

"Symbiotic mutants of Rhizobium meliloti which produce non-succinylated exopolysaccharide"
John A. Leigh 165

"Rhizobium mutants defective in lipopolysaccharide and infection"
K.D. Noel, P. Pachori, B. Kulpaca, K.A. Vandenbosch, B.A. Brink, and J.R. Cava 167

"Analysis of three Rhizobium phaseoli genes, psi, psr and pss, which affect exopolysaccharide synthesis and symbiotic nitrogen fixation and/or nodulation"
D. Borthakur, J.W. Lamb, and A.W.B. Johnston. 169

"Involvement of pSym nodulation genes in production of surface and extracellular components of Rhizobium trifolii which interact with white clover root hairs"
Frank B. Dazzo, Rawle I. Hollingsworth, Saleela Philip, Kathryn B. Smith, Margaret A. Welsch, Janet Salzwedel, Pamela Morris, and Lorna McLaughlin 171

Page

"Rhizobium exopolysaccharides are essential for the formation of nitrogen fixing nodules in the Rhizobium-legume symbiosis"
S.P. Djordjevic, H.C. Chen, J.X. Gray, J.J. Weinman, M.A. Djordjevic, J.W. Redmond, M. Batley, and B.G. Rolfe. 173

"Coinoculation with symbiotically defective mutants of Rhizobium meliloti"
S. Klein, A.M. Hirsch, C.A. Smith, and E.R. Signer. . . . 179

"Surface properties of Rhizobium meliloti associated with symbiosis"
Joseph Kieber, Ralph Clover, Turlough M. Finan, and Ethan R. Signer 182

"Degradative enzymes in Rhizobium meliloti"
Mary F. Lopez and Ethan R. Signer 185

"Identification of host specificity DNA regions determining the broad host range nodulation of Rhizobium strain NGR234"
Murali Nayudu, Greg L. Bender, Brant J. Bassam, Martha Sinclair, and Barry G. Rolfe 188

"Nif, Fix and Nod gene clusters in Bradyrhizobium japonicum, and nifA-mediated control of symbiotic nitrogen fixation"
H. Hennecke, H.-M. Fischer, S. Ebeling, M. Gubler, B. Thöny, M. Göttfert, J. Lamb, M. Hahn, T. Ramseier, B. Regensburger, A. Alvarez-Morales, and D. Studer. . . . 191

"Molecular genetics of nodulation of soybean by Bradyrhizobium japonicum"
G. Stacey, A.J. Nieuwkoop, Z. Banfalvi, J.-S. So, N. Deshmane, M.G. Schell, and D. Gerhold. 197

"Characterization of genes essential for symbiotic nitrogen fixation from Bradyrhizobium japonicum strain I110"
John D. Noti, Allen C. Yun, Otto Folkerts, Istvan Torok, and Aladar A. Szalay 202

"Nodulation genes of the stem nodulating Sesbania rostrata symbiont, strain ORS571"
M. Holsters, G. Van Den Eede, K. Goethals, M. van Montagu, and B. Dreyfus 208

"Nod-linked host specific gene for soybean (Peking) nodulation in Rhizobium fredii USDA193"
Neela Ramakrishnan and Alan G. Atherly. 211

Page

"Genomic organization of nodulation genes in Rhizobium phaseoli"
C. Quinto, J. Martinez, M.a. Cevallos, A. Davalos, and Y. Peralta 214

"Common and host specific nodulation genes in Rhizobium meliloti and their conservation in other Rhizobia"
A. Kondorosi, E. Kondorosi, B. Horvath, M. Gottfert, C. Bachem, F. Rodriguez-Quinones, Z. Banfalvi, P. Putnoky, Z. Gyorgypal, M. John, J. Schmidt, and J. Schell 217

"Host specific nodulation: effects of multiple nodD genes of Rhizobium meliloti"
Mary A. Honma and Frederick M. Ausubel. 223

"Nodulation genes of Rhizobium leguminosarum"
J.A. Downie, B.P. Surin, I.J. Evans, L. Rossen, J.L. Firmin, C.A. Shearman, and A.W.B. Johnston 225

"Interactions between Rhizobium mililoti and Rhizobium trifolii nodulation genes: what is the basis for dominance by R. mililoti?"
Roger Innes, Michael Djordjevic, Barry Rolfe, Jean De'Narie, and Peter Kuempel 229

"Multiple host-specificity loci in the broad host-range Rhizobium NGR234"
A. Lewin, C. Rosenberg, J. Stanley, D.N. Dowling, J.-F. Manen, F. Debelle, and W.J. Broughton 232

"Conserved nodulation genes are obligatory for non-legume nodulation"
Kieran F. Scott, Marlene Saad, G. Dean Price, Peter M. Gresshoff, Heather Kane, and Kaw Yan Chua 238

"Characterization of symbiotic genes and regulation of their expression in Rhizobium leguminosarum pre"
Jan Hontelez, Rene Klein Lankhorst, Jan-Dirk Jansma, Evert Jacobsen, Rommert C. van den Bos, and Ab van Kammen. 241

"Regulation of the promoters in the nodulation region of the symbiosis plasmid pRL1J1 of Rhizobium leguminosarum"
Herman P. Spaink, Rob J.H. Okker, Carel A. Wijffelman, Elly Pees, and Ben J.J. Lugtenberg. 244

"Narigenin induces the nodABC promotor of Rhizobium leguminosarum as well as Tsr factor production"
B.A.J. Zaat, A.A.N. van Brussel, C.A. Wijffelman, H.P. Spaink, R.H.J. Okker, E. Pees, and B.J.J. Lugtenberg. 247

Page

"An ntrC homologue in B. japonicum"
W. Szeto and F. Cannon. 250

"Glutamine synthetases of Rhizobium leguminosarum"
S. Colonna-Romano, R. Defez, M. Filser, M. Guida,
M. Iaccarino, A. Lamberti, and A. Riccio. 255

"Molecular analysis of a Fix cluster from Rhizobium meliloti"
D. Kahn, J. Batut, P. Boistard, M.L. Daveran, M. David,
O. Domergue, A.M. Garnerone, J. Ghai, C. Hertig,
D. Infante, and M.H. Renalier 258

"Regulation of the nitrogen fixation (Nif) genes in Rhizobium meliloti"
Shin-Ping Wang, Bin-Fu Shen, and San Chiun Shen 264

"The unusual symbiosis between the nitrogen fixing bacterium ORS571 and its host Sesbania rostrata: regulation of nitrogen fixation and assimilation genes in the free living versus symbiotic state"
F. De Bruijn, K. Pawlowski, P. Ratet, U. Hilgert,
and J. Schell 266

"Analysis of Azorhizobium sesbaniae ORS571 N_2 fixation genes"
Robert A. Ludwig, Robert G.K. Donald, Albert I. Loroch,
and David W. Nees 272

"Identification, characterisation and sequence analysis of the Rhizobium leguminosarum nifA gene"
S.S. Manian, P. Grönger, U.B. Priefer, and A. Pühler. . . 276

"Analysis of hup DNA and Hup host range of Rhizobium leguminosarum BIO"
H.V. Tichy, C. Schild, H.M. Ripke, L.M. Nelson, H. Fees,
and W. Lotz 279

"Bioluminescence in root nodules of soybean controlled by nitrogenase promoters"
Roman P. Legocki, Misuk Legocki, Thomas O. Baldwin,
and Aladar A. Szalay. 282

"In vivo cloning of genes from Bradyrhizobium japonicum"
Ketan S. Shah and L. David Kuykendall 288

"Genes for the catabolism and synthesis of a nodule-specific, opine-like compound are closely linked and on the Sym plasmid of Rhizobium meliloti"
Peter J. Murphy, Nina Heycke, Zsofia Banfalvi,
Adam Kondorosi, Jacques Tempé, and Jeff Schell. 292

Page

"Molecular biology of genes involved in carbon metabolism in *Rhizobium meliloti* and *Bradyrhizobium japonicum*"
F. O'Gara, B. Boesten, M. O'Regan, B. Kiely, B. Higgisson, C. Condon, K. Birkenhead, and S. Manian . . 295

"*Azorhizobium sesbaniae* ORS571 conducts synergistic N_2 fixation and nicotinic acid oxidation"
Robert A. Ludwig, Timothy T. Popin, and Christopher L. Kitts. 298

"At least three loci encode the leaf-curl phenotype in *Rhizobium* strain IC3342"
N.M. Upadhyaya, K.F. Scott, W.T. Tucker, and P.J. Dart. . 301

Section V: MOLECULAR GENETICS OF THE OTHER DIAZOTROPHIC ORGANISMS

"Use of heterologous hybridization in phylogenetic studies of symbiotic *Anabaena* strains"
C. Franche, B.E.G. Gunning, B.G. Rolfe, and J. Plazinski. 305

"*Chromobacterium lividum* NCTC 10590 is a nitrogen-fixing *Agrobacterium radiobacter*"
M.H. Soliman and G.R.K. Sastry. 307

"Studies on the diazotrophic nature of *Agrobacterium*"
Lalita Kanvinde, M.H. Soliman, H. Wardhan, Lise Nowell, D. Fox, and G.R.K. Sastry 309

"Developments in the genetic analysis of *Azospirillum*"
Mark Vanstockem, Kris Michiels, Maggi Maris, Jos Vanderleyden, and August P. Van Gool. 313

Section VI: SUPPLEMENTARY ARTICLES FOR SECTIONS I AND II

"Role of *Vir* genes in the excision of T-DNA from the ti-plasmid"
K. Veluthambi, R.K. Jayaswal, and S.B. Gelvin 319

"Cloning vectors for *Coryneform* bacteria"
Maya Kozlowski, Marta Srulovicz, and R. Wayne Davis . . . 325

"Cloning of *Serratia liquefaciens* chitinase gene(s)"
Sadhna Joshi and Maya Kozlowski 328

AUTHOR INDEX 331

SUBJECT INDEX. 335

CONTRIBUTORS

Alvarez-Morales, A.	Mikrobiologisches Institut, Eidgenössische Technische Hochschule, Eth-Zentrum, Universitätstrasse 2, CH-8092 Zürich, Switzerland
Arnold, W.	University of Bielefeld, W. Germany
Atherly, Alan G.	Department of Genetics, Ames, Iowa 50011 U.S.A.
Aubry, C.	Institut de Microbiologie, Université Paris-Sud, 91405 - Orsay Cedex, France
Ausubel, Frederick M.	Department of Molecular Biology, Massachusetts General Hospital, Boston, MA 02114 U.S.A.
Bachem, C.	Max-Planck-Institut fur Zuchtungsforschung, D-5000 Koln 30, Federal Republic of Germany
Bakker, Peter	Department of Molecular Cell Biology, Institute of Molecular Biology and Department of Phyto pathology, University of Utrecht, Utrecht, The Netherlands
Baldwin, Thomas O.	Department of Biochemistry and Biophysics, Texas A & M University, College Station, Texas 77843-2128 U.S.A.
Banfalvi, Z.	Department of Microbiology and Graduate Program of Ecology, The University of Tennessee, Knoxville, Tennessee, U.S.A.
Banfalvi, Z.	Institute of Genetics, Biological Research Center, Hung. Acad. Sci., P.O.B. 521 H-6701, Szeged, Hungary
Banfalvi, Zsofia	Institute of Genetics, Biological Research Center, Hungarian Academy of Science, H-6701 Szeged, P.O. Box 521, Hungary
Barran, Leslie R.	Plant Research Centre, Agriculture Canada, Ottawa, Ontario K1A 0C6 Canada
Bassam, Brant J.	Department of Genetics, Research School of Biological Sciences, Australian National University, P.O. Box 475, Canberra, Australia 2601
Batley, M.	School of Chemistry, Macquarie University, North Ryde, NSW, Australia 2113
Batut, J.	Laboratoire de Biologie Moleculaire, CNRS-INRA, B.P. 27, F 31326 Castanet-Tolosan, Cedex, France
Bender, Greg L.	Department of Genetics, Research School of Biological Sciences, Australian National University, P.O. Box 475, Canberra, Australia 2601
Bennetzen, J.L.	Department of Biological Sciences, Purdue University, West Lafayette, IN 47907 U.S.A.
Berger, Wendy H.	Institute for Molecular Biology, Department of Biology, Indiana University, Bloomington, IN 47405 U.S.A.
Beyou, A.	Institut de Microbiologie, Université Paris-Sud, 91405 - Orsay Cedex, France
Bhagwat, Arvind	Molecular Biology and Agriculture Division, Bhabha Atomic Research Centre, Bombay - 400085, India
Birkenhead, K.	Microbiology Department, University College, Cork, Ireland
Bisseling, Ton	Department of Moleculair Biology, Agricultural University, 6703 BC Wageningen, The Netherlands
Boesten, B.	Microbiology Department, University College, Cork, Ireland

Boistard, P.	Laboratoire de Biologie Moleculaire, CNRS-INRA, B.P. 27, F 31326 Castanet-Tolosan, Cedex, France
Borthakur, D.	John Innes Institute, Colney Lane, Norwich NR4 7UH, UK
Boulanger, F.	Institut de Microbiologie, Université Paris-Sud, 91405 - Orsay Cedex, France
Bradley, Desmond J.	John Innes Institute, Colney Lane, Norwich, NR4 7UH, UK
Bressan, R.A.	Department of Horticulture, Purdue University, West Lafayette, IN 47907 U.S.A.
Brewin, Nicholas J.	John Innes Institute, Colney Lane, Norwich, NR4 7UH, UK
Brink, B.A.	Marquette University, Milwaukee, WI U.S.A.
Brisson, N.	Departement de Biochimie, Université de Montréal, C.P. 6128, Succursale A, Montréal, Québec H3C 3J7 Canada
Brom, Susana	Centro de Investigacion Sobre Fijacion de Nitrogeno, Universidad Nacional Autonoma de Mexico, AP, Postal 565-A, Cuernavaca, Morelos, Mexico
Bromfield, Eden S.P.	Plant Research Centre, Agriculture Canada, Ottawa, Ontario K1A 0C6 Canada
Broughton, W.J.	Laboratoire de Biologie Moleculaire des Plantes Superieures, Universite de Geneve, 1 chemin de l'Imperatrice, 1292 Chambesy, Geneve, Suisse
Butcher, Geoffrey W.	AFRC Institute of Animal Physiology, Babraham, Cambridge, UK
Campos, F.	Depto. de Biologia Molecular de Plantas, Centro de Investigacion sobre Fijacion de Nitrogeno, UNAM, Apdo. Postal 565-A, Cuernavaca, Morelos, Mexico
Cannon, F.	BioTechnica International, Inc., 85 Bolton Street, Cambridge, MA 02140 U.S.A.
Carroll, Bernard J.	Botany Department, Australian National University, Canberra, ACT 2600, Australia
Cava, J.R.	Marquette University, Milwaukee, WI U.S.A.
Cevallos, Miguel Angel	Centro de Investigacion Sobre Fijacion de Nitrogeno, Universidad Nacional Autonoma de Mexico, AP, Postal 565-A, Cuernavaca, Morelos, Mexico
Charles, D.J.	Department of Horticulture, Purdue University, West Lafayette, IN 47907 U.S.A.
Chen, H.C.	Genetics Department, Research School Biological Sciences, Australian National University, Canberra, Australia 2601
Chiu, J.	Department of Horticulture, Purdue University, West Lafayette, IN 47907 U.S.A.
Chua, Kaw Yan	Centre for Recombinant DNA Research, Research School of Biological Sciences, Canberra, Australia
Close, T.J.	Department of Plant Pathology, University of California, Davis, CA 95616 U.S.A.
Clover, Ralph	Department of Biology, Massachusetts Institute of Technology, Cambridge, MA 02139 U.S.A.
Colonna-Romano, S.	International Institute of Genetics and Biophysics, CNR Via Marconi 10, 80125 Naples

Condon, C.	Microbiology Department, University College, Cork, Ireland
Crawford, Mark S.	Biotechnology Center, Rightmire Hall, Ohio State University, Columbus, Ohio 43210 U.S.A.
Dart, P.J.	Centre for Recombinant DNA Research and Department of Molecular Biology, Research School of Biological Sciences, The Australian National University, A.C.T. 2601, Australia
Daveran, M.L.	Laboratoire de Biologie Moleculaire, CNRS-INRA, B.P. 27, F 31326 Castanet-Tolosan, Cedex, France
David, M.	Laboratoire de Biologie Moleculaire, CNRS-INRA, B.P. 27, F 31326 Castanet-Tolosan, Cedex, France
Davila, Guillermo	Centro de Investigacion Sobre Fijacion de Nitrogeno, Universidad Nacional Autonoma de Mexico, AP, Postal 565-A, Cuernavaca, Morelos, Mexico
Day, David A.	Botany Department, Australian National University, Canberra, ACT 2600, Australia
Dazzo, Frank B.	Department of Microbiology & Public Health, Michigan State University, East Lansing, Michigan 48824 U.S.A.
De Bruijn, F.	Max Planck Institut für Züchtungsforschung, 5000 Köln 30, W. Germany
de Lajudie, P.	Laboratoire de Biologie des Sols, Orstom, BP1386, Dakar, Senegal
De'Narie, Jean	Laboratoire de Biologie Moleculaire CNRS-INRA, BP 27 F-31326, Castanet-Tolosan, Cedex, France
Debelle, F.	Laboratoire de Biologie Moleculaire, CNRS-INRA, B.P. 27, 31326 Castanet-Tolosan, Cedex, France
Defez, R.	International Institute of Genetics and Biophysics, CNR Via Marconi 10, 80125 Naples
Delves, Angela C.	Botany Department, Australian National University, Canberra, ACT 2600, Australia
Denny, Timothy P.	Department of Plant Pathology, University of Georgia, Athens, GA 30602 U.S.A.
Deshmane, N.	Department of Microbiology and Graduate Program of Ecology, The University of Tennessee, Knoxville, Tennessee, U.S.A.
Djordjevic, M.A.	Genetics Department, Research School Biological Sciences, Australian National University, Canberra, Australia 2601
Djordjevic, S.P.	Genetics Department, Research School Biological Sciences, Australian National University, Canberra, Australia 2601
Domergue, O.	Laboratoire de Biologie Moleculaire, CNRS-INRA, B.P. 27, F 31326 Castanet-Tolosan, Cedex, France
Donald, Robert G.K.	Thimann Laboratories, University of California, Santa Cruz, California 95064 U.S.A.
Donaldson, Pauline A.	Department of Biology and Institute of Biochemistry, Carleton University, Ottawa, Ontario K1S 5B6 Canada
Dow, J.M.	John Innes Institute, Colney Lane, Norwich, U.K. NR4 7UH

Dowling, D.N.	Laboratoire de Biologie Moleculaire des Plantes Superieures, Universite de Geneve, 1 chemin de l'Imperatrice, 1292 Chambesy, Geneve, Suisse
Downie, J.A.	CSIRO Division of Plant Industry, Canberra, ACT 2601, Australia
Dreyfus, B.	Laboratoire de Biologie des Sols, O.R.S.T.O.M., Dakar, Sénégal
Ebeling, S.	Mikrobiologisches Institut, Eidgenössische Technische Hochschule, Eth-Zentrum, Universitätstrasse 2, CH-8092 Zürich, Switzerland
Enriquez, C.	Depto. de Biologia Molecular de Plantas, Centro de Investigacion sobre Fijacion de Nitrogeno, UNAM, Apdo. Postal 565-A, Cuernavaca, Morelos, Mexico
Estramareix, C.	Institut de Microbiologie, Université Paris-Sud, 91405 - Orsay Cedex, France
Ettinger, William F.	Biotechnology Center, Rightmire Hall, Ohio State University, Columbus, Ohio 43210 U.S.A.
Evans, I.J.	John Innes Institute, Colney Lane, Norwich NR4 7UH, UK
Feenstra, Willem J.	Department of Genetics, University of Groningen, 9751 NN Haren, The Netherlands
Fees, H.	Institut fur Mikrobiologie und Biochemie, Universitat Erlangen-Nurnberg, D-8520 Erlangen, F.R.G.
Filser, M.	International Institute of Genetics and Biophysics, CNR Via Marconi 10, 80125 Naples
Finan, Turlough M.	Department of Biology, McMaster University, Hamilton, Ontario L8S 4K1 Canada
Firmin, J.L.	John Innes Institute, Colney Lane, Norwich NR4 7UH, UK
Fischer, H.-M.	Mikrobiologisches Institut, Eidgenössische Technische Hochschule, Eth-Zentrum, Universitätstrasse 2, CH-8092 Zürich, Switzerland
Fleury-Guerout, A.-M.	Institut de Microbiologie, Université Paris-Sud, 91405 - Orsay Cedex, France
Flores, Margarita	Centro de Investigacion Sobre Fijacion de Nitrogeno, Universidad Nacional Autonoma de Mexico, AP, Postal 565-A, Cuernavaca, Morelos, Mexico
Folkerts, Otto	Boyce Thompson Institute for Plant Research, Cornell University, Ithaca, N.Y. 14853 U.S.A.
Fortin, Marc G.	Centre for Plant Molecular Biology, Department of Biology, McGill University, 1205 Docteur Penfield Avenue, Montreal, Québec H3A 1B1 Canada
Fox, D.	Department of Genetics, University of Leeds, Leeds LS2 9JT, U.K.
Franche, C.	Departments of Developmental and Molecular Biology, R.S.B.S., A.N.U., Canberra City, A.C.T. 2601, Australia
Franssen, Henk	Department of Molecular Biology, Wageningen, The Netherlands
Frenk, Silvia	Centro de Investigacion Sobre Fijacion de Nitrogeno, Universidad Nacional Autonoma de Mexico, AP, Postal 565-A, Cuernavaca, Morelos, Mexico

Fuggi, A.	Dipartimento di Biologia Vegetale, University of Naples, Italy
Galfre, Giovanni	AFRC Institute of Animal Physiology, Babraham, Cambridge, UK
Garciarrubio, Alejandro	Centro de Investigacion Sobre Fijacion de Nitrogeno, Universidad Nacional Autonoma de Mexico, AP, Postal 565-A, Cuernavaca, Morelos, Mexico
Garnerone, A.M.	Laboratoire de Biologie Moleculaire, CNRS-INRA, B.P. 27, F 31326 Castanet-Tolosan, Cedex, France
Gerhold, D.	Department of Microbiology and Graduate Program of Ecology, The University of Tennessee, Knoxville, Tennessee, U.S.A.
Ghai, J.	Laboratoire de Biologie Moleculaire, CNRS-INRA, B.P. 27, F 31326 Castanet-Tolosan, Cedex, France
Giroux, H.	Departement de Biochimie, Université de Montréal, C.P. 6128, Succursale A, Montréal, Québec H3C 3J7 Canada
Gloudemans, Ton	Department of Molecular Biology, Wageningen, The Netherlands
Goethals, K.	Laboratorium voor Genetica, Rijksuniversiteit Gent, B-9000 Gent, Belgium
Goldthwaite, Jonathan	Department of Biology, Boston College, 140 Commonwealth Avenue, Newton, Massachusetts 02167 U.S.A.
Golinska, B.	Institute of Biochemistry, University of Agriculture, Wotynska 35, 60-637 Poznan, Poland Institute of Bioorganic Chemistry, Polish Academy of Sciences, Noskowskiego 12/14, 61-704 Poznan, Poland
Gonzalez, Victor	Centro de Investigacion Sobre Fijacion de Nitrogeno, Universidad Nacional Autonoma de Mexico, AP, Postal 565-A, Cuernavaca, Morelos, Mexico
Gottfert, M.	Institute of Genetics, Biological Research Center, Hung. Acad. Sci., P.O.B. 521 H-6701, Szeged, Hungary
Göttfert, M.	Mikrobiologisches Institut, Eidgenössische Technische Hochschule, Eth-Zentrum, Universitätstrasse 2, CH-8092 Zürich, Switzerland
Gottlob-McHugh, S.G.	Department of Biology, University of Ottawa, Ottawa, Ontario, Canada
Govers, Francine	Department of Molecular Biology, Wageningen, The Netherlands
Gray, J.X.	Genetics Department, Research School Biological Sciences, Australian National University, Canberra, Australia 2601
Gresshoff, Peter M.	Botany Department, Australian National University, Canberra, ACT, 2600
Gresshoff, Peter M.	Department of Botany, Australian National University, GPO Box 4, Canberra, ACT 2601, Australia
Grönger, P.	Lehrstuhl für Genetik, Fakultät für Biologie, Universität Bielefeld, D-4800 Bielefeld, F.R.G.

Gubler, M.	Mikrobiologisches Institut, Eidgenössische Technische Hochschule, Eth-Zentrum, Universitätstrasse 2, CH-8092 Zürich, Switzerland
Guida, M.	International Institute of Genetics and Biophysics, CNR Via Marconi 10, 80125 Naples
Gunning, B.E.G.	Departments of Developmental and Molecular Biology, R.S.B.S., A.N.U., Canberra City, A.C.T. 2601, Australia
Gyorgypal, Z.	Institute of Genetics, Biological Research Center, Hung. Acad. Sci., P.O.B. 521 H-6701, Szeged, Hungary
Hahn, M.	Mikrobiologisches Institut, Eidgenössische Technische Hochschule, Eth-Zentrum, Universitätstrasse 2, CH-8092 Zürich, Switzerland
Handa, A.K.	Department of Horticulture, Purdue University, West Lafayette, IN 47907 U.S.A.
Hanks, Joanna F.	Department of Biological Sciences, Wellesley College, Wellesley, Massachusetts 02181 U.S.A.
Hellmiss, Renate	Institute for Molecular Biology, Department of Biology, Indiana University, Bloomington, IN 47405 U.S.A.
Hertig, C.	Laboratoire de Biologie Moleculaire, CNRS-INRA, B.P. 27, F 31326 Castanet-Tolosan, Cedex, France
Heycke, Nina	Max-Planck-Institut für Züchtungsforschung, 5000 Köln 30, W. Germany
Higgisson, B.	Microbiology Department, University College, Cork, Ireland
Hilgert, U.	Max Planck Institut für Züchtungsforschung, 5000 Köln 30, W. Germany
Hirsch, Ann M.	Department of Biological Sciences, Wellesley College, Wellesley, Massachusetts 02181 U.S.A.
Hollingsworth, Rowle I.	Department of Microbiology & Public Health, Michigan State University, East Lansing, Michigan 48824 U.S.A.
Holsters, M.	Laboratorium voor Genetica, Rijksuniversiteit Gent, B-9000 Gent, Belgium
Honma, Mary A.	Department of Molecular Biology, Massachusetts General Hospital, Boston, MA 02114 U.S.A.
Hontelez, Jan	Department of Molecular Biology, Agricultural University, Wageningen, The Netherlands
Hooykaas, P.J.J.	Department of Biochemistry, University of Leiden, The Netherlands
Horvath, B.	Institute of Genetics, Biological Research Center, Hung. Acad. Sci., P.O.B. 521 H-6701, Szeged, Hungary
Hua, S.-S.T.	Western Regional Research Center, ARS, USDA, 800 Buchanan Street, Berkeley, CA 94710 U.S.A.
Huguet, T.	Laboratoire de Biologie Moléculaire des Relations Plantes-Microorganismes, INRA-CNRS, BP 27, Auzeville, 31326 Castanet-Tolosan, Cedex, France
Iaccarino, M.	International Institute of Genetics and Biophysics, CNR Via Marconi 10, 80125 Naples
Infante, D.	Laboratoire de Biologie Moleculaire, CNRS-INRA, B.P. 27, F 31326 Castanet-Tolosan, Cedex, France

Innes, Roger	Department of MCD-Biology, University of Colorado, Boulder, CO 80309 U.S.A.
Iyer, V.N.	Department of Biology and Institute of Biochemistry, Carleton University, Ottawa, Ontario K1S 5B6 Canada
Jacobs, F.	Centre for Plant Molecular Biology, Department of Biology, McGill University, 1205 Docteur Penfield Avenue, Montreal, Québec H3A 1B1 Canada
Jacobsen, Evert	Department of Genetics, University of Groningen, 9751 NN, Haren The Netherlands
Jansma Jan-Dirk	Department of Molecular Biology, Agricultural University, Wageningen, The Netherlands
Jayaswal, R.K.	Department of Horticulture, Purdue University, West Lafayette, IN 47907 U.S.A.
Ji, Joon M.	Institute for Molecular Biology, Department of Biology, Indiana University, Bloomington, IN 47405 U.S.A.
Jochimsen, Bjarne	Department of Molecular Biology and Plant Physiology, University of Aarhus, DK-8000 Aarhus, C., Denmark
Johansen, Eric	Agrigenetics Advanced Science Company, 5649 East Buckeye Road, Madison, Wisconsin 53716 U.S.A.
John, M.	Max-Planck-Institut fur Zuchtungsforschung, D-5000 Koln 30, Federal Republic of Germany
Johnson, D.A.	Department of Biology, University of Ottawa, Ottawa, Ontario, Canada
Johnston, A.W.B.	John Innes Institute, Colney Lane, Norwich NR4 7UH, UK
Kado, Clarence I.	Department of Plant Pathology, University of California, Davis, CA 95616 U.S.A.
Kahn, D.	Laboratoire de Biologie Moleculaire, CNRS-INRA, B.P. 27, F 31326 Castanet-Tolosan, Cedex, France
Kamp, R.	Institute of Biochemistry, University of Agriculture, Wotynska 35, 60-637 Poznan, Poland Institute of Bioorganic Chemistry, Polish Academy of Sciences, Noskowskiego 12/14, 61-704 Poznan, Poland
Kane, Heather	Centre for Recombinant DNA Research, Research School of Biological Sciences, Canberra, Australia
Kanvinde, Lalita	Department of Genetics, University of Leeds, Leeds LS2 9JT, U.K.
Kerr, Allen	Waite Agricultural Research Institute, University of Adelaide, Adelaide, South Australia
Kieber, Joseph	Department of Biology, Massachusetts Institute of Technology, Cambridge, MA 02139 U.S.A.
Kiely, B.	Microbiology Department, University College, Cork, Ireland
Kiss, György B.	Institute of Genetics, Biological Research Center, Hungarian Academy of Sciences, Szeged, P.O.B. 521, Hungary
Kitts, Christopher L.	Thimann Laboratories, University of California, Santa Cruz, California 95064 U.S.A.
Klein, S.	Department of Biology, Massachusetts Institute of Technology, Cambridge, MA 02139 U.S.A.

Kneen, B.E.	Boyce Thompson Institute, Ithaca, N.Y. 14853 U.S.A.
Kolattukudy, P.E.	Biotechnology Center, Rightmire Hall, Ohio State University, Columbus, Ohio 43210 U.S.A.
Kondorosi, Adam	Institute of Genetics, Biological Research Center, Hungarian Academy of Science, H-6701 Szeged, P.O. Box 521, Hungary
Kondorosi, E.	Institute of Biochemistry, Biological Research Center, Hung. Acad. Sci., P.O.B. 521 H-6701, Szeged, Hungary
Krotzky, Arno	Botany Department, Australian National University, Canberra, ACT 2600, Australia
Kuempel, Peter	Department of MCD-Biology, University of Colorado, Boulder, CO 80309 U.S.A.
Kulpaca, B.	Marquette University, Milwaukee, WI U.S.A.
Kuykendall, David	USDA, ARS, Nitrogen Fixation and Soybean Genetics Laboratory, Beltsville, Maryland 20705 U.S.A.
Laetsch, W.M.	Department of Botany, University of California, Berkeley, CA 94720 U.S.A.
Laliberte, J.-F.	Genetic Engineering Section, Plant Research Centre, Agriculture Canada, Ottawa, Canada
Lamb, J.W.	Mikrobiologisches Institut, Eidgenössische Technische Hochschule, Eth-Zentrum, Universitätstrasse 2, CH-8092 Zürich, Switzerland
Lamberti, A.	International Institute of Genetics and Biophysics, CNR Via Marconi 10, 80125 Naples
Lankhorst, Rene Klein	Department of Molecular Biology, Agricultural University, Wageningen, The Netherlands
Lara, M.	Centro de Investigacion Sobre Fijacion de Nitrogeno, Universidad Nacional Autonoma de Mexico, APDO, Postal 565-A, Cuernavaca, Mor. Mexico
Larsen, Knud	Department of Molecular Biology and Plant Physiology, University of Aarhus, DK-8000 Aarhus C., Denmark
Larue, T.A.	Boyce Thompson Institute, Ithaca, N.Y. 14853 U.S.A.
Lee, L.	Department of Biological Sciences, Purdue University, West Lafayette, IN 47907 U.S.A.
Lefebvre, D.D.	Genetic Engineering Section, Plant Research Centre, Agriculture Canada, Ottawa, Canada
Legocki, A.B.	Institute of Biochemistry, University of Agriculture, Wotynska 35, 60-637 Poznan, Poland Institute of Bioorganic Chemistry, Polish Academy of Sciences, Noskowskiego 12/14, 61-704 Poznan, Poland
Legocki, Misuk	Boyce Thompson Institute for Plant Research, Cornell University, Ithaca, N.Y. 14853 U.S.A.
Legocki, Roman P.	Boyce Thompson Institute for Plant Research, Cornell University, Ithaca, N.Y. 14853 U.S.A.
Leigh, John A.	Department of Microbiology SC-42, University of Washington, Seattle, WA 98195 U.S.A.
Lewin, A.	Laboratoire de Biologie Moleculaire des Plantes Superieures, Universite de Geneve, 1 chemin de l'Imperatrice, 1292 Chambesy, Geneve, Suisse

Lewis, D. Mark	Plant Research Centre, Agriculture Canada, Ottawa, Ontario K1A 0C6 Canada
Lopez, Mary F.	Department of Biology, Massachusetts Institute of Technology, Cambridge, MA 02139 U.S.A.
Loroch, Albert I.	Thimann Laboratories, University of California, Santa Cruz, California 95064 U.S.A.
Lotz, W.	Institut fur Mikrobiologie und Biochemie, Universitat Erlangen-Nurnberg, D-8520 Erlangen, F.R.G.
Ludwig, Robert	Thimann Laboratories, University of California, Santa Cruz, California 95064 U.S.A.
Lugtenberg, Ben J.J.	Department of Plant Molecular Biology, University of Leiden, Nonnensteeg 3, 2311 VJ Leiden, The Netherlands
Macol, Lisa A.	Department of Biological Sciences, Wellesley College, Wellesley, Massachusetts 02181 U.S.A.
Madrzak, C.J.	Institute of Biochemistry, University of Agriculture, Wotynska 35, 60-637 Poznan, Poland Institute of Bioorganic Chemistry, Polish Academy of Sciences, Noskowskiego 12/14, 61-704 Poznan, Poland
Manen, J.-F.	Laboratoire de Biologie Moleculaire des Plantes Superieures, Universite de Geneve, 1 chemin de l'Imperatrice, 1292 Chambesy, Geneve, Suisse
Manian, S.	Microbiology Department, University College, Cork, Ireland
Manian, S.S.	Lehrstuhl für Genetik, Fakultät für Biologie, Universität Bielefeld, D-4800 Bielefeld, F.R.G.
Marineau, C.	Departement de Biochimie, Université de Montréal, C.P. 6218, Succursale A, Montréal, Québec H3C 3J7 Canada
Maris, Maggi	F.A. Janssens Memorial Laboratory for Genetics, K.U. Leuven, Kardinaal Mercierlaan 92, B-3030 Heverlee, Belgium
Martinez, Esperanza	Centro de Investigacion Sobre Fijacion de Nitrogeno, Universidad Nacional Autonoma de Mexico, AP, Postal 565-A, Cuernavaca, Morelos, Mexico
Marugg, Joey	Department of Molecular Cell Biology, Institute of Molecular Biology and Department of Phyto pathology, University of Utrecht, Utrecht, The Netherlands
Mathews, Anne	Botany Department, Australian National University, Canberra, ACT, 2600
Matthysse, Ann G.	Department of Biology, University of North Carolina, Chapel Hill, NC 27514 U.S.A.
McLaughlin, Lorna	Department of Microbiology & Public Health, Michigan State University, East Lansing, Michigan 48824 U.S.A.
McLoughlin, Thomas J.	Agrigenetics Advanced Science Company, 5649 East Buckeye Road, Madison, Wisconsin 53716 U.S.A.
McQueen, D.A.R.	Departments of Plant Pathology and Biochemistry, University of Minnesota, St. Paul, MN 55108 U.S.A.
Merlo, Ann Owens	Agrigenetics Advanced Science Company, 5649 East Buckeye Road, Madison, Wisconsin 53716 U.S.A.

Michiels, Kris	F.A. Janssens Memorial Laboratory for Genetics, K.U. Leuven, Kardinaal Mercierlaan 92, B-3030 Heverlee, Belgium
Mignotte, C.	Institut de Microbiologie, Université Paris-Sud, 91405 - Orsay Cedex, France
Miller, K.L.	Western Regional Research Center, ARS, USDA, 800 Buchanan Street, Berkeley, CA 94710 U.S.A.
Moerman, Marja	Department of Molecular Biology, Wageningen, The Netherlands
Morris, Pamela	Department of Microbiology & Public Health, Michigan State University, East Lansing, Michigan 48824 U.S.A.
Murphy, Peter J.	Max-Planck-Institut für Züchtungsforschung, 5000 Köln 30, W. Germany
Nap, Jan-Peter	Department of Molecular Biology, Wageningen, The Netherlands
Nayudu, Murali	Department of Genetics, Research School of Biological Sciences, Australian National University, P.O. Box 475, Canberra, Australia 2601
Nees, David W.	Thimann Laboratories, University of California, Santa Cruz, California 95064 U.S.A.
Nelson, L.M.	Institut fur Mikrobiologie und Biochemie, Universitat Erlangen-Nurnberg, D-8520 Erlangen, F.R.G.
Nieuwkoop, J.	Department of Microbiology and Graduate Program of Ecology, The University of Tennessee, Knoxville, Tennessee, U.S.A.
Noel, K.D.	Marquette University, Milwaukee, WI U.S.A.
Noti, John D.	Boyce Thompson Institute for Plant Research, Cornell University, Ithaca, N.Y. 14853 U.S.A.
Nowell, Lise	Department of Genetics, University of Leeds, Leeds LS2 9JT, U.K.
Nuti, M.P.	Istituto di Chimica e Industrie agrarie, Universita di Padova, Italy
O'Gara, F.	Microbiology Department, University College, Cork, Ireland
Okker, Rob H.J.	Department of Plant Molecular Biology, University of Leiden, Nonnensteeg 3, 2311 VJ Leiden, The Netherlands
O'Regan, M.	Microbiology Department, University College, Cork, Ireland
Olsson, Jane E.	Botany Department, Australian National University, Canberra, ACT 2600, Australia
Ophel, Kathy	Waite Agricultural Research Institute, University of Adelaide, Adelaide, South Australia
Ortega, J.L.	Centro de Investigacion Sobre Fijacion de Nitrogeno, Universidad Nacional Autonoma de Mexico, APDO, Postal 565-A, Cuernavaca, Mor. Mexico
Pachori, P.	Marquette University, Milwaukee, WI U.S.A.
Padilla, J.	Depto. de Biologia Molecular de Plantas, Centro de Investigacion sobre Fijacion de Nitrogeno, UNAM, Apdo. Postal 565-A, Cuernavaca, Morelos, Mexico

Palacios, Rafael	Centro de Investigacion Sobre Fijacion de Nitrogeno, Universidad Nacional Autonoma de Mexico, AP, Postal 565-A, Cuernavaca, Morelos, Mexico
Pawlowski, K.	Max Planck Institut für Züchtungsforschung, 5000 Köln 30, W. Germany
Pees, Elly	Department of Plant Molecular Biology, University of Leiden, Nonnensteeg 3, 2311 VJ Leiden, The Netherlands
Peralta, Ernest G.	Genentech Inc., 460 Point San Bruno Blvd., South San Francisco, CA 94080 U.S.A.
Pinero, Daniel	Centro de Investigacion Sobre Fijacion de Nitrogeno, Universidad Nacional Autonoma de Mexico, AP, Postal 565-A, Cuernavaca, Morelos, Mexico
Philip, Saleela	Department of Microbiology & Public Health, Michigan State University, East Lansing, Michigan 48824 U.S.A.
Plazinski, J.	Departments of Developmental and Molecular Biology, R.S.B.S., A.N.U., Canberra City, A.C.T. 2601, Australia
Popin, Timothy T.	Thimann Laboratories, University of California, Santa Cruz, California 95064 U.S.A.
Postma, Jenne G.	Department of Genetics, University of Groningen, 9751 NN, Haren The Netherlands
Price, G. Dean	Department of Botany, Australian National University, GPO Box 4, Canberra, ACT 2601, Australia
Priefer, U.B.	Lehrstuhl für Genetik, Fakultät für Biologie, Universität Bielefeld, D-4800 Bielefeld, F.R.G.
Pühler, A.	Lehrstuhl für Genetik, Fakultät für Biologie, Universität Bielefeld, D-4800 Bielefeld, F.R.G.
Putnoky, P.	Institute of Genetics, Biological Research Center, Hung. Acad. Sci., P.O.B. 521 H-6701, Szeged, Hungary
Quinto, Carmen	Centro de Investigacion Sobre Fijacion de Nitrogeno, Universidad Nacional Autonoma de Mexico, AP, Postal 565-A, Cuernavaca, Morelos, Mexico
Ramakrishnan, Neela	Department of Genetics, Ames, Iowa 50011 U.S.A.
Ramseier, T.	Mikrobiologisches Institut, Eidgenössische Technische Hochschule, Eth-Zentrum, Universitätstrasse 2, CH-8092 Zürich, Switzerland
Ratet, P.	Max Planck Institut für Züchtungsforschung, 5000 Köln 30, W. Germany
Ream, Walt	Institute for Molecular Biology, Department of Biology, Indiana University, Bloomington, IN 47405 U.S.A.
Redmond, J.W.	School of Chemistry, Macquarie University, North Ryde, NSW, Australia 2113
Reed, Jason	Department of Biology, Massachusetts Institute of Technology, Cambridge, MA 02139 U.S.A.
Regensburger, B.	Mikrobiologisches Institut, Eidgenössische Technische Hochschule, Eth-Zentrum, Universitätstrasse 2, CH-8092 Zürich, Switzerland

Renalier, M.H.	Laboratoire de Biologie Moleculaire, CNRS-INRA, B.P. 27, F 31326 Castanet-Tolosan, Cedex, France
Reyes, O.	Institut de Microbiologie, Université Paris-Sud, 91405 - Orsay Cedex, France
Riccio, A.	International Institute of Genetics and Biophysics, CNR Via Marconi 10, 80125 Naples
Richaud, F.	Institut de Microbiologie, Université Paris-Sud, 91405 - Orsay Cedex, France
Ripke, H.M.	Institut fur Mikrobiologie und Biochemie, Universitat Erlangen-Nurnberg, D-8520 Erlangen, F.R.G.
Roberts, Daniel P.	Departments of Microbiology and Plant Pathology, University of Georgia, Athens, GA 30602 U.S.A.
Rodriguez-Quinones, F.	Institute of Genetics, Biological Research Center, Hung. Acad. Sci., P.O.B. 521 H-6701, Szeged, Hungary
Rogowsky, P.	Department of Plant Pathology, University of California, Davis, CA 95616 U.S.A.
Rolfe, Barry G.	Department of Genetics, Research School of Biological Sciences, Australian National University, P.O. Box 475, Canberra, Australia 2601
Romero, David	Centro de Investigacion Sobre Fijacion de Nitrogeno, Universidad Nacional Autonoma de Mexico, AP, Postal 565-A, Cuernavaca, Morelos, Mexico
Rosenberg, C.	Laboratoire de Biologie Moleculaire, CNRS-INRA, B.P. 27, 31326 Castanet-Tolosan, Cedex, France
Rossen, L.	John Innes Institute, Colney Lane, Norwich NR4 7UH, UK
Saad, Marlene	Centre for Recombinant DNA Research, Research School of Biological Sciences, Canberra, Australia
Salzwedel, Janet	Department of Microbiology & Public Health, Michigan State University, East Lansing, Michigan 48824 U.S.A.
Sanchez, F.	Depto. de Biologia Molecular de Plantas, Centro de Investigacion sobre Fijacion de Nitrogeno, UNAM, Apdo. Postal 565-A, Cuernavaca, Morelos, Mexico
Sastry, G.R.K.	Department of Genetics, University of Leeds, Leeds LS2 9JT, U.K.
Schell, J.	Max Planck Institut für Züchtungsforschung, 5000 Köln 30, W. Germany
Schell, Jeff	Max-Planck-Institut für Züchtungsforschung, 5000 Köln 30, W. Germany
Schell, M.G.	Department of Microbiology and Graduate Program of Ecology, The University of Tennessee, Knoxville, Tennessee, U.S.A.
Schell, Mark A.	Departments of Microbiology and Plant Pathology, University of Georgia, Athens, GA 30602 U.S.A.
Schild, C.	Institut fur Mikrobiologie und Biochemie, Universitat Erlangen-Nurnberg, D-8520 Erlangen, F.R.G.
Schippers, Bob	Department of Molecular Cell Biology, Institute of Molecular Biology and Department of Phyto pathology, University of Utrecht, Utrecht, The Netherlands

Schmidt, J.	Max-Planck-Institut fur Zuchtungsforschung, D-5000 Koln 30, Federal Republic of Germany
Schottel, J.L.	Departments of Plant Pathology and Biochemistry, University of Minnesota, St. Paul, MN 55108 U.S.A.
Schuller, Kathryn A.	Botany Department, Australian National University, Canberra, ACT 2600, Australia
Scofield, G.	John Innes Institute, Colney Lane, Norwich, U.K. NR4 7UH
Scott, K.F.	Centre for Recombinant DNA Research and Department of Molecular Biology, Research School of Biological Sciences, The Australian National University, A.C.T. 2601, Australia
Sebastian, Joseph	Biotechnology Center, Rightmire Hall, Ohio State University, Columbus, Ohio 43210 U.S.A.
Segovia, Lorenzo	Centro de Investigacion Sobre Fijacion de Nitrogeno, Universidad Nacional Autonoma de Mexico, AP, Postal 565-A, Cuernavaca, Morelos, Mexico
Shah, Ketan S.	USDA, ARS, Nitrogen Fixation and Soybean Genetics Laboratory, Beltsville, Maryland 20705 U.S.A.
Shaw, Joe J.	Department of Plant Pathology, University of California, Davis, CA 95616 U.S.A.
Shearman, C.A.	John Innes Institute, Colney Lane, Norwich NR4 7UH, UK
Shen, Bin-Fu	Institute of Plant Physiology, Department of Molecular Genetics, 300 Fonglin Road, Shanghai 200032, Peoples Republic of China
Shen, San Chiun	Institute of Plant Physiology, Department of Molecular Genetics, 300 Fonglin Road, Shanghai 200032, Peoples Republic of China
Signer, Ethan R.	Department of Biology, Massachusetts Institute of Technology, Cambridge, MA 02139 U.S.A.
Sikorski, M.	Institute of Biochemistry, University of Agriculture, Wotynska 35, 60-637 Poznan, Poland Institute of Bioorganic Chemistry, Polish Academy of Sciences, Noskowskiego 12/14, 61-704 Poznan, Poland
Simpson, Robert B.	Molecular Biology Group, ARCO Plant Cell Research Institute, 6560 Trinity Court, Dublin, CA 94568 U.S.A.
Sinclair, Martha	Department of Genetics, Research School of Biological Sciences, Australian National University, P.O. Box 475, Canberra, Australia 2601
Smith, C.A.	Department of Biological Sciences, Wellesley College, Wellesley, MA 02181 U.S.A.
Smith, Kathryn B.	Department of Microbiology & Public Health, Michigan State University, East Lansing, Michigan 48824 U.S.A.
So, J.-S.	Department of Microbiology and Graduate Program of Ecology, The University of Tennessee, Knoxville, Tennessee, U.S.A.
Soliman, M.H.	Department of Genetics, Faculty of Agriculture, Cairo University, Cairo, A.R.E.

Spaink, Herman P.	Department of Plant Molecular Biology, University of Leiden, Nonnensteeg 3, 2311 VJ Leiden, The Netherlands
Spielmann, Albert	Molecular Biology Group, ARCO Plant Cell Research Institute, 6560 Trinity Court, Dublin, CA 94568 U.S.A.
Squartini, A.	Istituto di Chimica e Industrie agrarie, Universita di Padova, Italy
Stacey, G.	Department of Microbiology and Graduate Program of Ecology, The University of Tennessee, Knoxville, Tennessee, U.S.A.
Stanley, J.	Laboratoire de Biologie Moleculaire des Plantes Superieures, Universite de Geneve, 1 chemin de l'Imperatrice, 1292 Chambesy, Geneve, Suisse
Strozycki, P.	Institute of Biochemistry, University of Agriculture, Wotynska 35, 60-637 Poznan, Poland Institute of Bioorganic Chemistry, Polish Academy of Sciences, Noskowskiego 12/14, 61-704 Poznan, Poland
Studer, D.	Mikrobiologisches Institut, Eidgenössische Technische Hochschule, Eth-Zentrum, Universitätstrasse 2, CH-8092 Zürich, Switzerland
Surin, B.P.	CSIRO Division of Plant Industry, Canberra, ACT 2601, Australia
Szalay, Aladar A.	Boyce Thompson Institute for Plant Research, Cornell University, Ithaca, N.Y. 14853 U.S.A.
Szeto, W.	BioTechnica International, Inc., 85 Bolton Street, Cambridge, MA 02140 U.S.A.
Szybiak-Strozycka, U.	Institute of Biochemistry, University of Agriculture, Wotynska 35, 60-637 Poznan, Poland Institute of Bioorganic Chemistry, Polish Academy of Sciences, Noskowskiego 12/14, 61-704 Poznan, Poland
Tempé, Jacques	Institute de Microbiologie, Bat. 409, Université de Paris-Sud, F-91405 Orsay, France
Thomas, Joseph	Molecular Biology and Agriculture Division, Bhabha Atomic Research Centre, Bombay - 400085, India
Thöny, B.	Mikrobiologisches Institut, Eidgenössische Technische Hochschule, Eth-Zentrum, Universitätstrasse 2, CH-8092 Zürich, Switzerland
Tichy, H.V.	Institut fur Mikrobiologie und Biochemie, Universitat Erlangen-Nurnberg, D-8520 Erlangen, F.R.G.
Torok, Istvan	Institute of Biochemistry, Hungarian Academy of Sciences, H-6701, Szeged, Hungary
Tucker, W.T.	Centre for Recombinant DNA Research and Department of Molecular Biology, Research School of Biological Sciences, The Australian National University, A.C.T. 2601, Australia
Upadhyaya, N.M.	Centre for Recombinant DNA Research and Department of Molecular Biology, Research School of Biological Sciences, The Australian National University, A.C.T. 2601, Australia

Valderrama, B.	Centro de Investigacion Sobre Fijacion de Nitrogeno, Universidad Nacional Autonoma de Mexico, APDO, Postal 565-A, Cuernavaca, Mor. Mexico
Vam Vikites, D.	Boyce Thompson Institute, Ithaca, N.Y. 14853 U.S.A.
van Brussel, Anton A.N.	Department of Plant Molecular Biology, University of Leiden, Nonnensteeg 3, 2311 VJ Leiden, The Netherlands
van den Eede, G.	Laboratorium voor Genetica, Rijksuniversiteit Gent, B-9000 Gent, Belgium
van der Hofstad, Gerard	Department of Molecular Cell Biology, Institute of Molecular Biology and Department of Phyto pathology, University of Utrecht, Utrecht, The Netherlands
van Gool, August P.	F.A. Janssens Memorial Laboratory for Genetics, K.U. Leuven, Kardinaal Mercierlaan 92, B-3030 Heverlee, Belgium
van Kammen, Albert	Department of Molecular Biology, Wageningen, The Netherlands
van Montagu, M.	Laboratorium voor Genetica, Rijksuniversiteit Gent, B-9000 Gent, Belgium
Vandenbosch, K.A.	University of Wisconsin, Madison, WI U.S.A.
Vanderleyden, Jos	F.A. Janssens Memorial Laboratory for Genetics, K.U. Leuven, Kardinaal Mercierlaan 92, B-3030 Heverlee, Belgium
Vanstockem, Mark	F.A. Janssens Memorial Laboratory for Genetics, K.U. Leuven, Kardinaal Mercierlaan 92, B-3030 Heverlee, Belgium
Vazquez, M.	Depto. de Biologia Molecular de Plantas, Centro de Investigacion sobre Fijacion de Nitrogeno, UNAM, Apdo. Postal 565-A, Cuernavaca, Morelos, Mexico
Vegh, Zoltan	Institute of Genetics, Biological Research Center, Hungarian Academy of Sciences, Szeged, P.O.B. 521, Hungary
Verma, Desh Pal S.	Centre for Plant Molecular Biology, Department of Biology, McGill University, 1205 Docteur Penfield Avenue, Montreal, Québec H3A 1B1 Canada
Vincze, Eva	Institute of Genetics, Biological Research Center, Hungarian Academy of Sciences, Szeged, P.O.B. 521, Hungary
Vreeland, V.J.	Department of Botany, University of California, Berkeley, CA 94720 U.S.A.
Walker, Graham C.	Department of Biology, Massachusetts Institute of Technology, Cambridge, MA 02139 U.S.A.
Wang, Shin-Ping	Institute of Plant Physiology, Department of Molecular Genetics, 300 Fonglin Road, Shanghai 200032, Peoples Republic of China
Wardhan, H.	Department of Genetics, University of Leeds, Leeds LS2 9JT, U.K.
Weinman, J.J.	Genetics Department, Research School Biological Sciences, Australian National University, Canberra, Australia 2601

Weisbeek, Peter	Department of Molecular Cell Biology, Institute of Molecular Biology and Department of Phyto pathology, University of Utrecht, Utrecht, The Netherlands
Welsch, Margaret A.	Department of Microbiology & Public Health, Michigan State University, East Lansing, Michigan 48824 U.S.A.
Wijffelman, Carel A.	Department of Plant Molecular Biology, University of Leiden, Nonnensteeg 3, 2311 VJ Leiden, The Netherlands
Wittmann-Liebold, B.	Institute of Biochemistry, University of Agriculture, Wotynska 35, 60-637 Poznan, Poland Institute of Bioorganic Chemistry, Polish Academy of Sciences, Noskowskiego 12/14, 61-704 Poznan, Poland
Wood, Elizabeth A.	John Innes Institute, Colney Lane, Norwich, NR4 7UH, UK
Yun, Allen C.	Boyce Thompson Institute for Plant Research, Cornell University, Ithaca, N.Y. 14853 U.S.A.
Zaat, Bas A.J.	Department of Plant Molecular Biology, University of Leiden, Nonnensteeg 3, 2311 VJ Leiden, The Netherlands
Zhang, M.	Centre for Plant Molecular Biology, Department of Biology, McGill University, 1205 Docteur Penfield Avenue, Montreal, Québec H3A 1B1 Canada

Section I

MOLECULAR GENETICS OF *AGROBACTERIUM* AND PLANT TRANSFORMATION

ECOLOGY OF AGROBACTERIUM: PLASMIDS AND BIOVARS

KATHY OPHEL AND ALLEN KERR

Waite Agricultural Research Institute, University of Adelaide
Adelaide, South Australia

1. INTRODUCTION

At present, there are three recognized chromosomal forms(or biovars) of Agrobacterium tumefaciens. The differentiating characteristics of the three biovars have been described elsewhere(Bergey's manual,1985). In nature, these chromosomal forms have distinct host ranges. Biovar 1 is a ubiquitous soil organism, with pathogenic and nonpathogenic forms found on a wide range of dicotyledonous hosts whereas biovars 2 and 3 have specific host associations. Biovar 2 are found in association with stonefruit and biovar 3 are associated virtually exclusively with grapevine. In artificial inoculations, biovar 2 pathogens have a wide host range and, although some biovar 3 are limited to grapevine, many biovar 3 are pathogenic on a wider range of host plants in glasshouse inoculations. Although pathogenicity and host range determinants are coded for by the Ti-plasmid, it is apparent that the chromosome plays a role in the specificity of root colonization. The root colonization of biovars 2 and 3 on vines and almonds was studied over a 12-month period. Reciprocal plasmid transfers were made between biovars 2 and 3 and studies on root colonization patterns of these transconjugants are underway. All strains used in this study are described in Table 1.

TABLE 1. Legend of strains

AGROBACTERIUM:

Designation	Biovar	Plasmids	Opine Type	Source
K27	2	pTiK27; cryptic	nopaline	peach gall
K128	2	cryptic	–	peach gall
K309	3	pTiK309; cryptic	octopine	vine gall
K377	3	pTiK377	nopaline	vine gall

E.COLI:

Designation	Strain	Plasmid	Resistance
K382	C600 rif^r	RP4	km cb tc
K1001	HB101 rif^r	pDP35::Kpn 1 fragment of A6(INC region)	km cb tc
K1002	HB101 rif^r	pPH1JI	gm cm sm sp

2. RESULTS

2.1. Reciprocal plasmid transfers between biovars 2 and 3

2.1.1. Transfer of biovar 2 Ti-plasmid to biovar 3 background. All attempts to transfer a biovar 2 Ti-plasmid to a Ti-plasmid containing biovar 3 background were unsuccessful. Therefore, the wide host range plasmid pDP35 containing the cloned Kpn 1 fragment of pTiA6(which contains the INC region of A6) was used to eliminate, by conjugation, the biovar 3 Ti-plasmid pTiK377 from strain K377. The INC-containing pDP35 was then further eliminated by pPH1JI, another IncP wide host range plasmid. It was then possible to mobilize the biovar 2 Ti-plasmid pTiK27 into the Ti-plasmidless K377. By conjugation, RP4 was then transferred to K27 and K27::RP4 was then further mated with K377 pTi$^-$. Transconjugants were selected on minimal media containing 0.2% nopaline plus 2% NaCl(selection for biovar 3 background). All plasmid transfers and eliminations were confirmed by agarose gel electrophoresis, pathogenicity tests on tomato stems and ability to utilize opines.

2.1.2. Transfer of biovar 3 Ti-plasmid to biovar 2 background. The broad host range conjugative plasmid RP4 was transferred to biovar 3 strain K309 by conjugation. K309::RP4 was then further mated with a biovar 2 recipient K128chl^r. Transfer of pTiK309 to K128 was confirmed by selection on minimal media containing 0.2% octopine plus chloramphenicol, agarose gel electrphoresis and pathogenicity tests on tomato stems.

2.2 Root colonization studies

2.2.1. Root colonization of almonds by wild-type biovars 2 and 3. Fig. 1 shows mean populations(expressed as colony-forming units or cfu per cm root section) of biovars 2(K27) and 3(K309) over a 12-month period. Unwounded almond seedlings were dipped in a 10^8cells/ml bacterial suspension of K27 or K309 before planting; pots were grown outdoors in large pots containing non-sterile soil in a randomized block design. Populations of biovar 2 were consistently higher(by 10- to 100-fold) during the first 6 months of the sampling period,when the initial galls were being formed. Galls were formed only on almonds inoculated with biovar 2.

2.2.2. Root colonization of grapevines by wild-type biovars 2 and 3. Fig. 2 shows mean populations of biovars 2(K27) and 3(K309) over a 10-month period. Unwounded 1-year Cabernet Sauvignon rootlings were dipped in 10^8cells/ml bacterial suspensions before planting, in a duplicate of the experiment described in 2.2.1. Populations of biovar 2 were somewhat higher than biovar 3 populations but not to the same extent as on almond roots. Stem populations (below-ground) were generally higher than root populations (data not shown) but the trend in colonization patterns between biovars on roots and stems was comparable. However, when samples were taken from surface-sterilized sections of stem (8-10 cm above soil level), only biovar 3 was found.

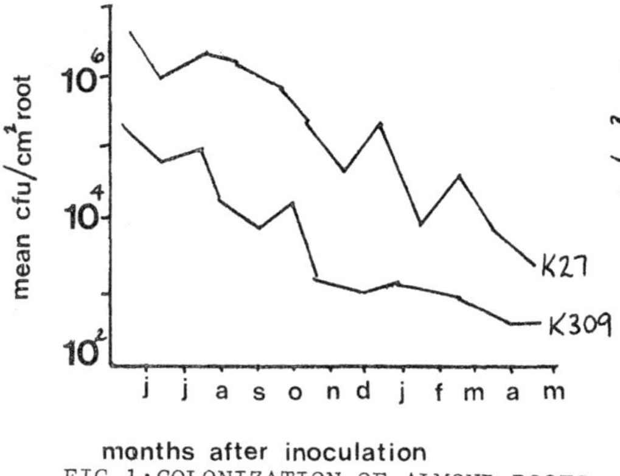

FIG.1:COLONIZATION OF ALMOND ROOTS

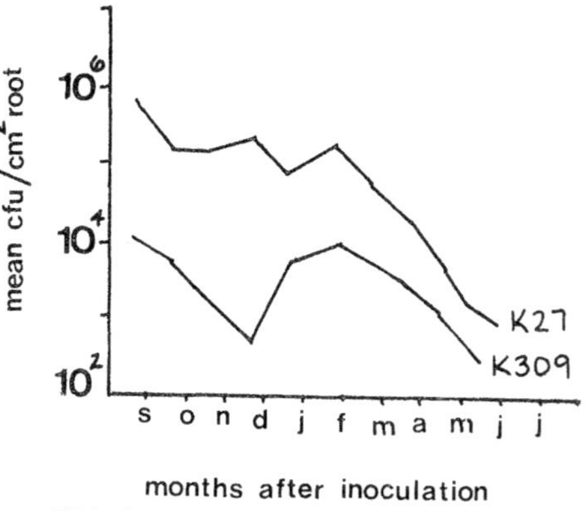

FIG.2:COLONIZATION OF VINE ROOTS

2.2.3. Root colonization by constructed strains. Preliminary results, after 3 months of sampling, when the biovar 2/biovar3 reciprocal transconjugants described in 2.1.1 and 2.1.2 and their wild-type parental strains, were inoculated onto almond roots, indicate that establishment and initial colonization of almond roots is a function of the chromosomal background of the strain. Both transconjugants have maintained populations on almond roots and stem which are virtually identical to those of their wild-type background strains. Similiar trials have been initiated on grapevines.

3. DISCUSSION

The ability of *Agrobacterium* biovars 2 and 3 to differentially colonize roots and stem of almond and grapevine appears to be a function of the chromosomal background. Biovar 2 is an extremely efficient colonizer of almonds and biovar 3, though a poor colonizer of vine roots, appears to have the ability to move into the vascular system of the grapevine. It is also interesting to note the inability of biovar 2 plasmids to establish in a biovar 3 background when a biovar 3 Ti-plasmid is present. Perhaps the transfer of plasmids between biovars is less frequent than is commonly assumed and there is a tighter relationship between plasmid type and chromosome.

Initial establishment on roots appears to be independent of the plasmid component. It is possible to speculate that plasmid type may play a role in determining the level of colonization after the formation of galls; this is being investigated at present. It seems evident that the determination of natural host range is not only coded for by Ti-plasmid determinants but that there is a large contribution of the chromosome, especially in the early stages of the infection process.

ACKNOWLEDGEMENTS

Special thanks to S.K. Farrand for the use of the cloned INC fragment of A6.

THE AGROBACTERIUM RHIZOGENES ROOT-INDUCING SYSTEM.

F. RICHAUD, C. AUBRY, A. BEYOU, F. BOULANGER, C. ESTRAMAREIX, A.-M. FLEURY-GUEROUT, C. MIGNOTTE and O. REYES.

Institut de Microbiologie, Université Paris-Sud, 91405 - Orsay Cedex, (France).

1. INTRODUCTION

In sensitive dicotyledoneous plants, *Agrobacterium rhizogenes* inoculation in wounds result first in the formation of a small callus that resembles the lesions induced in the same plant hosts by the related pathogen *Agrobacterium tumefaciens*. The fundamental difference between both pathogens is that in the former, a population of callus cells promptly differenciates in root-like tissue (hairy root), which accounts for most of the cell proliferation in advanced stages of the infection. The hairy-root tissue of many plants can be propagated in axenic medium for long periods. In these conditions, portions of the cultured hairy roots often differenciate in shoots, from which whole plants can be regenerated. Molecular biology studies show that the cultured hairy roots and the regenerated plants carry DNA sequences (T-DNA or transferred DNA) corresponding to an *A. rhizogenes* large plasmid (the Ri plasmid) in their genome (Chilton *et al.*, 1982 ; White *et al.*, 1982 ; Willmitzer *et al.*, 1982).

The conversion of *A. rhizogenes*-infected cells into calli and hairy roots is determined by genes encoded in the T-DNA regions of different types of Ri plasmids. The T-DNA sequences found both in regenerated plants and in cultured root-tumors of related plant varieties infected with the same type of *A. rhizogenes* may differ, which suggests that cell transformation by T-DNA, and/or their subsequent clonal selection, depends on factors that determine differences between species, (Huffman *et al.*, 1984 ; Taylor *et al.*, 1985 ; R. Peerbolte, personal communication). On the other hand, many genes of the plant genomic T-DNA are specifically expressed in certain organs of regenerated plants (Durand-Tardif *et al.*, 1985 ; Ooms *et al.*, 1986), thus suggesting that these genes may be sensitive to some of the plant mechanisms that control and maintain differenciation. An alternative and complementary approach is the study of how mutations on the T-DNA affect root-tumor formation on explants or whole plants before extensive clonal selection takes place.

Mutations affecting the T-DNA loci of *A. rhizogenes* A4 (essentially identical to 1855) and 8196 involved in the production of hairy root have been isolated and characterized (Cardarelli *et al.*, 1985 ; White *et al.*, 1985 ; Plessis *et al.*, 1985 ; Boulanger *et al.*, 1986 ; Estramareix *et al.*, 1986). The tumor inducing phenotypes of insertion mutants and deletions in the T-DNA can be conveniently observed by infecting *Kalankoe diagremontiana* and *Daucus carota*, two plants known to show characteristic responses to different *A. rhizogenes* types.

2. PROCEDURE

Materials and methods

Bacterial strains, transposons and techniques were described in Ratet & Richaud 1986, Boulanger *et al.*, 1986 and Estramareix *et al.*, 1986.

3. RESULTS AND DISCUSSION.

3.1. Results

3.1.1. A description of typical mutant hairy-root induction phenotypes observed is presented in Table 1.

3.1.2. TABLE 1

Hairy-Root Phenotypes Induced by *A. rhizogenes* Strains.

	K. diagremontiana	*D. carota* slices apical	basal
A4 wild-type	callus ; type 1 and 2 roots (stems and leaves)	callus ; type 2 roots	callus ; type 2 roots
rolA	abundant type 2 roots (leaves)	?	callus ; roots
Δ116	same	?	callus ; roots
rolB	no response (leaves)	* callus ; type 2 roots	callus ; type 2 roots
rolC	callus (leaves)	?	callus ; roots
rolD	callus (leaves)	?	callus ; roots
ΔCM3 *	no tumor response type 3 roots (stems)	no response	no response
Δ100	same (stems)	?	callus ; roots
ΔHM7 *	few type 2 roots	callus ; type 2 roots	no response
Δ102	no response (leaves)	?	no response
rolB/ΔHM7*	no response (leaves)	no response	no response
8196 wild-type	few type 2 roots	callus ; type 2 roots	no response
IB5	no response	no response	no response

Root type:
1 : curly, thick roots (Fig.1)
2 : straight, thin, highly ramified roots (Fig. 1, 2 and 3)
3 : straight, thin, poorly ramified, arising borderline to the wound roots (Fig. 4)

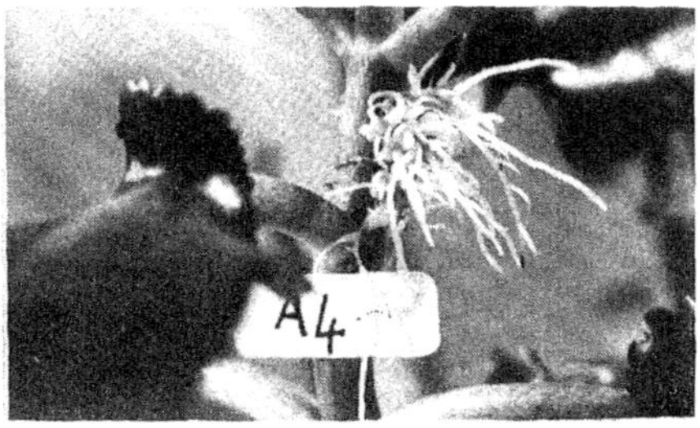

Fig. 1

Fig. 2

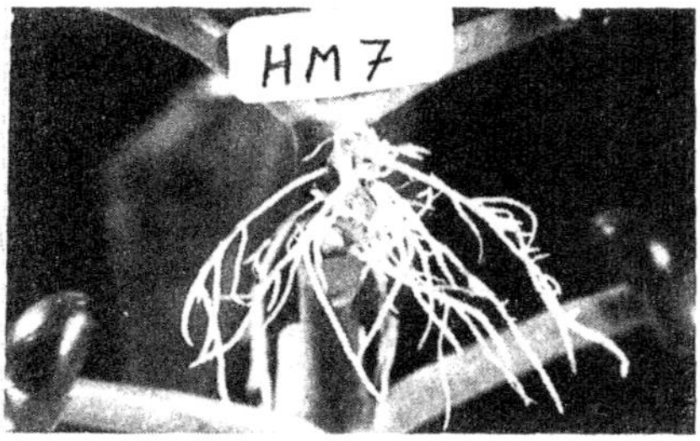

Fig. 3

Fig. 4

3.1.3. Figure 5
Location of mutations in *A. rhizogenes* A4 T1-DNA

```
       3          6   8  10     12 13 14                  tms1
-----------------------------------------------------//-----------
1 2  4  5      7   9          11          15 16 17 18   tms2
                T1-DNA                                  Tr-DNA
rol                     A  B  C        D
ΔCM3*           ≡≡≡≡≡≡≡≡≡≡≡≡≡≡≡≡≡≡≡≡≡
Δ100                 ≡≡≡≡≡≡≡≡≡≡≡≡≡≡≡≡≡≡≡≡≡≡≡≡
Δ116            ≡≡≡≡≡≡≡≡≡
ΔHM7*,   Δ102                                          ≡≡≡≡≡≡≡≡≡
```

Notes: Table 1 and Figure 5 have been compiled using data from White *et al.* (1985); Boulanger *et al.*, (1986) and Estramareix *et al.*, (1986). Data and mutations not described by White *et al.*, (1985) are identified by * characters. Δ characters stand for deletions. The extent of the deletions is indicated by ≡ characters. Numbers from 1 to 18 stand for T1-DNA *orf*'s in the A4 system have been described by Slightom *et al.*, (1986). The positions of loci *rolA*, *rolB*, *rolC* and *rolD* of White *et al.*, (1985) relative to the *orfs* are shown. The A4 T-DNA is transferred to the plant genome in two separate regions called T1-DNA and Tr-DNA.

3.1.4. Hairy-Roots Phenotypes Among A4 T-DNA Mutants. Table 1 lists only mutant loci with an observable hairy-root induction phenotype. The most striking characteristics of the A4 T-DNA system are: (i) the inactivation of single loci abolishes hairy-root induction in some, but not in all, the plant targets considered (no mutant presenting such a phenotype in the apical side of carrot slices was found in a sizable collection of single-insertion mutants at different sites in the A4 core T-DNA region) ; (ii) hairy-root induction in all plant systems considered can be achieved through the inactivation of multiple T-DNA loci. This is illustrated by the examples below. Single insertions or deletions at the *tms* loci abolish hairy-root induction in both *K. diagremontiana* leaves and in the basal side, but not in the apical side, of *D. carota* slices.

Double *rolB/tms* mutations completely abolish rhizogenesis in this system (Estramareix *et al.*, 1986), which suggests that in *tms* mutants, the *rolB* locus is also needed for hairy-root induction in carrots. A further example of this is the observation that the ΔCM3 deletion abolishes root induction in carrot slices, though the deletions Δ116 and Δ100 do not. Thus either of the overlapping gene clusters removed in Δ116 (orf6 through orf9/rolA) and in Δ100 (*orf8/rolA/rolB/rolC/orf13/orf14/rolD/orf16/orf17*) is dispensable for hairy-root induction in carrot slices provided that genes of the other cluster are present.

The simplest interpretation of these observations is that hairy-root induction is redundantly determined in the T-DNA. Such redundancy can be visualized (i) as a unique hairy-root-induction pathway where some or all of the steps involved are encoded more than once, or (ii) as different parallel hairy-root-induction pathways not sharing any critical step. In the later case, the additional assumption that a given pathway can be functional or indispensable in some plant tissues -or species- and not in others would explain the various phenotypes presented by the same mutant allele in different plan test-systems. The phenotypes observed in the A4 system would thus be accounted for by two parallel rhizogenetic pathways encoded in the T1-DNA defined respectively by the genes removed in the Δ116 and the Δ100 deletions (collectively designated *har*, for "hairy root"). Both pathways would require auxine (the product of the Tr-DNA genes *tms1* and *tms2* (White *et al.*, 1985). Either pathway would account for hairy-root induction in *D. carota*, but not in *K. diagremontiana* leaves, which requires *rolB*, a gene contained in the Δ100 deletion. Still a third pathway, operative in root-induction in *K. diagremontiana* stems by the ΔCM3 mutant, is defined by the Tr-DNA (*aux*) genes and the T1-DNA genes remaining in ΔCM3.

The concept of parallel multiple rhizogenetic pathways is also consistent with the morphology described by White *et al.*, (1985) for the roots induced by A4 mutants in *K. diagremontiana* (Table 1). Mutants affected in *rolA* produce abundant thin, filiform, straight roots in infected leaves. Mutants that have large deletions affecting the *rol* region, such as Δ100, induce abundant long thin straight roots in the tissue surrounding the wounds of infected stems ; no tissue proliferation is seen in the wound itself. This is also the case for the ΔCM3 deletion (Boulanger *et al.*,1986). Hairy roots induced by the *tms* deletion mutant ΔHM7 in *K. diagremontiana* stems resemble those induced by the ΔCM3 and Δ100 mutants (long, filiform, straight roots). However the ΔHM7-induced roots branch more often and sprout from the infected wound, not from the surrounding tissue. This is also true for the roots induced by ΔHM7 in the apical surface of carrot slices. Some roots of similar morphology often appear after infection with wild-type A4, mixed with the more abundant short curly roots, which may be related to occasional transformation of some target cells by T1-DNA only.

3.1.5. Hairy root phenotypes among *A. rhizogenes* 8196 mutants. Recent evidence suggests that the T-DNA of pRi8196 (Lahners *et al.*, 1984) presents DNA homologies with the T1-DNA of pRi A4 (A. Combard, manuscript in preparation). These extend to most of the A4 T1-DNA core, but do not include the *rolD* or the *tms* region. Phenotypically 8196-induced hairy roots are more similar to those induced by A4 *aux* mutants than to those induced by an A4 *rolD* mutant (fex long, branched, thin, straight roots, but little callus development in *K. diagremontiana* leaves; thin root development only on the apical side of carrot slices,(Ryder *et al.*, 1985). Physiologically, the 8196 T-DNA also behaves as A4 *aux* mutants. Coinfection of 8196 with ΔCM3 induce hairy-root development in the basal side of carrot slices, a tissue where both mutants are defective in single infection. This suggests that 8196 pathogenicity can be reinforced by the *tms* genes of the ΔCM3 mutant. On the other hand, 8196 mutants carrying single insertions and completely defective for hairy-root induction in carrots are found (Boulanger *et al.*, 1986 ; C. Aubry, unpublished). Indeed, one of these (IB5) maps in the homology between 8196 and the *rolA, B* region of A4 T-DNA's. Thus, the available evidence is consistent with the notion that 8196 and A4 have evolved from a common ancestor.

3.1.6. Discussion. Hairy-root induction by A4 may result from redundant, parallel, cooperating rhizogenetic pathways (Fig. 6). These (at least 3) pathways are such that inactivation of any of the three does not abolish completely hairy-root, while only the *aux* pathway is sufficient on choosen organs. The number and organization of the pathways should be found by examining systematically pairwise combinations of already available insertion and deletion mutants for their root-induction phenotype of low stringency plant test systems, as the apical surface of *D. carota* slices. Some interesting regions were few single site mutations occur (as the *orf5/orf9* region, defined by the deletions Δ100 and ΔCM3) should be explored by site-directed insertion mutagenesis.

Figure 6 : Model for hairy-root development

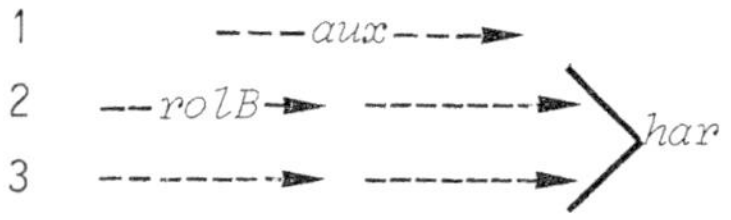

REFERENCES

Boulanger F. *et al.*, 1986, Plant Mol. Biol. 6, 271-279.
Cardarelli M. *et al.*, 1985, Plant Mol. Biol. 5, 385-391.
Chilton M.-D. *et al.*, 1982, Nature, 295, 432-434.
Durand-Tardif M. *et al.*, 1985, J. Mol. Biol., 186, 557-564.
Estramareix C. *et al.*, 1986, Plasmid, 15, 245-247.
Huffman G.A. *et al.*, 1984, J. Bacteriol., 157, 269-276.
Lahners . *et al.*, 1984, Plasmid, 11, 130-140.
Ooms G. *et al.*, 1986, Plant Mol. Biol. 6, 321-330.
Plessis A. *et al.*, 1985, Plasmid, 14, 17-27.
Ratet P. & Richaud F., 1986, Gene, 42, 185-192.
Ryder M.H. *et al.*, 1985, Plant Physiol. 77, 215-221.
Slightom J. *et al.*, 1986, J. Biol. Chem. 261, 108-121.
Taylor B. *et al.*, 1985, Mol. Gen. Genet. 201, 554-557.
White F.F. *et al.*, 1982, Proc. Natl. Acad. Sci. USA, 79, 3193-3197.
White F.F. *et al.*, 1985, J. Bacteriol., 164, 33-44.
Willmitzer L. *et al.*, 1982, Mol. Gen. Genet., 186, 16-22.

EFFECT OF THE PRESENCE OF THE PLASMID pSA AND OF AUXIN ON THE ATTACHMENT OF AGROBACTERIUM TUMEFACIENS TO PLANT HOST CELLS

ANN G. MATTHYSSE

1. INTRODUCTION

Tumor formation by Agrobacterium tumefaciens involves the transfer of plasmid DNA from the bacteria to the plant host cell. One of the earliest steps in tumor formation appears to be the attachment of the bacteria to the surface of the plant cell. In general bacterial attachment does not appear to require the active participation of the plant host cell. Bacteria attach with only slightly altered kinetics to plant cells which have been killed by treatment with heat or glutaraldehyde(6). In contrast the bacteria appear to play an active role in the attachment process; bacteria which have been killed with heat or glutaraldehyde do not attach to plant cells(4). Bacterial attachment is required for virulence; bacterial mutants which fail to attach to plant cells are avirulent(2,4).

2. THE EFFECT OF THE PLASMID pSA ON VIRULENCE OF AGROBACTERIUM

When the wide host range R plasmid pSA is introduced into A. tumefaciens, the bacteria become avirulent(3). This loss of virulence is not due to any irreversible alteration in the bacteria since bacteria which are cured of pSA simultaneously recover virulence(3). Bacteria containing pSA can also be restored to virulence by the addition of auxin to the inoculated wound sites in the host plant(1).

Observations made by New et al(7) showed that strains of Agrobacterium carrying pSA fail to interfere with the ability of virulent strains to cause tumors on bean leaves. This interference, which is normally observed with the same strains lacking pSA is believed to reflect the occupation of binding sites on the plant cell surface by the interfering strain. Thus the presence of pSA could be presumed to prevent the binding of Agrobacterium to plant host cells and consequently to prevent bacterial virulence.

3. BINDING OF AGROBACTERIUM TO PLANT CELLS

Virulent strains of Agrobacterium bind to carrot suspension culture cells. This binding can be observed directly in the microscope.

Alternatively binding can be measured indirectly by separating plant cells with their attached bacteria from the free bacteria by filtration and then determining the number of free and bound bacteria by viable cell counts or by the use of radioactively labelled bacteria(5).

A. tumefaciens strains 1D1, Ce-12, and C58 containing pSA all show binding to carrot cells which is indistinguishable from the binding of the same strains lacking pSA when the binding is measured in Murashige and Skoog tissue culture medium. However, when the bacteria are grown in medium containing no auxin, and the plant cells are depleted of auxin before they are incubated with the bacteria in medium without auxin, then only the wild type strains bind to the plant cells. Strains containing pSA do not bind to carrot cells in the absence of auxin. In these experiments bacterial binding was determined in the light microscope and by viable cell counts of free and attached bacteria.

Surprisingly, the effect of the hormone appears to be on the bacteria rather than on the plant cells. Wild type A. tumefaciens bound to dead, as well as to living, plant cells. The binding of bacteria containing pSA to heat-killed auxin-depleted carrot cells was dependent on the presence of auxin in the incubation medium.

Wild type A. tumefaciens bind to carrot cells with only a very short lag time , less than 2 min. Bacteria containing pSA and grown in the absence of auxin showed a lag time of about 30 min. before the beginning of bacterial attachment. Wild type bacteria bind to plant cells when inhibitors of bacterial protein synthesis are included in the incubation medium along with the plant cells(4). Attachment to carrot cells of bacteria containing pSA is prevented by the inclusion of tetracycline in the incubation medium. Thus bacterial protein synthesis appears to be required for the attachment of bacteria carrying pSA incubated in the presence of auxin.

4. THE NATURE OF THE DEFECT IN AGROBACTERIA CARRYING pSA

The avirulence of A. tumefaciens carrying pSA can be reversed by the introduction of auxin into the wound site along with the bacteria(1). When the amount of IAA released into the culture medium during the growth of strains 15955 and 15955(pSA) was compared, the strain carrying pSA produced only about one fourth as much IAA as the wild type strain(1). Thus the presence of pSA seems to inhibit the production of IAA by the bacterium. The mechanism of this inhibition remains unknown.

Binding of A. tumefaciens to plant cells appears to involve bacterial surface proteins and lipopolysaccharide. New et al(7) found that LPS from strain 15955 but not from 15955(pSA) inhibited tumor formation in wounded bean leaves. However, LPS from either strain 15955 or 15955(pSA) inhibited the binding of virulent strain A6 to carrot cells with approximately equal efficiency. The reason for this discrepancy between the results in wounded bean leaves and in suspension cultures is unknown.

One class of avirulent bacterial mutants which fails to bind to carrot cells appears to be lacking one or more surface polypeptides(4). When surface proteins extracted from Agrobacterium were examined using PAGE some bands were observed which were present in wild type bacteria grown with and without auxin, but which were present in strains carrying pSA only if the bacteria were grown with auxin. The possible role of these proteins in bacterial binding remains to be determined.

5. CONCLUSION

The effect of the plasmid pSA on the virulence and binding of Agrobacterium to plant cells appears to represent a requirement for bacterial auxin production for both of these processes. One of the interesting aspects of this auxin requirement is that the effect of the plant hormone appears to be on the bacteria rather than on the plant cells.

This research was supported by grant 85-CRCR-1-1902 from USDA.

REFERENCES

1. Chernin LS, Lobanok EV, Fomicheva VV, Kartel NA: Crown gall-supressive IncW R plasmids cause a decrease in auxin production by *Agrobacterium tumefaciens*. Mol Gen Genet 195:195-199, 1984.

2. Douglas CJ, Halperin W, Nester EW: *Agrobacterium tumefaciens* mutants affected in attachemnt to plant cells. J Bacteriol 152:1265-1271, 1982.

3. Farrand SK, Kado CI, Ireland CR: Supression of tumorigenicity by the IncW R plasmid pSA. Mol Gen Genet 181:44-51, 1981.

4. Matthysse AG: Mutants of *Agrobacterium tumefaciens* which fail to bind to plant host cells. Submitted for publication.

5. Matthysse AG, Holmes KV, Gurlitz RHG: Elaboration of cellulose fibrils by *Agrobacterium tumefaciens* during attachment to carrot cells. J Bacteriol 145:583-595, 1981.

6. Matthysse AG, Holmes KV, Gurlitz RHG: Binding of Agrobacterium tumefaciens to carrot protoplasts. Physiol Plant Pathol 20:27-33, 1982.

7. New PB, Scott JJ, Ireland CR, Farrand SK, Lippincott BB, Lippincott JA: Plasmid pSA causes loss of LPS-mediated adherance in *Agrobacterium*. J Gen Microbiol 129:3657-3660, 1983.

DUAL REGULATION OF VIRULENCE GENES OF AGROBACTERIUM PLASMID pTiC58

P. ROGOWSKY, T. J. CLOSE, AND C. I. KADO

Department of Plant Pathology, University of California, Davis, California 95616

1. INTRODUCTION

Agrobacterium tumefaciens plasmid pTiC58 contains a cluster of genes spanning approximately 30 kb that are essential for virulence (Vir). The Vir region of pTiC58 is composed of at least six complementation groups much like the Vir region of octopine Ti plasmids (Klee et al., 1983; Hille et al., 1984; Hooykaas et al., 1984, Lundquist et al, 1984). These groups are designated *virA*, *virB*, *virG*, *virC*, *virD*, and *virE*, and occur in the same order in octopine and nopaline Ti plasmids (Fig. 1). Each

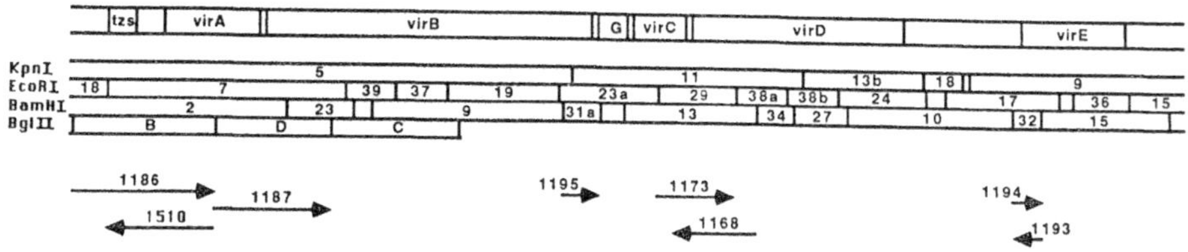

complementation group seems so far to be organized as an operon, and genes that have been compared at the level of DNA sequence or antibody cross reaction show considerable homologies (Hagiya, et al., 1985; Hirooka & Kado, 1986; Close et al., in press; Yanofsky et al., in press; Winans et al., in press; Hooykaas et al., unpublished). One function of the Vir genes is to process a section of the pTi plasmid that is transferred and integrated into plant genomes (T-DNA) (Alt-Moerbe et al., 1986).

Two types of regulation of the Vir genes have been previously reported. A negative control is represented by ros, the chromosomal mutant, which permits the expression of *virC* and *virD* from both octopine and nopaline type Ti plasmids (Close et al., 1985), regardless of growth conditions. Positive regulation, involving the *virA* and *virG* loci, was shown in octopine type Ti plasmid to be responsible for the induction of *virB*, *virG*, *virC*, *virD*, and *virE* by plant phenolic compounds such as acetosyringone (Stachel et al., 1985; Bolton et al., 1986). We have used fusions of Vir promoters to a promoterless *Vibrio fischeri* luciferase operon (lux) to measure the level of expression of Vir genes from the nopaline type Ti plasmid pTiC58 over the course of induction. Light production was monitored in free living *A. tumefaciens* cultures that were induced by acetosyringone, and in *A. tumefaciens* that were in contact with freshly sliced carrot tissue. Both systems gave the same overall conclusion that luminescence faithfully represents Vir gene expression.

Studies on T-DNA processing have so far utilized indirect assays that involve *E. coli* for the detection of intermediates (Koukolikova-Nicola et al., 1985; Alt-Moerbe et al., 1986), or were carried out directly in A. tumefaciens but relied on cocultivation and subsequent separation of bacteria and plant cells (Albright et al., in press). We were able to investigate the processing directly in *A. tumefaciens* using Southern blots by utilizing the same conditions of acetosyringone induction that were used in our luminescence monitoring assays. Analysis of processing in Vir- mutants enabled us to verify that the *virD* locus is involved in T-DNA processing. We have observed these intermediates in the absence of a plant inducer by using the *Agrobacterium* Ros mutant (Close et al., submitted).

2. PROCEDURE

Bacterial strains and media.

A. tumefaciens LBA4301 Rec-, pTi-, Rif^r (Klapwijk et al., 1979) and the Ros mutant were maintained in medium 523 at 30o C. Escherichia coli HB101 F-, pro, leu, thi, lacY, Str^r, r-m-, EndoI-, recA- (Boyer and Roulland-Dussoix, 1969) was grown in LB medium at 30o C. Acetosyringone inductions were carried out in Murashige-Skoog medium supplemented with 12.5 mM potassium phosphate buffer, pH 5.7.

Plasmids and Genetic Analysis.

The lux promoter proficient vector pUCD607 was described previously (Shaw and Kado, 1986). The lux promoter probe vector pUCD615 is described elsewhere (Hirooka et al., submitted; Rogowsky et al., in preparation).

Bioluminescence assay.

Light emitted by the induction of each lux-vir gene fusion was measured by an end-on photometer arranged in a light tight box. The emitted photons were translated into light units directly by a microprocessor. This "luminometer" was developed in collaboration with Beckman Instruments, Inc., Fullerton, California. Measurements were made under the same geometry using induced (by the addition of 100 uM acetosyringone) and uninduced cultures in 250 ml nepheloflasks. Light emmitted by bacteria on freshly cut carrot root slices was recorded photographically using high speed color film.

3. RESULTS

Organization of the Vir region

The Vir region of pTiC58 contains six operons currently designated *virA*, *virB*, *virG*, *virC*, *virD*, and *virE* (Fig. 1). A locus *tzs* involved in *transzeatin biosynthesis* (Barry et al., 1986) is located to the left of *virA*, and a recently identified locus *vraA* is adjacent to the *virE* operon (Hirooka et al., submitted). These latter loci are not absolutely required for tumorigenesis.

Regulation of the Vir region

All of the operons in the vir region were analyzed for their ability to respond to acetosyringone by measuring the amount of light produced by A. tumefaciens 9 hours after induction. Each Vir operon promoter was fused to a promoterless lux cassette on plasmid vector pUCD615 and introduced by conjugal mating using pRK2013 into A. tumefaciens LBA4301 containing pTiC58. As shown in Table 1, there are four levels of induction: 1) virB, virE, and tzs are induced to levels comparable to the level of a highly constitutive tet promoter in pUCD607; 2) virA and virD reach a level of expression that is about ten fold lower (note that virA is constitutively expressed at a low level); 3) virC reaches the lowest level of all inducible genes, although the fold increase from an extremely low basal level is considerable; and 4) virG and vraA promoters are not induced by acetosyringone, a pattern that is similar to the constitutive tet promoter in pUCD607 and the vector itself (pUCD615).

virA and virG are required for positive regulation of Vir

To determine if each Vir operon can be induced with acetosyringone independently of other Vir loci, the fusion plasmids were also introduced into LBA4301 without pTiC58. In none of these cases could induction be achieved with acetosyringone in the absence of pTiC58. We also found that pTiC58 containing inactivating mutations in either virG or virA prevented the induction of virB and virE with acetosyringone. Based on this study, virG and virA are both required for the regulation of these Vir operons. However, we have found that amplification of virG on a multicopy plasmid vector leads to elevated expression of virB and virE even in the absence of virA or acetosyringone.

virD operon is involved in T-DNA processing

Since the right border of the T-DNA has been implicated as part of the recognition region for processing, we have used a restriction fragment containing the right portion of the T-DNA, including the right border, of pTiC58 as a probe in Southern blots. DNA from wild-type A. tumefaciens and the Ros mutant, both containing pTiC58 under acetosyringone induced and uninduced conditions, revealed a small population of T-DNA intermediates represented by fragments derived from double strand cleavage at the right border among a larger population of the full size fragment containing the border. The amount of T-DNA intermediates was in the following order: wild-type $\ll$ Ros $<$ wild-type + acetosyringone $<$ Ros + acetosyringone, which suggested a role of virC or virD in T-DNA processing. To determine which Vir genes are involved in T-DNA processing, we assayed for the T-DNA intermediates in the presence of individual Tn5 insertional mutations in each Vir operon. We found that mutations in virB, virC and virE do not affect T-DNA intermediate formation in an induced Ros mutant. virA mutants gave low levels of intermediates formation similar to that of an intact pTiC58 in an uninduced Ros mutant. On the other hand, a mutation early in virD completely prevented the formation of T-DNA intermediates, whereas a mutation in a late virD gene allow normal processing. Thus, the more proximal genes in virD seem to be involved in this aspect of T-DNA processing, and the most distal virD genes are not.

TABLE 1. Induction of vir Genes with Acetosyringone

Plasmid in LBA4301(pTiC58)	Promoter fusion	Light/Cell* [quanta/min x 10-6] +AS	-AS	Fold Increase
pUCD1186	virA	1598	69	23
pUCD1187	virB	14838	13	1141
pUCD1168	virC	138	3.5	39
pUCD1173	virD	1629	23	71
pUCD1194	virE	13206	41	322
pUCD1195	virG	56	156	0.3
pUCD1510	tzs	24112	113	213
pUCD1193	vraA	15	52	0.2
pUCD607	tet	19969	27744	0.7
pUCD615	none**	16	17	0.9

* 9 h after induction with acetosyringone (AS).

** Basal level is due to a weak unidentified promoter activity in the vector part of the plasmid.

4. DISCUSSION

Positive regulation of genes in the pTiC58 Vir region involves the interaction of inducers, such as acetosyringone, with the products of *virA* and *virG*, as has been shown for octopine type Ti plasmids. We observed that increasing the amount of *virG* product is sufficient to cause a considerable increase in the basal expression of other Vir genes, but *virA* is necessary for a response to acetosyringone. Thus, we believe that *virG* protein may serve as the positive regulatory element for switching on *virB*, *virC*, *virD* and *virE* (Fig. 2), and *virA* may be responsible for

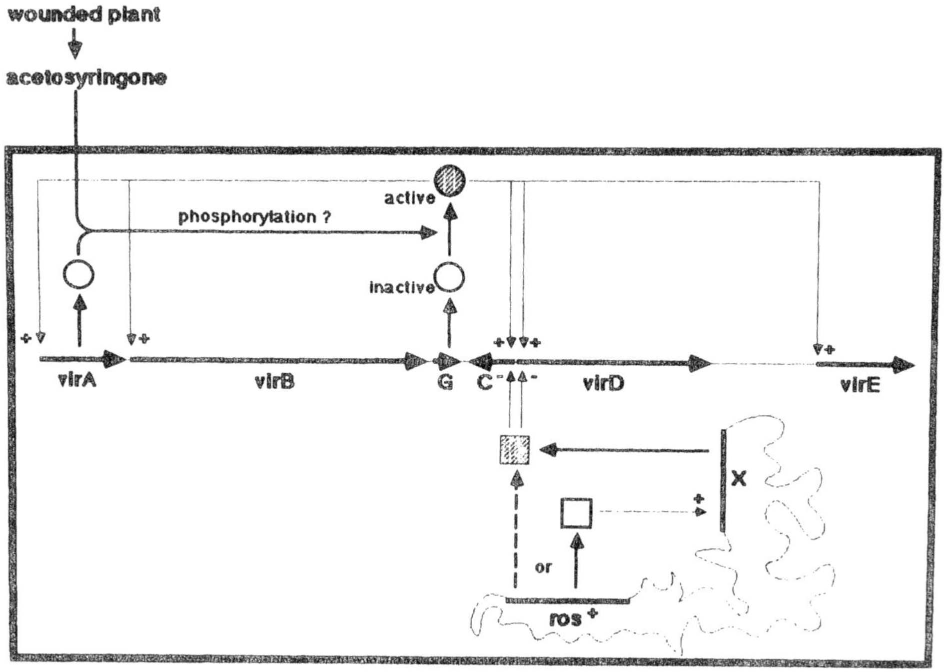

transducing the signal to *virG*. Another aspect of Vir gene regulation is represented by the *ros* chromosomal gene, which negatively regulates *virC* and *virD*. It is noteworthy that the same set of regulatory mechanisms are operating in octopine and nopaline type Ti plasmids.

Though there is striking similarity in the regulation of octopine and nopaline Ti plasmids, we have also discovered some differences. pTiC58 *virA* can be induced to a much larger extent than that reported for the homologous octopine Ti plasmid gene. In addition, the pTiC58 *virG* promoter is not induced. It seems possible, however, that transcription of *virG* increases by readthrough from the inducible *virB* promoter to achieve autoregulation of nopaline Ti plasmids as well.

Through the use of acetosyringone induced cultures, we have been able to identify the particular Vir locus involved in the generation of T-DNA intermediates. We observed the highest levels of T-DNA right border cleavage in the Ros mutant, a strain maximizes the level of virC and virD expression. Our results indicate double stranded cleavage of intermediates during T-DNA processing but do not rule out the formation of single-stranded T-DNA intermediates. The presence of single-stranded molecules is mechanistically attractive since classical bacterial conjugative functions seem to involve linear single-stranded DNA. Whatever the processing mechanism may be should prove interesting.

5. ACKNOWLEDGEMENTS

This work was supported by NIH research grant CA-11526 from the National Cancer Institute, DHHS.

6. REFERENCES

1. Alt-Moerbe, J., B. Rak, and J. Schroder. 1986. EMBO J.5: 1129-1135.
2. Barry, G. F., S. G. Rogers, M. B. Hein, J. G. Niedermeyer, N. L. Hoffmann, L. M. Blatt, R. T. Fraley, C. R. Sharp, and R. B. Horsch. Curr. Top. Plant Biochem. Physiol. 4: 101-110. (D. D. Randall, D. G. Blevins, R. L. Larson & T. Tagawa, eds.) University of Missouri, Columbia, MO.
3. Bolton, G. W., E. W. Nester, and M. P. Gordon. 1986. Science 232: 983-985.
4. Close, T. J., R. C. Tait, and C. I. Kado. 1985. J. Bacteriol. 164: 774-781.
5. Hagiya, M., T. J. Close, R. C. Tait, and C. I. Kado. 1985. Proc. Natl. Acad. Sci. USA 82: 2669-2673.
6. Hirooka, T., and C. I. Kado. 1986. J. Bacteriol. in press.
7. Hooykaas, P. J. J., M. Hofker, H. Den Dulk-Ras, and R. A. Schilperoort. 1984. Plasmid 11: 195-205.
8. Klee, H. J., F. F. White, V. N. Iyer, M. P. Gordon, and E.W. Nester. 1983. J. Bacteriol. 153: 878-883.
9. Koukolikova-Nicola, Z., R. D. Shillito, B. Hohn, K. Wang, M. Van Montagu and P. Zambryski. 1985. Nature 313:191-196.
10. Lundquist, R. C., T. J. Close, and C. I. Kado. 1984. Mol. Gen. Genet. 193: 1-7.
11. Shaw, J. J. and C. I. Kado. 1986. Bio/Technology 4: 560-564.
12. Stachel, S. E., E. W. Nester, and P. C. Zambryski. 1986. Proc. Nat. Acad. Sci. USA 83: 379-383.

OVERDRIVE, A T-DNA TRANSMISSION ENHANCER ON THE A. TUMEFACIENS TUMOR-INDUCING PLASMID

ERNEST G. PERALTA[1], RENATE HELLMISS, JOON M. JI, WENDY H. BERGER, AND WALT REAM

Institute for Molecular Biology, Dept. of Biology, Indiana University, Bloomington, IN 47405, USA; [1] Present Address: Genentech Inc, 460 Point San Bruno Blvd., South San Francisco, CA 94080

1. INTRODUCTION

Agrobacterium tumefaciens incites crown gall tumors on many dicotyledonous plants when viable bacteria infect wounded tissue (2, 22). Virulent strains contain a 190 kilobase (kb) tumor-inducing (Ti) plasmid that carries genes essential for tumorigenesis (8, 38). During tumorigenesis a specific segment of the Ti plasmid, the T-DNA, integrates into plant nuclear DNA (5, 19). Plant tumor cells express T-DNA genes responsible for tumorous growth, but T-DNA transmission does not require tumorigenesis or T-DNA encoded proteins (20, 29). Similar 23 bp direct repeats lie at both ends of four different T regions (1, 33, 40), and these repeats signal the T-DNA borders since T-DNA ends occur in or near these repeats in several different tumors (11, 32, 34, 41). Deletions that remove the T region right border severely attenuate virulence on most plants even though the tumor maintenance genes remain intact (11, 15, 24, 26-27, 31, 39). Such deletion mutants provide an assay for right border function: we can reintroduce T-DNA borders to the right of the T region and measure their ability to restore virulence. T-DNA transmission requires only a right border repeat in *cis* (26-27, 39), but Ti plasmid sequences that lie to the right of the right border repeat greatly stimulate its function (13, 25-27).

In this study we identified a specific sequence flanking a right border repeat that stimulated T-DNA transmission. This flanking sequence, which we call *overdrive*, lies to the right of the TL right border repeat in the octopine-type plasmid pTiA6NC. Related sequences occur to the right of three other right border repeats (1, 6, 33). We constructed a series of deletions that extend into *overdrive* and demonstrated that efficient T-DNA transmission required *overdrive*. To determine whether *overdrive* functions as a separate element, we synthesized this sequence and cloned it to the right of three different border repeats. Although right border repeats function best in one orientation (26-27, 39), *overdrive* functioned in either orientation. *overdrive* also retained its activity when relocated, in either orientation, to the left of the border repeat. *overdrive* function did not depend on its exact distance from the border repeat; *overdrive* retained its activity when moved either 10 bp closer to or 433 bp farther from the border repeat. Thus, *overdrive* formed a discrete element distinct from the border repeat. These results indicate *overdrive* may distinguish right and left T-DNA borders. The relationship of T-DNA right borders to *overdrive* resembles that of inverted repeats to recombinational enhancers in several site-specific inversion systems (12, 14, 16, 28).

2. MATERIALS AND METHODS

We used methods and strains described earlier (25-27).

3. RESULTS

3.1. Border Assay System

To identify sequences required for T-DNA transmission, we developed an assay system for border function (26). The octopine-type Ti plasmid pTiA6NC contains two adjacent but noncontiguous T regions designated TL and TR; TL, the left T-DNA, contains all the genes required for tumor maintenance (24, 37). A deletion mutant (pWR113 in WR3095) of pTiA6NC lacking the three border repeats that lie to the right of TL (without affecting the tumor maintenance genes) renders WR3095 essentially avirulent on all plants tested. A wide host range shuttle vector, pWR64, contains sequences from EcoRI fragment 1 (7) to the right of TL in pWR113 (26). To assay the border function of restriction fragments containing border repeats, we inserted them into the shuttle vector at the single HindIII site, introduced the border fragments into pWR113 by homologous recombination in A. tumefaciens (30), and tested their abililty to restore virulence. Since deletions that remove EcoRI fragment 1 do not affect virulence (24), our reintroduced constructs did not interfere with other sequences important for T-DNA transmission. The new location of the fragments did not adversely affect efficient right border function. The right border functioned poorly when we reduced the normal righthand flanking sequences to 4 bp (26). Thus, our previous work indicates that sequences normally lying to the right of the border repeat greatly stimulate its function.

3.2. Deletion Analysis of the T-DNA Transmission Enhancer

To identify the Ti plasmid sequences that stimulate border repeat activity, we constructed a series of deletions that extended into sequences present to the right of the TL right repeat (Fig. 1). Strains containing the TL right border repeat flanked on the right by at least 40 bp of Ti sequences exhibited full virulence (strains WR1101, WR1103). An otherwise identical strain (WR1102) that contained the TL right repeat flanked on the right by only 25 bp of Ti sequences exhibited greatly reduced virulence. Strains with the normal right border replaced by a synthetic right repeat from TL of pTiA6 (octopine-type) or pTiT37 (nopaline-type) or the TL left repeat (on a 767 bp HindIII-HincII fragment) showed similar weak virulence (26). Therefore, three different border repeats exhibited similar weak activity, and efficient T-DNA transmission required sequences located within 40 bp to the right of the TL right repeat; the sequences within 25 bp to the right of the repeat did not include all of the stimulatory sequences. A conserved sequence, 5′ TAAPuTPy-CTGTPuT-TGTTTGTTTG 3′, begins 17 bp to the right of the pTiA6NC TL right border repeat and 16 bp to the right of the TR right repeat (1). Sequences sharing 75 to 100 % homology with an 8 bp core sequence (5′ TGTTTGTT 3′; underlined above) lie to the right of T-DNA right border repeats in the nopaline-type Ti plasmid pTiT37 (6) and the A. rhizogenes Ri plasmid pRiA4 (33). The deletion which removed 15 bp (including the core sequence) from the right end of the conserved sequence also eliminated the stimulatory effect of flanking sequences on TL right border repeat function (Fig. 1).

3.3. Synthetic T-DNA Transmission Enhancer: overdrive

To determine whether the conserved flanking region identified by deletion analysis contained all the sequences required to fully stimulate T-DNA border repeat function, we synthesized an oligonucleotide comprising the conserved flanking sequence found to the right of the pTiA6NC TL right border repeat. We inserted this sequence in its normal orientation only 6 bp to the right of a synthetic pTiT37 right border repeat. The resulting strain exhibited much greater virulence than the parental strain which contained only the synthetic pTiT37 right border repeat. This synthetic oligonucleotide enhanced the activity of a heterologous border repeat when positioned 10 bp closer than normal to a right border repeat. Thus, this conserved region contained the sequences needed to stimulate border repeat function, and the sequence retained its function even when moved 10 bp closer than normal to the border repeat. We call this conserved flanking sequence overdrive (25).

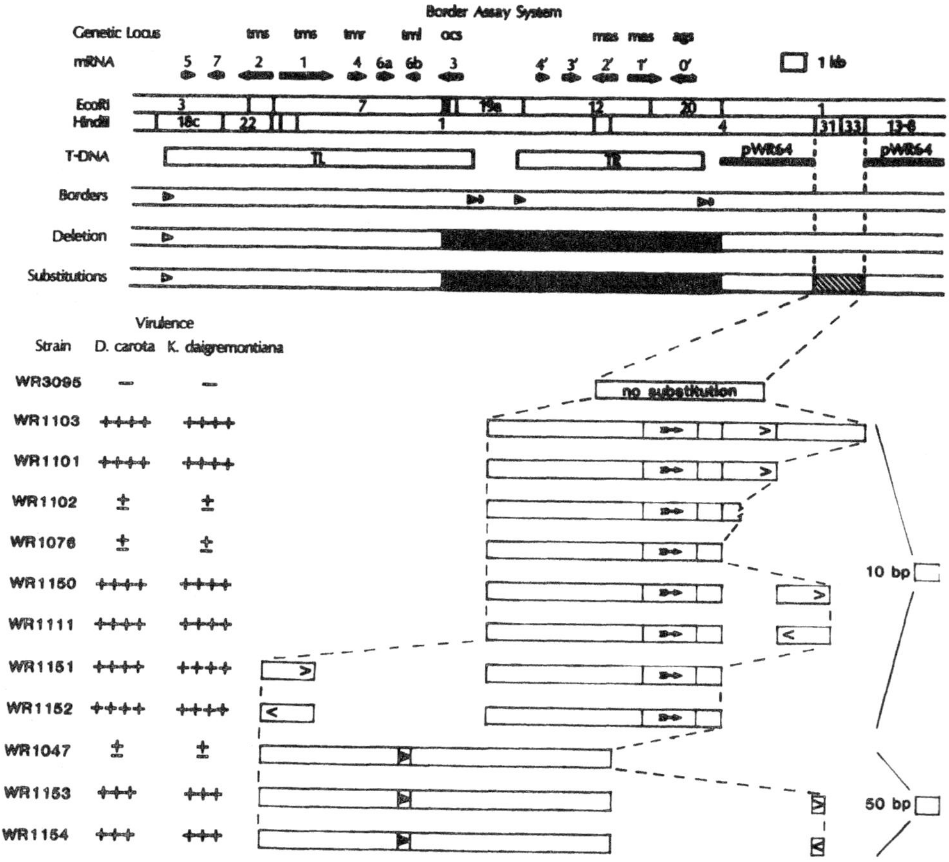

Figure 1. Border Assay System. The symbols indicate: ▶ or ▶ or ➤➤ = border repeat (▶ = TL left repeat and ➤➤ = TL right repeat in the expanded region);) or [>] = overdrive; [<] = inverted overdrive; ++++ = fully virulent; +++ = almost fully virulent; +/- = very weakly virulent; - = avirulent.

3.4. Orientation and Position Independence of overdrive

To determine whether its orientation affects the ability of overdrive to stimulate T-DNA transmission, we inserted the synthetic overdrive in each orientation 23 bp to the right of the pTiA6NC TL right border repeat (WR1111 & WR1150; Fig. 1). To determine whether the order of the border repeat and overdrive affects the efficiency of T-DNA transmission, we inserted the synthetic overdrive in each orientation 76 bp to the left of the pTiA6 TL right border repeat (WR1151 & WR1152; Fig. 1). All four strains exhibited much greater virulence than the parental strain (WR1076) which lacked overdrive. Thus, overdrive stimulated T-DNA transmission equally when placed, in either orientation, to the left or the right of the border repeat, and overdrive retained its function when positioned 60 bp farther than normal from the border repeat.

3.5. Effect of Distance on overdrive Activity

To test whether overdrive could act over a substantial distance to stimulate T-DNA transmission, we increased the distance between overdrive and a border repeat. In this instance, we used the border repeat which normally lies at the left end of TL in pTiA6 in place of the T-DNA right border repeat. This border repeat promoted T-DNA transmission inefficiently without overdrive (WR1047; see 3.2. & Fig. 1), but overdrive greatly stimulated T-DNA transmission when placed, in either orientation, 448 bp to the right of this border repeat. Thus, overdrive retained its function when moved 432 bp farther than normal from the border repeat, and overdrive stimulated T-DNA transmission from a properly-repositioned copy of a left border repeat.

4. DISCUSSION

Our experiments demonstrated that efficient T-DNA transmission requires two discrete sequences, the 23 bp T-DNA right border repeat and overdrive. Border repeats and overdrive apparently constitute separate elements that interact to promote efficient T-DNA transmission. These elements presumably do not form a single contiguous site because: (1) a border repeat can function inefficiently without overdrive, (2) the exact spacing and sequence between a border repeat and overdrive can vary without reducing border function, and (3) overdrive functions fully in either orientation whereas the right border repeat functions best in its wild-type orientation.

The specific sequence of a border repeat did not restrict it to function as either a "left" or "right" border repeat. The three border repeats tested (the pTiA6NC TL left and right border repeats and the pTiT37 right border repeat) all promoted T-DNA transmission at low efficiency when placed in the active orientation (26, 39) to the right of the T region in our assay system. Each of these border repeats promoted efficient T-DNA transmission when located near a synthetic overdrive sequence. Thus, a border repeat promoted T-DNA transmission as a function of its location and orientation with respect to the T-DNA; the flanking overdrive sequence strongly influenced the efficiency with which a repeat promoted T-DNA transmission.

Our 24 bp overdrive oligonucleotide contains all the sequences required for wild-type overdrive activity in our assay system, but full activity may require only a portion of this 24 bp sequence. We compared the sequences that flank the four known T-DNA right border repeats and found the 8 bp overdrive core sequence (5′ TGTTTGTT 3′) at slightly different locations to the right of each border repeat. Two mismatches occur within the core sequence that flanks the pTiT37 right border repeat, but the other three right borders (TL and TR of pTiA6NC and TL of pRiA4) each contain the exact core sequence. Comparisons between any two putative overdrive sequences reveal regions of homology longer than the 8 bp core sequence. For example, the pTiA6NC TL and TR right borders share homology in 19 of 24 positions in the overdrive region (5′ TAA-T--CTGT-T-TGTTTGTTTG 3′; core sequence underlined), the TL right borders of pTiA6NC and pRiA4 share homology in 15 of 17 positions in the overdrive region (5′ ATGTTTGTT--ATTGTT 3′), and the right borders of pTiT37 and TR in pTiA6NC share homology in 9 of 10 positions in the overdrive region (5′ ATTTGT-TGT 3′). However, the 8 bp core sequence constitutes the region most highly conserved in all four potential overdrive regions. overdrive activity may require only the 8 bp core sequence, or larger overdrives with somewhat different sequences may function in each plasmid due to variations in vir-encoded (virulence) proteins. The two pTiA6NC overdrives share the greatest homology possibly because both interact with identical vir-encoded proteins. Alternatively, some vir proteins may promote efficient T-DNA transmission without an overdrive.

Our results differ from those of Wang et al. (39): in the nopaline-type plasmid pTiC58, the nopaline-type right border repeat alone promotes efficient T-DNA transmission, but in the octopine-type Ti plasmid pTiA6NC, we observe only partial activity with either octopine or nopaline-type right border repeats alone (25-26). Jen and Chilton (13) identified a region to the right of the nopaline-type (pTiT37) right border repeat which exhibited overdrive activity when an octopine-type Ti plasmid supplied the vir functions. The region with overdrive activity includes the core sequence identified in our study, and the nopaline-type repeat exhibits only weak activity in their system. Thus, efficient T-DNA transmission may require overdrive when an octopine-type Ti plasmid provides the vir genes but not when a nopaline-type Ti plasmid supplies the vir functions. Further mutagenesis will define the sequences important for overdrive activity in pTiA6NC, and complementation tests with vir genes from pTiC58 (a nopaline-type Ti plasmid) will determine whether these vir genes can render the pTiA6NC right border repeat fully active without overdrive.

Our experiments, together with data from other studies, suggest a possible function for overdrive. Efficient T-DNA transmission requires only the right border repeat in its wild-type orientation (26, 39) and the overdrive sequence in cis (in pTiA6NC; 25-26). overdrive itself does not promote T-DNA transmission in the absence of a right border repeat (31), but overdrive greatly stimulates T-DNA transmission. Apparently, a directional T-DNA transmission process begins at the right border and moves leftward through the T region. Presumably, some of the Ti plasmid-carried vir genes (18), induced by exudates from wounded plant cells (23, 35-36), encode proteins which act at the right border repeat to initiate T-DNA transmission. Since both left and right border repeats (without overdrive) exhibit weak unidirectional right border activity in our assay, we propose that overdrive greatly enhances interaction between the right border and the appropriate vir proteins, perhaps through a mechanism similar to that envisioned for eukaryotic transcriptional enhancers (42). Therefore, most T-DNA transmission events will initiate at the right border and move leftward through the T-DNA.

Other site-specific recombination systems resemble the T-DNA border repeat-overdrive system. Several genetic recombination pathways, for example phage lambda integration (21) and Tn3 resolution (17), require specific flanking sequences in addition to repeat sequences directly involved in a crossover site. Site-specific inversions that mediate tail fiber switching in bacteriophages (e.g. Mu, P1, P7) and variation of Salmonella typhimurium flagellar antigens require inverted repeat sequences and specific flanking sequences in cis for efficient inversion (4, 12, 14, 16, 28). These flanking sequences, called recombinational enhancers, stimulate site-specific inversion independent of their position and orientation relative to the inverted repeats. We will continue to test whether the overdrive sequence can also enhance T-DNA transmission in a similar fashion.

5. ACKNOWLEDGEMENTS

We thank Drs. M.-D. Chilton, E. Bach, and A. Montoya for providing the pTiT37 right border repeat, Drs. M.- D. Chilton and G. Jen for communicating similar results prior to publication, T. Lynch for expert technical assistance, and Dr. T. Alton for comments on the manuscript. A grant from the National Science Foundation (PCM8316006) supported this research. E. Peralta received a Graduate and Professional Opportunity Program Fellowship (USDEG-83-453) and a Bayard Franklin Floyd Memorial Fellowship.

6. REFERENCES

1. Barker R, Idler K, Thompson D, Kemp J (1983). Plant Mol Biol 2:335.
2. Bevan MW, Chilton M-D (1982). Annu Rev Genet 16:357.
3. Caplan AB, Van Montagu M, Schell J (1985). J Bacteriol 161:655.
4. Craig NL (1985). Cell 4:649.
5. DeBeuckeleer M, Lemmers M, DeVos M, Willmitzer L, Van Montagu M, Schell J (1981). Mol Gen Genet 183:283.
6. Depicker A, Stachel S, Dhaese P, Zambryski P, Goodman H (1982). J Mol Appl Genet 1:561.
7. DeVos G, DeBeuckeleer M, Van Montagu M, Schell J (1981). Plasmid 6:249.
8. Garfinkel D, Simpson R, Ream L W, White F, Gordon M, Nester E (1981). Cell 27:143.
9. Hepburn A, White J (1985). Plant Mol Biol 5:3.
10. Holsters M, Silva B, Van Vliet F, Genetello C, DeBlock M, Dhaese P, Depicker A, Inze D, Engler G, Villarroel R, Van Montagu M, Schell J (1980). Plasmid 3:212.
11. Holsters M, Villarroel R, Gielen J, Seurinck J, DeGreve H, Van Montagu M, Schell J (1983). Mol Gen Genet 190:35.
12. Huber H, Iida S, Arber W, Bickle T (1985). Proc Natl Acad Sci USA 82:3776.
13. Jen G, Chilton, M-D (1986). in preparation.
14. Johnson RC, Simon MI (1985). Cell 4:781.
15. Joos H, Inze D, Caplan A, Sormann M, Van Montagu M, Schell J (1983). Cell 32:1057.
16. Kahmann R, Rudt F, Koch C, Mertens G (1985). Cell 41:771.
17. Kitts P, Symington L, Dyson P, Sherratt D (1983). EMBO J 2:1055.
18. Klee HJ, White F, Iyer V, Gordon M, Nester E (1983). J Bacteriol 153:878.
19. Lemmers M, DeBeuckeleer M, Holsters M, Zambryski P, Depicker A, Hernalsteens J, Van Montagu M, Schell J (1980). J Mol Biol 144:353.
20. Leemans J, Deblaere R, Willmitzer L, DeGreve H, Hernalsteens J, Van Montagu M, Schell J (1982). EMBO J 1:147.
21. Nash HA (1981). Annu Rev Genet 15:143. 31.
22. Nester E, Gordon M, Amasino R, Yanofsky M (1984). Annu Rev Plant Physiol 35:387.
23. Okker R, Spaink H, Hille J, van Brussel T, Lugtenberg B, Schilperoort RA (1984). Nature 312:564.
24. Ooms G, Hooykaas P, Van Veen R, Van Beelen P, Regensburg-Tunik T, Schilperoort RA (1982). Plasmid 7:15.
25. Peralta E, Hellmiss R, Ream W (1986). EMBO J in press.
26. Peralta EG, Ream LW (1985). Proc Natl Acad Sci USA 82:5112.
27. Peralta E, Ream LW (1985). In Szalay A, Legocki R (eds.): "Advances in the Molecular Genetics of the Bacteria-Plant Interaction," Ithaca: Cornell Univ. Publishers,p 124.
28. Plasterk R, van de Putte P (1985). EMBO J 4:237.
29. Ream LW, Gordon M, Nester E. (1983). Proc Natl Acad Sci USA 80:1660.

30. Ruvkun GB, Ausubel FM (1981). Nature 289:85.
31. Shaw CH, Watson M, Carter G, Shaw C (1984). Nucleic Acids Res 12:6031.
32. Simpson RB, O'Hara P, Kwok W, Montoya A, Lichtenstein C, Gordon M, Nester E (1982). Cell 29:1005.
33. Slightom JL, Durand-Tardif M, Jouanin L, Tepfer D (1986). J Biol Chem 261:108.
34. Slightom JL, Jouanin L, Leach F, Drong RF, Tepfer D (1985). EMBO J 4:3069.
35. Stachel S, An G, Flores C, Nester E (1985). EMBO J 4:891.
36. Stachel S, Messens E, Van Montagu M, Zambryski P (1985). Nature 318:624.
37. Thomashow MF, Nutter R, Montoya A, Gordon M, Nester E (1980). Cell 19:729.
38. Van Larebeke N, Engler G, Holsters M, Van den Elsacker S, Zaenen I, Schilperoort RA, Schell J (1974). Nature 252:169.
39. Wang K, Herrera-Estrella L, Van Montagu M, Zambryski P (1984). Cell 38:455.
40. Yadav N, Vanderleyden J, Bennet D, Barnes W, Chilton M-D (1982). Proc Natl Acad Sci USA 79:6322.
41. Zambryski P, Depicker A, Kruger K, Goodman H (1982). J Mol Appl Genet 1:361.
42. DeFranco D, Yamamoto K (1986). Mol Cell Biol 6:993.

PHYSICAL STRUCTURE AND GENETICS OF THE T-DNA IN PLANTS TRANSFORMED BY AGROBACTERIUM TUMEFACIENS

Albert Spielmann and Robert B. Simpson
Molecular Biology Group
ARCO Plant Cell Research Institute
6560 Trinity Court, Dublin, CA 94568

1. INTRODUCTION

Agrobacterium tumefaciens strains incite crown gall tumors when inoculated on many dicotyledonous plants (1,2). A portion of the endogenous Ti plasmid, called the T-DNA, is transferred and stably incorporated in the nuclear DNA of cells transformed during the infection process. Imperfect 25 basepair direct repeat sequences flank the T-DNA and are required for the transfer (7), whereas none of the genes present on the T-DNA are essential for the transfer. A region outside the T-DNA, called the vir region, is also required for the T-DNA transfer, although the functions of vir region genes are currently unknown. When the T-DNA and the vir region are present on the same plasmid, the term "unitary" vectors will be used in contrast to "binary" vectors, where the T-DNA and the vir region are on two separate plasmids.

During the past few years, this natural transformation system has been modified in order to make it suitable as a tool to introduce any desirable genes into plant cells without interfering with regeneration of normal plants. Analyses of the inserted T-DNA have shown that one to several copies of the T-DNA are stably integrated in the plant genome and maintained through meiosis, that the integration sites are random and that T-DNA copies are sometimes incomplete or rearranged.

2. RESULTS

To use this transformation system in genetic engineering, it is desirable to be able to predict the copy number and the number of locations in the plant genome (loci), and to reduce the number of aberrant copies. In attempt to identify factors responsible for the variation in these characteristics, we have compiled in Table 1 the T-DNA structures found in transgenic plants transformed either by a unitary (A) or binary (B) vector system. Since only the plants where the loci number was known are listed, this Table represents a sub-population of transformed plants described in the literature. The number of copies of the T-DNA per haploid genome was determined by Southern blot analysis while "loci" refers to the number of genetic loci determined by analysis of progeny of transgenic plants for T-DNA encoded traits. To be able to make a direct comparison between unitary and binary system (such as average copy number per plant, average loci number per plant and average copy number per locus), only the plants where the T-DNA copy number was accurately investigated were taken into consideration (* in Table 1).

The data summarized in table 1C show that the average copy number per locus is very similar for each system (1.3 and 1.5). However, the copy number per plant in the binary system is twice as much as in the unitary system (3.1 and 1.4, respectively). Furthermore, the loci number per plant in the binary system is also twice as much as in the unitary system (2.1 and 1.1, respectively). Figure 1 shows the distribution of loci number per plant, both for the unitary (shaded) and the binary system (white). The unitary system produced a majority of transformed

TABLE 1: T-DNA structures in transgenic plants where the loci number is known.

	Number of plants	Number of copies	Number of loci	References
A. UNITARY				
Nicotiana	1	1*	1	3,4,5
	1	2*	2	3
	2	nd	1	6
	3	1-5	1	7
	1	<10	2	7
	1	nd	1	8,9
Petunia	2	nd	1	10
	1	nd	1	11
	1	1*	1	12
	2	2*	1	12,13
	3	1*	1	13
	6	1*	1	14
	1	2*	2	14
Tomato	1	1*	1	10
	1	3*	1	15
	1	~5	2	15
	-	-	-	
Total	33		37	
Total (*)	21	30	23	

(*) Including only plants where the copy number is accurately known

	Number of plants	Number of copies	Number of loci	References
B. BINARY				
Nicotiana	1	2	1	10
	2	2	2	10
	1	4	2	10
	1	6	2	10
	1	4	4	10
	1	2	2	17
	-	-	-	
Total	7	20	13	

C. SUMMARY

	Plants	Copies	Loci	Copies/ plant	Copies/ locus	Loci/ plant
Unitary	21	30	23	1.4	1.3	1.1
Binary	7	20	13	3.1	1.5	2.1
Total	28	50	36	1.8	1.4	1.3

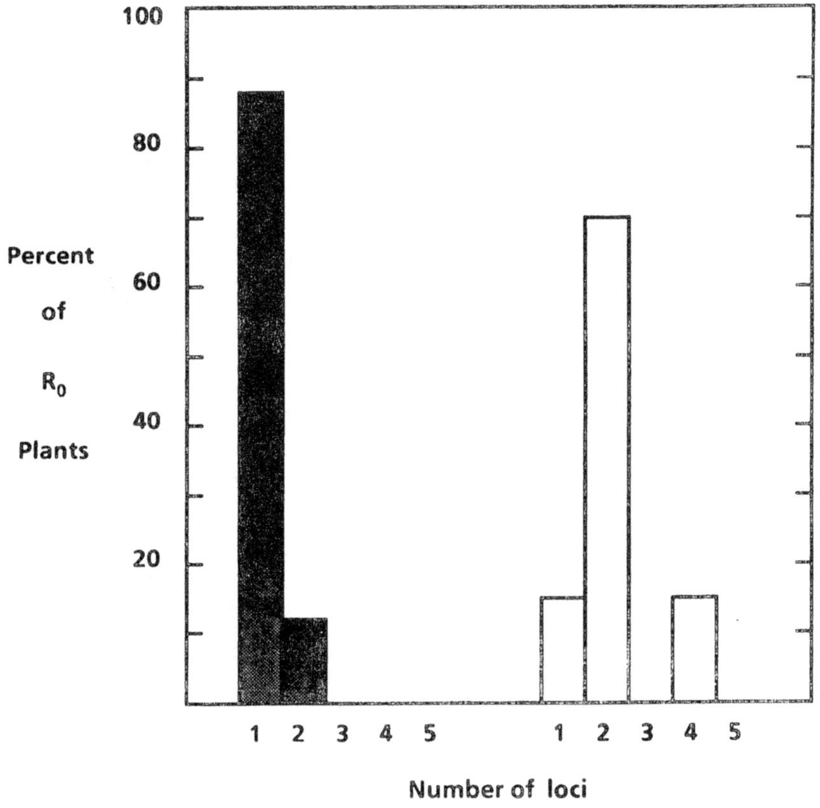

Figure 1. Distribution of T-DNA loci number per plant. Shaded box: "unitary" vector system; white box: binary vector system.

plants having only one locus, whereas plants arising from the binary system frequently showed two or more loci. Unpublished data from our laboratory extend this observation for the binary system (17 of 32 transformed tobacco plants have T-DNA copies at two or more loci).

An interesting common feature between the two systems is that, in several cases, DNA analysis demonstrates the presence of multiple copies (characterized by their different junction fragments with plant DNA), whereas genetics analysis shows a fewer number of loci. This observation suggests that two or more T-DNA copies are genetically linked (but not necessary joined together directly), or some copies may be phenotypically silent, due to mutation, methylation or chromosomal position. In fact, we have recently demonstrated (unpublished data) the presence of a silent T-DNA copy in transformant XSO-11 (16). Another common feature of both systems is that T-DNA copies are frequently aberrant, as indicated by Southern blot analysis (see references in Table 1).

3. DISCUSSION AND PROSPECTS

A priori, the striking difference, both in copy number per plant and in loci number per plant between unitary and binary systems could be attributed to several other factors, other than the type of vector used. Such factors include the bacterial strain, the Ti plasmid family, the plant species, the type of plant tissue infected, the infection procedure, and if a selectable marker was used, the level of antibiotic used to select the transformed plant cells.

Based on the limited amount of data, the only good correlation apparent was the type of vector used. One possibility is that the copy number of the plasmid in bacterial cell is the important difference between the two systems. Based on the parent plasmid RK2, the copy number in Agrobacteria for the binary vectors varies from 4 to 7, whereas for the unitary system the copy number is one. If productive attachment of the bacteria to the plant cell is the limiting step in T-DNA transfer (19), then the number of copies stably integrated in a plant cell could well depend on the number of T-DNA copies present in the few bacteria bound productively to that plant cell. We are currently testing the hypothesis that the copy number of the binary vector plasmid in the bacteria determines the copy number of the T-DNA in the plant cell.

The integration of foreign DNA at multiple loci in a single transgenic plant opens new approaches in plant genetic engineering. One obvious advantage of having multiple copies of a particular gene lies in the possibility of high levels of gene expression. This can be useful in a variety of experiments, such as altering plant phenotype by overproduction of a particular enzyme or compound, or by the use of antisense RNA to create "down" mutations in trans. Foreign DNA sequences integrated randomly at many loci in a particular genome may be useful as an insertion mutagen for gene isolation, much as some transposable elements have been used in animals and plants (21). Also, the segregation in the progeny of transgenic plants of two or more T-DNAs makes possible the co-transfer of selected and unselected T-DNA to unlinked loci in the same plant cell. Progeny could be identified that contain only the unselected T-DNA and thus are hosts for a second round of transformation using the same selectable marker. In a similar way, the combination of a binary vector system and multiple transformation may allow the transfer of a total genomic library made in Escherichia coli to plant cells via Agrobacterium tumefaciens, a first step in isolating genes through functional complementation. Finally, the inserted T-DNA can be used as a selectable tag for monitoring the inheritance of one or several chromosomes in progeny.

REFERENCES

1. Gheysen G, Dhaese P, Van Montagu M and Schell J: DNA flux across genetic barriers: The crown gall phenomenon. In: Hohn B and Dennis ES (ed): Genetic flux in plants. Springer-Verlag Wien New York, 1985.
2. Fraley RT, Rogers SG and Horsch RB: Genetic transformation in higher plants. Crit. Rev. Plant Sci. 4:1-46, 1986.
3. De Block M, Herrera-Estrella L, Van Montagu M, Schell J and Zambryski P: Expression of foreign genes in regenerated plants and in their progeny. EMBO J. 3:1681-1689, 1984.
4. Otten L, De Greve H, Hernalsteens JP, Van Montagu M, Schieder O, Straub J, and Schell J: Mendelian transmission of genes introduced into plants by the Ti plasmids of Agrobacterium tumefaciens. Mol. Gen. Genet. 183:209-213, 1981.
5. Memelink J, Wullems GJ and Schilperoort RA: Nopaline T-DNA is maintained during regeneration and generative propagation of transformed tobacco plants. Mol. Gen. Genet. 190:516-522, 1983.
6. Horsch RB, Fraley RT, Rogers SG, Sanders PR, Lloyd A and Hoffmann N: Inheritance of functional foreign genes in plants. Science 223:496-498, 1984.

7. Abel PP, Nelson RS, De B, Hoffmann N, Rogers SG, Fraley RT and Beachy RN: Delay of disease development in transgenic plants that express the tobacco mosaic virus coat protein gene. Science 232:738-743, 1986.
8. Wostemeyer A, Otten L and Schell JS: Sexual transmission of T-DNA in abnormal tobacco regenerants transformed by octopine and nopaline strains of Agrobacterium tumefaciens. Mol. Gen. Genet. 194:500-507, 1984.
9. Sengupta-Gopalan C, Reichert NA, Barker RF, Hall TC and Kemp JD: Developmentally regulated expression of the bean B-phaseolin gene in tobacco seed. Proc. Natl. Acad. Sci. USA 82:3320-3324, 1985.
10. Horsch RB, Fry JE, Hoffmann NL, Eichholtz D, Rogers SG and Fraley RT: A simple and general method for transferring genes into plants. Science 227:1229-1231, 1985.
11. Nagy F, Morelli G, Fraley RT, Rogers SG and Chua N-H: Photoregulated expression of a pea rbsS gene in leaves of transgenic plants. EMBO J. 1985
12. Fraley RT, Rogers SG, Horsch RB, Eichholtz DA, Flick JS, Fink CL, Hoffmann L and Sanders PR: The SEV system: A new disarmed Ti plasmid vector system for plant transformation. Biotechnology 3:629-635, 1985.
13. Wallroth M, Gerats AGM, Rogers SG, Fraley RT and Horsch RB: Chromosomal localization of foreign genes in Petunia hybrida. Mol. Gen. Genet. 202:6- 15, 1986.
14. Rogers SG, Bisaro DM, Horsch RB, Fraley RT, Hoffmann NL, Brand L, Elmer JS and Lloyd AM: Tomato golden mosaic virus A component DNA replicates autonomously in transgenic plants. Cell, 45:593-600, 1986.
15. McCormick S, Niedermeyer J, Fry J, Barnason A, Horsch R and Fraley R: Leaf disc transformation of cultivated tomato (L. esculentum) using Agrobacterium tumefaciens. Plant Cell Reports 5:81-84, 1986.
16. Spielmann A and Simpson RB: T-DNA structure in transgenic tobacco plants with multiple independent integration sites. Mol. Gen. Genet., in press.
17. de Framond AJ, Back EW, Chilton WS, Kayes L and Chilton MD: Two unlinked T-DNAs can transform the same tobacco plant cell and segregate in the F1 generation. Mol. Gen. Genet. 202:125-131, 1986.
18. Van Lijsebettens M, Inze D, Schell J and Van Montagu M: Transformed cell clones as a tool to study T-DNA integration mediated by Agrobacterium tumefaciens. J. Mol. Biol. 188:129-145, 1986.
19. Depicker A, Herman L, Jacobs A, Schell J and Van Montagu M: Frequencies of simultaneous transformation with different T-DNAs and their relevance to the Agrobacterium/plant cell interaction. Mol. Gen. Genet. 201:477-484, 1985.
20. Petit A, Berkaloff A and Tempe J: Multiple transformation of plant cells by Agrobacterium may be responsible for the complex organization of T-DNA in crown gall and hairy root. Mol. Gen. Genet. 202:388-393, 1986.
21. Fedoroff NV, Furtek DB and Nelson Jr OE: Cloning of the bronze locus in maize by a simple and generalizable procedure using transposable controlling element activator (Ac). Proc. Natl. Acad. Sci. USA 81:3825-3829, 1984.

MAMMALIAN METALLOTHIONEIN FUNCTIONS IN PLANTS

D.D. LEFEBVRE and J.-F. LALIBERTE
Genetic Engineering Section, Plant Research Centre, Agriculture Canada, Ottawa, Canada.

1. INTRODUCTION

Mammalian metallothioneins (MT) are cysteine-rich proteins which bind Cu, Zn, and their toxic analogue Cd (1). Their primary role in metabolism is uncertain but they are thought to be involved in copper and zinc homeostasis (2). MT is a small protein (M.W. 6,000) for which cDNA clones have been isolated (3-6).

To test for mammalian metallothionein function in plants we infected Brassica campestris c.v. "Just Right" with cauliflower mosaic virus harbouring in its genome a cDNA clone of Chinese hamster metallothionein-II (CHMT-II). In this report we show that transfection results in a high level of expression of CHMT-II and that the protein binds high levels of cadmium within the plant cells. This is probably instrumental in imparting Cd-resistance to these infected plants.

2. PROCEDURES

2.1 Vector construction and plant infection

The cloned cauliflower mosaic virus pCa-BB1 which has a unique XhoI site replacing the non-essential open reading frame II (ORF II) was used as an expression vector (7). In this unique site was placed a Chinese hamster metallothionein-II cDNA clone (6). This viral vector, coined pCa-MT-II was used to infect plants. Twenty ug of pCa-MT-II were digested with SalI to release the viral genome from its bacterial plasmid. This was mixed with 50 ul TE buffer containing 2 mg celite and rubbed on the leaves of Brassica campestris c.v. "Just Right". After 1-2 weeks first signs of infection were seen and systemic infection occured within 3-4 weeks.

2.2 DNA preparation

Viral DNA was purified according to the protocol of Gardner and Shepherd (8). The DNA was then restricted and run on a 1.2% agarose gel and analyzed by Southern blot (9).

2.3 Protein preparation and immunoblotting

0.5 g of fresh leaves were ground in 5 ml of ice cold 5mM Tris-HCl pH 8.6, 1mM phenylmethylsulphonylfluoride and were centrifuged at 105,000 Xg for 90 min. The soluble proteins were carboxymethylated with iodoacetamide and separated by SDS polyacrylamide electrophoresis. Western blot (Fig. 1) was performed by probing the nitrocellulose transferred proteins with rabbit antiserum against rat metallothionein-I which reacted equally well with MT-II. Detection was performed by using anti-rabbit antiserum coupled to biotin, then ^{35}S-streptavidin followed by autoradiography (10).

2.4 Cd-binding and Cd-resistance studies

Leaf disks were exposed to ^{115}Cd-labelled 1.0mM $CdCl_2$ in 1.0mM K_2SO_4, 1.0mM $CaCl_2$, 2% sucrose for 18 h at 21°C under constant light. The disks were freeze-dried, ground in KMT buffer (10mM KCl, 1mM $MgCl_2$, 10mM Tris-HCl pH7.6) containing 10% (w/v) Bio Beads SM-4 and centrifuged at 10,000 Xg for 15 min. The supernatant was analyzed for bound and total Cd, and for protein content. Chelex-100 was used to remove non-bound cadmium leaving the protein associated Cd in solution. Cd resistance was tested for by placing petioles of intact leaves into the same solution as used for labelling leaf disks except that the Cd concentration was increased to 10mM.

3. RESULTS AND DISCUSSION

The cauliflower mosaic virus was chosen as an expression vector for MT because of its high transcription and translation levels (7). Gene expression can be ascertained 4 weeks after the initial infection which is easily performed with cloned naked viral DNA. In addition infection and therefore foreign gene expression takes place immediately in mature leaf tissue unlike in other forms of plant transformation which must await plant regeneration from protoplasts or calli (11-13).

Infection of *B. campestris* by CaMV was not impaired by positioning CHMT-II into the ORFII region. The viral DNA of systemically infected leaves was purified and restriction enzyme analysis showed the insert to be stable for at least 3 cycles of reinfection.

CaMV infection results in a high level of transcription. CHMT-II was placed under the control of the very strong 35S promotor in the ORF-II position. If no regulation exists limiting the translation of CaMV transcripts, a high level of metallothionein should be produced. The synthesis of MT protein in *B. campestris* leaves was measured by using an antiserum made against rat metallothionein I (14). Figure 1 shows that a protein in infected leaves gave an antigenic response (lane C) which co-migrated with MT-II (upper of two bands) of Chinese hamster liver (lane A). This signal was not observed in the control leaves (lane B) nor in leaves infected with wild-type CaMV (not shown). This co-migration indicates that the protein made in the plant is not a fusion product with other viral proteins and if post-translational modifications took place,

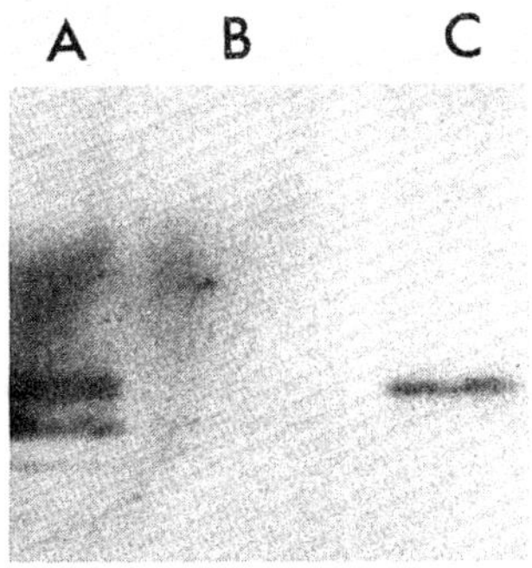

Figure 1. Metallothionein immunoblot of soluble proteins from A. Chinese hamster liver. B. *Brassica campestris* control leaf. C. *B. campestris* Ca-MT-II infected leaf.

they did not alter the electrophoretic mobility of the protein with respect to the liver isolate. Finally, since the immunoblot analysis was performed with plants showing signs of infection for over 2 weeks, it appears that the CHMT-II protein, although foreign to the plant cells, is not significantly degraded. The CHMT-II content of the leaf is estimated at approximately 0.5% of the leaf protein.

Although expression of CHMT-II was high it remained to be seen if protein function was normal. In vivo Cd-binding studies showed that control leaf tissue exposed to 1.0mM Cd contained 3 times the bound Cd of Ca-BB1 (wild-type virus) infected leaves but only two thirds that of the Ca-MT-II infected leaves. Ca-MT-II leaves contained no free Cd whereas control and Ca-BB1 leaves had 44.0 and 31.0 nmol free Cd per mg leaf protein respectively. It appears that normal viral infection slows Cd absorption whereas CHMT-II produced in Ca-MT-II leaves acts as a cytoplasmic sink overcoming the virally associated reduction in Cd-uptake.

Resistance to Cd was assessed by placing intact leaf petioles with their xylem streams carefully maintained into 10mM $CdCl_2$ solutions. Preliminary experiments of this nature demonstrated that acute exposures to high Cd levels were necessary since excised leaves remained alive for only 5 to 7 days. In 10mM Cd control leaves wilted and died within 24 h whereas both Ca-BB1 and Ca-MT-II infected leaves remained turgid and alive for 3 days. The viral background therefore complicates Cd-tolerance analysis probably because absorption is retarded. It is, however, apparent that even where cytoplasmic content of the Ca-MT-II leaves far surpassed that of the control leaf (1.4 times) there was no serious effect on the Ca-MT-II leaves while the control leaves quickly scenesced.

In summary mammalian MT functions normally in the plant cytoplasm. Although the magnitude of Cd absorption is affected by viral infection, MT does seem to increase tolerance to this toxic element.

REFERENCES

1. Kagi JHR and Nordberg M: Metallothionein. Basel: Brikhaser, 1979.
2. Brady FO: Trends Biochem Sci 7: 143-145, 1982.
3. Durnam DM, Perrin F, Gannon F and Palmiter RD: Proc Natl Acad Sci USA 77(11): 6511-6515, 1980.
4. Karin M and Richards RI: Nature 299: 797-802, 1982.
5. Andersen RD, Birren BW, Ganz T, Piletz JE and Herschman HR: DNA 2(1): 15-21, 1983.
6. Griffith BB, Walters RA, Enger MD, Hildebrand CE, and Griffith JK: Nucleic Acids Res. 11(3): 901-910, 1983.
7. Brisson N, Paskowsky J, Penswick JR, Gronenborn B, Potrykus I and Hohn T: Nature 310: 511-514, 1984.
8. Gardner RC and Shepherd RJ: Virology 106: 159-161, 1980.
9. Maniatis C, Fritsch EF and Sambrook J: Molecular Cloning: A Laboratory Manual. Cold Spring Harbor Lab, Cold Spring Harbor, New York, 1982.
10. Brower MS, Brakel CL and Garry K: Anal Biochem 147: 382-386, 1985.
11. Schell J and Van Montagu M: Bio/Technology 1(2): 175-180, 1983.
12. Potrykus I, Shillito RD, Saul MW and Paszkowski J: Plant Molec Biol Rep 3(3): 117-128, 1985.
13. Reich TJ, Iyer VN, Scobie B, Miki B: Can J Bot 64: in press.
14. Vander Mallie RJ and Garvey JS: J Biol Chem 254 (17): 8416-8421, 1979.

Tumorigenesis and root nodulation by *Agrobacterium tumefaciens* carrying Rhizobium sym plasmids.

Pauline A. Donaldson and V.N. Iyer, Department of Biology and Institute of Biochemistry, Carleton University, Ottawa, Ontario, Canada K1S 5B6

Introduction

Agrobacterium and *Rhizobium* are closely related bacterial genera in the family Rhizobiceae. *Agrobacterium* and *Rhizobium* genes contributing to the induction of, respectively, plant tumors and root-nodules, are encoded on large plasmids. *Agrobacterium tumefaciens* is well known for its ability to cause tumorous growth of plant cells by transferring a segment of its own DNA (T-DNA) into the nuclear genome of plants. The mechanism of T-DNA transfer and integration is still unknown. Ti plasmid genes (other than T-DNA) required for these events have been identified within the virulence (*vir*) region, which encodes at least 6 complementation groups (1). In addition, *A. tumefaciens* chromosomal virulence genes are required. The Ti plasmid *vir* genes act *in trans* with respect to T-DNA (2) and many are conserved among various types of Ti plasmids: both those conferring a wide and a narrow host range for tumor formation. In addition, DNA homology with the *vir* region (but not necessarily functional homology) has been demonstrated for some sym plasmids of *Rhizobium* (3). The implication of this DNA homology remains unclear. *Rhizobium* genes determining root-nodule development (*ndv*) have both structural and functional homology with chromosomal *vir* genes (*chv A* and *chv B*) of *Agrobacterium*. We are interested in possible similarities between mechanisms of infection by *Agrobacterium* and *Rhizobium*. In *Rhizobium*, large plasmids (pSym) encode genes which contribute to root-nodule formation. Typically, *Rhizobium* penetrate root hairs via an invagination of the plant cell wall and are released into the cytoplasm of newly divided cortical cells, where they participate in nitrogen fixation. The mechanism by which *Rhizobium* stimulates cortical cell division is unknown. In spite of the reported DNA-DNA homology between parts of *vir* and parts of some sym plasmids, Hoekema *et al.* (4) have shown that in *Rhizobium*, sym plasmids do not encode a complete system of virulence. The experiments described here were designed to determine if *Rhizobium* sym plasmids encode functional counterparts to any Ti plasmid *vir* genes. We introduced sym plasmids into *A. tumefaciens* with Ti plasmid Vir- mutations. We then tested the constructed strains for restoration of wild-type tumor-forming ability. The tumor-forming and nodulating ability of control *A. tumefaciens* strains carrying both a wild type Ti plasmid and a sym plasmid were also tested to determine if the presence of the two plasmids in one bacterium caused inhibition of either nodulation or tumorigenesis.

Procedures

The bacterial strains used in this study are described in Table 1. The self-transmissible and transposon Tn5-labelled plasmids ptr5a and pIJ1016 were conjugatively transferred to the *A. tumefaciens* strains by bacterial filter matings. Following overnight growth on non-selective medium, transconjugants were selected on medium supplemented with nalidixic acid and kanamycin. *R. leguminosarum* PD102 transferred the non-self-transmissible *R. phaseoli* *sym* plasmid pPD102 (=p3622::Tn5.mob) to the *A. tumefaciens* strains during triparental matings in which the self-transmissable plasmid pRK2013 was used as a helper. *R. leguminosarum* PD102 was isolated following mobilisation of pPD102, from pools of *R. phaseoli* PD101::Tn5.mob derivatives to *R. leguminosarum* B151. *R. phaseoli* PD101::Tn5.mob derivatives were constructed using the suicide plasmid vehicle pGS39 (described by G. Selvaraj, Ph.D. thesis, Carleton University, 1983). The plasmid content of the *Agrobacterium* x *Rhizobium* transconjugants was examined by a modified Eckhardt-type gel electrophoresis procedure. Also, transconjugants from matings involving the vir::Tn3HoHo1 mutants were screened for resistance to Cb, the Ti plasmid marker, to determine if the Ti plasmid was still present. The nodulating ability of the

TABLE 1: *Agrobacterium* and *Rhizobium* strains

Strain	Relevant Characteristics	Reference
A. tumefaciens		
A348	Nal resistant, carries w.t. pTiA6NC	
A136	as above, no Ti plasmid	
Sm240	same as A348 with pTiA6NC virA::Tn3HoHo1	
Sm243	pTiA6NC *virB*::Tn3HoHo1	(6)
Sm365	pTiA6NC *virC*::Tn3HoHo1	(6)
Sm311	pTiA6NC *virD*::Tn3HoHo1	(6)
Sm355	pTiA6NC *virD*::Tn3HoHo1	(6)
Sm341	pTiA6NC *virE*::Tn3HoHo1	(6)
Sm361	pTiA6NC *virE*::Tn3HoHo1	(6)
R. trifolii		
LPR5035	Rif resistant, carries ptr5a (pSym::Tn5)	(7)
R. phaseoli		
RCR3622	natural isolate, p3622b = pSym	(3)
PD101	as above but Sp[1] resistant	this study
B316	*R. phaseoli* 8401 with pIJ1016[2]	(8)
R. leguminosarum		
B151	pSym- derivative of 128C53	(9)
PD102	B151 with pPD102 (=p3622b::Tn5.mob)	this study

Nal=nalidixic acid, Rif=rifampicin, Sp=spectinomycin, w.t. wild-type

1 an *R. phaseoli* strain cured of its indigenous Sym plasmid

2 pIJ1016 = pRL5JI::Tn5, an *R. leguminosarum* Sym plasmid; *mob* = *oriT* of RK2

bacterial strains was tested on sterile seedlings of Phaseolus vulgaris L. cv. Kentucky Wonder (bean), Pisum sativum L. cv. Novella (pea), and Trifoliium repens L. cv. White Dutch (white clover). For tumorigenesis assays, Kalanchoe daigremontiana leaves, Lycopersicum esculentum (tomato), and Nicotiana tabacum (tobacco) stems (all of greenhouse grown plants) were wounded with sterile toothpicks and inoculated with liquid suspensions of bacteria. Octopine synthase activity in some tumor tissue was examined using a procedure recently described by Holbrook, Haffner, and Miki (5).

Results

The constructed strains stably maintained both the Ti and the introduced Sym plasmid. Transconjugants derived from the vir::Tn3HoHo1 mutants all maintained resistance to carbenicillin, and those derived from the wild-type A. tumefaciens A348 maintained the ability to produce tumors on at least some host plants (data included in Table 3). Results of Eckhardt-type gel analysis of these A. tumefaciens transconjugants also indicated that plasmids corresponding in size to the respective sym plasmid and the Ti plasmid were present.

In the A. tumefaciens strains, the Sym plasmids conferred the ability to form ineffective nodules and root deformations on the respective legume host. This provided additional evidence that a Sym plasmid has been acquired by the strains. On 25-50% of pea plants inoculated (10 plants were tested per strain), A. tumefaciens carrying pIJ1016 formed small white nodules and root deformations such as swellings, thick short roots, or wavy roots. A. tumefaciens strains carrying pPD102 formed a small number of ineffective white nodules on a portion (20-50%, depending on the strain inoculated) of inoculated bean plants. The A. tumefaciens carrying ptr5a were able to induce root deformations and nodules on roots of clover. In no case was effective nodulation conferred by ptr5a in A. tumefaciens. Only root swellings or small, rounded, black (or occasionally white) nodules were observed. Results of nodulation assays with these strains suggest that the nodulating ability of A. tumefaciens PD61 (the pTi- strain carrying ptr5a) is enhanced in comparison with that of A. tumefaciens PD60 (A348 carrying ptr5a). Thus, future analysis of nodulation phenotypes conferred by Sym plasmids (or by cloned nodulation genes from Rhizobium) in A. tumefaciens must consider the effect of the indigenous Ti plasmid on nodulating ability. Table 2 summarizes the nodulating and root-deforming abilities of A. tumefaciens PD60 and PD61. The majority of the deformations induced by A. tumefaciens PD60 were small swellings or root thickenings which began to appear about 10 days later than those induced by A. tumefaciens PD41 (the latter were observed as early as 7 days after inoculation). A higher frequency of nodulation occurred following inoculation of PD61 than following inoculation of PD60, and the number of nodules formed on each plant was significantly higher for PD61 than for PD60 inoculated plants.

TABLE 2: Clover-nodulating ability of A. tumefaciens/ptr5a

Strain	# of plants[1] with root deformations[2]	nodules	Total # of nodules (av./plant)
PD60 (A348/ptr5a)	11 (31%)	3 (8%)	3 (1)
PD61 (A136/ptr5a)	21 (60%)	11 (31%)	79 (7)

[1] total # of plants inoculated per strain = 35
[2] includes root-swellings and nodules

The presence of Sym plasmids did not restore wild-type tumor-forming ability to any of the avirulent (Vir A-,B-, and D-) or altered-host-range mutants (Vir E- and Vir C-). A. tumefaciens A136 (pTi-), Sm 243 (Vir B-), Sm 311 (Vir D-), and Sm 355 (Vir D-) were all avirulent on each host plant tested. The introduction of either pPD102, ptr5a, or pIJ1016 into A136, Sm 243, Sm 311, or Sm 355 did not alter the virulence phenotype of these strains. Neither pPD102 nor ptr5a increased the virulence of the Vir A- mutant on any of the plant hosts. We were unable to construct a Vir A- strain carrying the sym plasmid pIJ1016. Table 3 summarizes results of virulence assays for A. tumefaciens A348, Sm 341, Sm 361 (Vir E-) and their corresponding pSym+ derivatives. The virulence of the Vir C- altered-host-range mutant was unchanged except that tumor-size on both K. daigremontiana and tomato was reduced in the presence of ptr5a. The Vir E- mutants formed small tumors on tomato but were avirulent on carrot, tobacco, and Kalanchoe. However, PD46 (Sm 341 carrying pPD102) formed small, slow-growing tumors at approximately 30% of K. daigremontiana leaf wounds; these tumors formed about 1.5 months after inoculation.

TABLE 3: Tumor-formation by A. tumefaciens A348 and Vir E- mutants in the presence and absence of pSym

Strain	plasmids present	Tumors[1] formed on Kalanchöe	tomato
A348	pTiA6NC w.t.	52/52	10/10
PD40	A348/pPD102	20/20	10/10
PD50	A348/pIJ1016	9/10M	10/10S
PD60	A348/ptr5a	10/10M	10/10M
Sm341	pTiA6NC virE::Tn3HoHol	0/42	10/10S
PD46	Sm341/pPD102	11/36S	10/10S
PD56	Sm341/pIJ1016	0/10	5/10S
PD66	Sm341/ptr5a	0/10	8/10S
Sm361	pTiA6NC virE::Tn3HoHol	0/34	10/10S
PD48	Sm361/pPD102	3/34	10/10S
PD58	Sm361/pIJ1016	0/10	4/10S
PD68	Sm361/ptr5a	0/10	8/10S

[1] shown as the fraction of inoculated sites that gave a response
M= tumors smaller than wild-type, S= very weak response relative to wild-type

Octopine synthase activity was not detected in these tumors, suggesting that T-DNA transfer may not have occurred. Octopine synthase activity was detected in tumors formed on this host by A. tumefaciens A348 or PD40 (A348 carrying pPD102). Co-inoculation of equal quantities of A. tumefaciens Sm341 and R. leguminosarum PD102 on leaves of K. daigremontiana also led to the formation of small, slow-growing tumors, at 5/20 inoculation sites even though, alone, neither strain could induce tumors on Kalanchoe. On Kalanchoe and tomato, A. tumefaciens A348 carrying either ptr5a or pIJ1016 formed considerably smaller tumors than did the parental A. tumefaciens strain. Virulence on tobacco was also significantly decreased. Depending on the age of tobacco inoculated the strain carrying pIJ1016 was either avirulent or attenuated in its virulence phenotype.

Discussion

Our findings suggest that functional counterparts to vir are not expressed by Sym plasmids in A. tumefaciens. 'Vir-like' genes may be present, but regulated in a different way - perhaps by plant factors, as in the case of some Ti plasmid vir genes and some nodulation genes. It is not clear whether sym plasmid sequences showing DNA homology with vir encode essential genes in Rhizobium, and so the implications of the reported DNA-DNA homology remain uncertain.

Small, slow-growing tumors were formed by A. tumefaciens PD45, and by simultaneous inoculation of A. tumefaciens Sm341 and R. leguminosarum PD102 (carrying the R. phaseoli pSym pPD102. This suggests that the effect of pPD102 may be due to a diffusible factor or to a factor acting directly at the wound site. Similarly, inter-strain complementation of mutated vir E loci by the wild-type locus has been reported, but this complementation requires the presence of all the other vir loci in the complementing bacterium, and it leads to complete restoration of virulence (10). The cellular proliferations we observed after inoculation of Kalanchoe with PD46 may have occurred in the absence of T-DNA transfer. In the absence of T-DNA transfer, 'artificially induced proliferations' of cells on potato discs are only formed by A. tumefaciens vir- mutants (including a Vir E- mutant) when an exogenous source of the auxin NAA is present (11). Possibly, the presence of the sym plasmid mimics the effect of the exogenous application of NAA, thereby allowing the Vir E- /pSym strain to induce the small calli we observed. There are reports that sym plasmids, at least in Rhizobium, contribute partially to the production of auxin by the bacteria (12).

We observed attenuation of wild-type Agrobacterium A348 virulence in the presence of ptr5a or pIJ1016. Hooykaas et al. (13) have reported that Ti plasmids confer only an attenuated tumor-forming phenotype to R. leguminosarum and R. phaseoli. In those studies, the Rhizobium strains carried their own indigenous sym plasmid. Our observations raise the possibility that the presence of the sym plasmid, as opposed to the heterologous chromosomal background, is responsible for the attenuated tumor phenotype. Thus, the Agrobacterium and the Rhizobium chromosomal backgrounds (with the possible exception of R. meliloti) may be equally suitable for Ti plasmid directed tumor formation.

Acknowledgements

We are grateful to N. Brewin, P. Hooykaas, E. Nester, S. Stachel, G. Selvaraj and C. Wijfellman for some of the strains and plasmids. The study was supported by research grants from the Natural Sciences and Engineering Council of Canada (NSERC). PD was the recipient of an NSERC Graduate Student Scholarship.

References

1 Stachel, S.E., E.W. Nester, and P.C. Zambryski (1986) Proc. Natl. Acad. Sci. (USA) 83: 379-383.
2 Hoekema, A., P.R. Hirsch, P.J.J. Hooykaas, R.A. Schilperoort (1983) Nature (London) 303: 179-180.
3 Prakash, R.K. and R.A. Schilperoort (1982) J. Bacteriol. 149: 1129-1134.
4 Hoekema, A., P.J.J. Hooykaas and R.A. Schilperoort (1984) J. Bacteriol. 158: 383-385.
5 Holbrook, L.A., M. Haffner, B.L.A. Miki (1986) Biochem. Cell. Biology 64: 126-132.
6 Horsch, R.B., H.J. Klee, S. Stachel, S.C. Winans, E.W. Nester, S.G. Rogers, and R.T. Fraley (1986) Proc. Natl. Acad. Sci. (USA) 83: 2571-2575.
7 Hooykaas, P.J.J., A.A.N. VanBrussel, H. den Dulk-Ras, G.M.S. Van Slogteren, R.A. Schilperoort (1981) Nature (London) 291; 351-353.
8 Gortz, R., I.J. Evans, J.A. Downie and A.W.B. Johnston (1985) Mol. Gen. Genet. 201: 296-300.
9 Brewin, N.J., E.A. Wood, A.W.B. Johnston, N.J. Dibb and G. Hombrecher (1982) J. Gen. Microbiol. 128: 1817-1827.
10 Otten, L., H. De Greve, J. Leemans, R. Hain, P. Hooykaas, J. Schell (1984) Mol. Gen. Genet. 195: 159-163.
11 Joos, H., B. Timmerman, M. Van Montagu, and J. Schell (1983) EMBO J. 2: 2151-2160.
12 Wang, T.L., E.A. Wood and N.J. Brewin (1982) Planta 155: 345-349.
13 Hooykaas, P.J.J., P.M. Klapwijk, M.P. Nuti, R.A. Schilperoort, and A. Rorsch (1977) J. Gen Microbiol. 98: 477-484.

Section II

MOLECULAR GENETICS OF PHYTOPATHOGENIC BACTERIA AND FUNGI

CUTINASE AND PECTINASE IN HOST-PATHOGEN AND PLANT-BACTERIAL INTERACTION

P. E. KOLATTUKUDY[1], JOSEPH SEBASTIAN, WILLIAM F. ETTINGER, and MARK S. CRAWFORD.

1. INTRODUCTION

The aerial parts of plants are covered by the cuticle which forms the boundary layer at which microbes come into contact with the plant. The cuticle is composed of an insoluble structural polymer, cutin, which is embedded in a complex mixture of soluble lipids collectively called wax (1,2). Cutin is composed of interesterified hydroxy and hydroxyepoxy fatty acids primarily derived from palmitic, oleic, and linoleic acids (Fig. 1). The component fatty acids are held together by ester bonds between the primary as well as secondary hydroxy groups and the carboxyls. The cutin polymer constitutes the major physical barrier to penetration of fungi into the aerial parts of plants (3). The cutin barrier is attached to a pectinaceous polymeric layer which may also serve as a barrier to penetration by pathogens. In this paper, we will briefly review progress recently made in our understanding of the role of fungal cutinases and pectinases in the interaction between pathogenic fungi and their host plants. We also describe a recent discovery that a phyllospheric bacterium which cohabits with an apparently nitrogen-fixing bacterium generates an extracellular cutinase and thus can provide a carbon source while receiving fixed N_2 from the cohabiting partner.

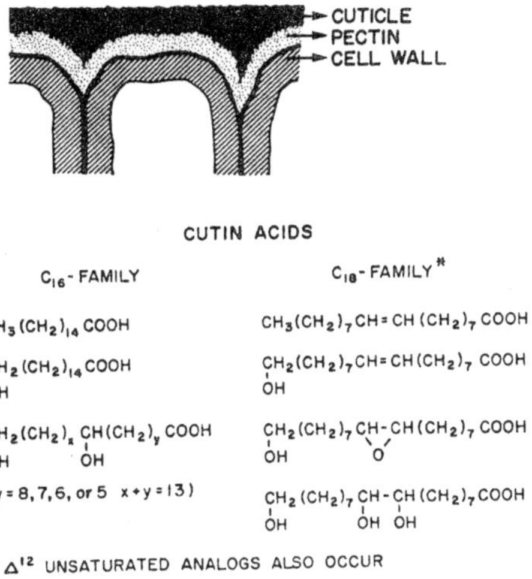

FIGURE 1. Top: schematic representation of the cuticle. Bottom: structure of the major monomers of cutin.

---

[1]Address all correspondence to P. E. Kolattukudy, Biotechnology Center, Rightmire Hall, Ohio State University, Columbus, Ohio 43210. This work was supported by NSF Grant DMB 8306835.

2. FUNGAL CUTINASE

2.1 Properties

Extracellular cutinases produced by a number of fungi have been purified and characterized (2,4). All of these fungal enzymes show similar properties. They are 25 kDa monomeric proteins containing a few percent O-glycosydically attached carbohydrates. They have similar amino acid compositions. They all have one or two histidines, one methionine, one tryptophane, and two disulphide bridges with no free sulfhydryl groups. However, immunological cross reactivity is limited to a few closely related species within the same genus.

The catalytic properties of cutinases produced by various fungi are also similar. The cutinases show specificity for the primary alcohol esters; secondary alcohol esters are hydrolyzed at much lower rates. The fungal enzymes all have a high (9 or above) pH optimum. Catalysis by fungal cutinases involves the classical catalytic triad of serine, histidine and a carboxyl group. Thus the enzymes are highly sensitive to inhibition by active serine-directed reagents such as diisopropylfluorophosphate and other organic phosphates.

2.2 Role in Pathogenicity

The role of cutinase production in cuticular penetration by a pathogen was studied first by electron microscopy of sections of pea stem inoculated with *Fusarium solani* f. sp. *pisi* spores and treated with ferritin conjugated anticutinase IgG. In these studies ferritin was associated with the cuticle only in the region where the germinating spore penetrated the cuticle (2,3). Similar results were obtained with *Colletotrichum gloeosporioides* on papaya fruit (M. B. Dickman, S. Patil, C. Soliday, P. E. Kolattukudy, unpublished). Thus, germinating spores penetrating the cuticle were found to secrete cutinase.

To see whether cutinase was necessary for penetration and thus for infection, the effect of specific inhibitors of cutinase on the penetration process was determined. Cutinase inhibitors were included in spore suspension drops which were then placed on appropriate host plants. To make sure that the test compound prevented infection solely by preventing penetration rather than by being toxic to the fungus, controls were run in which the cuticle-wall barrier under the spore suspension was breached with a pin. Inclusion of anticutinase IgG or organophosphates in spore suspensions of *F. solani* f. sp. *pisi* prevented lesion formation on pea stems with intact cuticles, but lesion formation was normal on pea stems in which the cuticle-wall barrier was mechanically breached (2). Similar results were obtained using *C. gloeosporioides* on papaya fruit, *C. graminicola* on corn seedlings, and *Venturia inequalis* on apple seedlings. Thus, cutinase inhibitors effectively prevent several pathogenic fungi from penetrating into their hosts, protecting the plant against infection. These results suggested that cutinase inhibition might be a plausible method to control fungal infections in the field. Papaya plots were sprayed biweekly with an organophosphate inhibitor of cutinase throughout an entire season and the fruits were screened for lesions after harvest. Treatment with the cutinase inhibitor protected the fruits from infection by *C. gloeosporioides*, although not as effectively as a classical fungicide (M. B. Dickman, W. Nishijima, S. S. Patil, P. E. Kolattukudy, manuscript submitted). Under field conditions fungi infect the fruit not only by hydrolyzing the intact cuticular barrier but also through pre-existing breaks in the cuticle. Therefore, antipenetrants alone would not be as effective as fungicides, but the use of antipenetrants with fungicides could minimize the need for the latter.

2.3 Cloning and Sequencing of Cutinase cDNA

cDNA libraries were prepared from poly $(A)^+$ RNA isolated from glucose-grown cultures of *F. solani* f. sp. *pisi* and *C. capsici* induced to produce cutinase by the addition of cutin hydrolysate. The *Fusarium* library (in pBR322) was screened by differential hybridization to ^{32}P labeled cDNA prepared from poly $(A)^+$ RNA isolated from induced and uninduced cultures. The clones which appeared to be unique to induced cultures were tested for the presence of cutinase cDNA by hybrid selected translation. An 850 bp cutinase cDNA was sequenced. The amino acid sequence deduced from the nucleotide sequence was verified by direct amino acid sequencing of nearly 40% of the protein (5). The *Colletotrichum* cDNA library was made in the expression vector λgt11 and screened directly with antiserum to the *Colletotrichum* cutinase. An 820 bp cDNA was isolated and sequenced. Again, the amino acid sequence deduced from the nucleotide sequence was verified by direct amino acid sequencing of approximately 20% of the protein (William F. Ettinger and P. E. Kolattukudy, manuscript in preparation).

2.4 Structure

The N terminus of the mature enzyme isolated from *Fusarium* has been shown to have a unique glucuronic acid attached to the N-terminal glycine in amide linkage (2,4). The first glycine in the open reading frame deduced from the cDNA sequence shows this residue to be at position number 32. This indicates that a 31 amino acid leader peptide is removed during post-translational modification of the enzyme. The active serine of the catalytic triad was identified by treating the enzyme with radioactive diisopropylfluorophosphate, followed by isolation and sequencing of a tryptic peptide containing the modified residue. An essential carboxyl group was located by coupling a carbodiimide activated carboxyl group with radioactive glycine ethyl ester followed by isolation and amino acid analysis of a tryptic peptide (William F. Ettinger and P. E. Kolattukudy, unpublished results). The histidine residue involved in catalysis was readily identified as the sole histidine in the enzyme.

The residues identified as being involved in the catalytic triad are located far apart in the primary structure. These residues must be held in juxtaposition by secondary and tertiary structure in order to be catalytically active. The two disulphide bridges present in the fungal cutinase are essential to maintain the catalytic activity of the enzyme; reduction of the disulphide bridges leads to complete inactivation of the enzyme (Wolfram Köller and P. E. Kollatukudy, manuscript in preparation). Presumably because of the functional significance of these disulphide bridges, the location and spacing of the cysteine residues is highly conserved between the *Fusarium* and the *Colletotrichum* enzymes.

A comparison of the primary structure of the enzymes from *Fusarium* and *Colletotrichum* shows conservation of sequence between the two enzymes (Fig. 2). About 50% of the amino acids are directly conserved. The longest unbroken stretch of complete homology is a seven residue long group surrounding the active serine. The relative positions of the four cysteines and the histidine, serine, and aspartate of the catalytic triad are strongly conserved.

2.5 Structure of Cutinase Genes

Recently, the *Fusarium* (C. L. Soliday and P. E. Kolattukudy, manuscript submitted) and *Colletotrichum* (William F. Ettinger and P. E. Kolattukudy, unpublished) genes were cloned and sequenced. The open reading frame in the genomic sequences is interrupted at the same location by 52 bp and 57 bp introns in *Fusarium* and *Colletotrichum*, respectively. These short introns show the usual junction and internal consensus sequences found in other fungal introns.

C. capsici and F. solani sequence homology

C. capsici .... asp glu leu glu ser gly ser ser ser
F. solani .... gly arg thr thr arg asp asp leu ile asn gly asn ser

ser asp cys pro lys val ile tyr ile phe ala arg ala ser thr glu pro gly asn met
ala ser cys arg asp val ile phe ile tyr ala arg gly ser thr glu thr gly asn leu

gly ile ser ala gly pro ile val ala asp ala leu glu ser arg tyr gly ala ser gln
gly ...... thr leu gly pro ser ile ala ser asn leu glu ser ala phe gly lys asp gly

val trp val gln gly val gly gly pro tyr ser ala asp leu ala ser asn phe ile ile
val trp ile gln gly val gly gly ala tyr arg ala thr leu gly asp asn ala leu ....

pro glu gly thr ser arg val ala ile asn glu ala lys arg leu phe thr leu ala asn
pro arg gly thr ser ser ala ala ile arg glu met leu gly leu phe gln gln ala asn

thr lys cys pro asn ser ala val val ala gly gly tyr ser gln gly thr ala val met
thr lys cys pro asp ala thr leu ile ala gly gly tyr ser gln gly ala ala leu ala

ala ser ser ile ser glu leu ser ser thr ile gln asn gln ile lys gly val val leu
ala ala ser ile glu asp leu asp ser ala ile arg asp lys ile ala gly thr val leu

ser ala ile thr lys asn leu gln asn leu gly arg ile pro asn phe ser thr ser lys
phe gly tyr thr lys asn leu gln asn arg gly arg ile pro asn tyr pro ala asp arg

thr glu val tyr cys ala leu ala asp ala val cys tyr gly thr leu phe ile leu pro
thr lys val phe cys asn thr gly asp leu val cys thr gly ser leu ile val ala ala

ala his phe leu tyr gln ala asp ala ala thr ser ala pro arg phe leu ala ala arg
pro his leu ala tyr gly pro asp ala arg gly pro ala pro glu phe leu ile glu lys

ile gly
val arg ala val arg gly ser ala

FIGURE 2. Comparison of the amino acid sequences of cutinase from *Colletotrichum capsici* and *Fusarium solani* f. sp. *pisi*. Regions of complete homology are underlined. Arrows indicate the members of the catalytic triad (Asp, Ser, His) and the cysteines. The vertical bars indicate the location of the intron which interrupts the coding sequences in the genes.

The 5' flanking sequence of the gene is of considerable interest since it may contain regulatory sequences. However, classical TATA and CAT sequences in the 5' flanking region were not found, although TAATAT and a closely associated CAAG sequence were found, 111 to 117 nucleotides from the ATG of the *Fusarium* gene. A functional analysis of these regions will be necessary to determine the significance of these sequences.

2.6 Cutinase Induction

Spores of *F. solani* f. sp. *pisi* were chosen as a model system to study how a fungal spore might sense that it is on the surface of a plant and initiate production of cutinase. *F. solani* f. sp. *pisi* spores contain a low level of cutinase associated with the spore wall (6). Further cutinase production is induced in spores by either the cutin polymer or chemically prepared cutin hydrolysate (7). Monomers purified from cutin hydrolysate were also tested as inducers. The most effective inducers were dihydroxy-C_{16} acid and trihydroxy-C_{18} acid (Fig. 3); both are components unique to cutin.
To test whether the induction of the enzyme involved regulation of the level of cutinase mRNA, non-germinating spores were suspended in a solution containing either cutin or cutin monomers.
At various times the spores were removed and the RNA extracted was probed with ^{32}P-labeled cutinase cDNA. This analysis showed that within 15 minutes after the spores came into contact with either cutin or cutin monomers, cutinase mRNA was detectable and the level increased for the next few hours. The level of cutinase in the culture fluid began to increase rapidly about 45 minutes after the spores were exposed to cutin or cutin monomers. This increase was inhibited by addition of cyclohexamide (Fig. 3). These observations have led to the hypothesis that the small amount of cutinase originally present in the spore wall is

responsible for producing cutin monomers when a spore is in contact with the surface of a plant; these monomers then induce further production of cutinase leading to penetration of the cuticle.

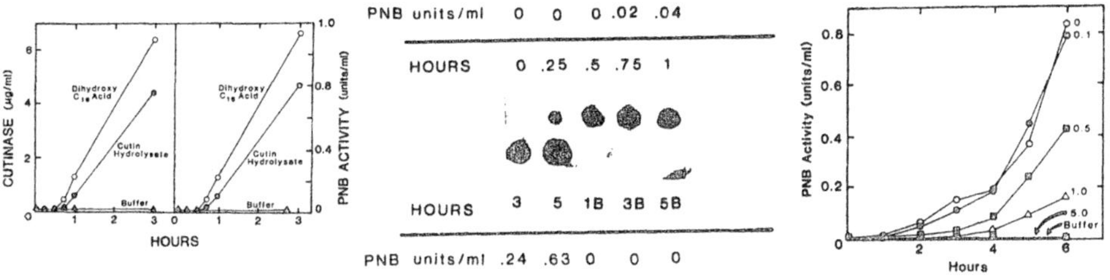

FIGURE 3. A) Time course of induction of cutinase activity (right) and cutinase protein (left), by cutin hydrolysate and dihydroxy-C_{16} acid in spore suspensions of *F. solani* f. sp. *pisi*. B) RNA dot blot probed with ^{32}P-labeled cutinase cDNA. The duration (hours) after exposure to cutin is indicated by the numbers. Numbers followed by B are controls incubated with no cutin. *p*-Nitrophenylbutyrate hydrolase activity (PNB) found in the extracellular fluid is also indicated. C) Effect of cyclohexamide on the induction of cutinase activity in spores. Cyclohexamide concentrations (µg/ml) are indicated on each line. Cutin (0.5 mg/ml) was added at time 0.

3. FUNGAL PECTINASES

Once a pathogen breaches the cutin barrier, it encounters the underlying carbohydrate polymers such as pectin. Further ingress of the pathogen into the plant would require disruption of pectin layers. Although many microorganisms are known to produce pectinases, the role of these enzymes in the fungal penetration into plants has not been elucidated. Spores and mycelia of *F. solani* f. sp. *pisi* can be induced to produce pectinases when pectin is the sole carbon source in the media (Mark S. Crawford and P. E. Kolattukudy, unpublished). Spores in media containing pectin produced polygalacturonase which reached a maximal level in about 8 h; this period corresponded with germination. After about 24 h pectate lyase appeared in the germination medium (Fig. 4).

Three pectinases produced by *F. solani* f. sp. *pisi* were purified to homogeneity: a 58 kDa cationic exopolygalacturonase, a 45 kDa anionic endopolygalacturonase, and a 26 kDa pectate lyase. Rabbit antibodies prepared against the pectate lyase were used to test whether the lyase is essential for infection. Specific inhibition of the lyase caused by the inclusion of antilyase IgG in a spore suspension drop placed on pea stem sections reduced lesion formation by approximately 50%.

Poly(A)$^{+}$ RNA isolated from *F. solani* f. sp. *pisi* induced to synthesize pectate lyase, when translated *in vitro*, generated a 29 kDa protein which immunologically cross reacted with antilyase IgG (Mark S. Crawford and P. E. Kolattukudy, unpublished). Since the mature lyase is a 26 kDa protein the primary translation product must be a precursor

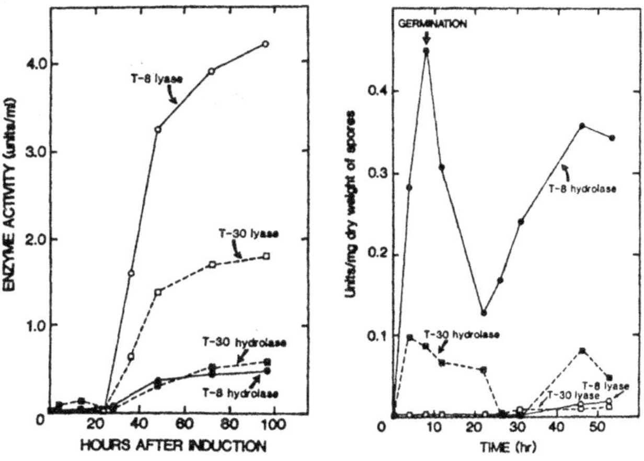

FIGURE 4. Induction of pectate lyase and hydrolase (polygalacturonase) activity in mycelia (left) and germinating spores (right) of *F. solani* f. sp. *pisi*. T-8 and T-30 are two different isolates of the fungus.

containing a leader sequence which is removed during the post-translational processing of the enzyme. cDNA prepared with the poly(A)$^{+}$ RNA isolated from the induced cultures was cloned in the expression vector λgt11. Screening of the recombinants with antilyase IgG revealed plaques containing the lyase cDNA. The availability of the cloned cDNA will allow a molecular approach to elucidate the role of the lyase gene in pathogenesis.

4. BACTERIAL CUTINASE

4.1 Discovery of a Cutinase Producing *Pseudomonas putida* in Association with an Apparently Nitrogen Fixing *Corynebacterium*.

It has been suggested that a well established and mixed population of microorganisms in the phyllosphere could make a substantial contribution to the nitrogen requirements of vegetation (8,9). Some nitrogen fixing organisms have been isolated from the leaf surface. These nitrogen fixers, when sprayed on the crop plants, increased the crop yield compared to crops which received no nitrogen fertilizer (10). Proposed sources of nutrients include leaf exudates or organic deposits on the leaf surface. Whether these nitrogen fixers are capable of utilizing any of the leaf surface components had not been determined. If the microorganisms are to utilize the insoluble polymer, cutin, on the leaf surface, these organisms would have to produce an extracellular cutinase. To test whether the apparently nitrogen fixing phyllospheric bacteria could use the plant cuticular polyester as a carbon source for growth, seven different bacterial strains isolated from plant leaves were tested for their ability to grow on a nitrogen free medium with cutin as the sole carbon source (Joseph Sebastian, A. K. Chandra, and P. E. Kolattukudy, manuscript submitted). Of these seven strains only one (designated BS2) survived under such conditions. When this bacterial isolate was transferred and grown on agar plates containing nutrient broth and yeast extract two morphologically distinct colonies were visible. They were identified as a fluorescent *Pseudomonas putida* and a *Corynebacterium* sp. In order to determine whether both species fixed N_2 and degraded cutin, the pure cultures of each were grown alone on nitrogen-free medium and in cutin-containing medium. *P. putida* strain was found to degrade cutin whereas *Corynebacterium* was able to grow on nitrogen-free medium. A series of co-culture experiments showed that the *Corynebacterium* can provide fixed N_2 while *P. putida* can provide carbon source. Thus the two together can grow in cutin-containing nitrogen-free *Azotobacter* mineral medium.

4.2 Induction of Cutinase in *P. putida*.

Since the *P. putida* strain showed ability to grow on cutin as the sole source of carbon, it was suspected that this bacterium produced an extracellular cutin degrading enzyme. The conditions of the growth of the bacterium for the optimum production of cutinase were determined. Since cutinases from fungi and pollen are known to catalyze hydrolysis of *p*-nitrophenyl esters of short chain fatty acids, the extracellular fluid was assayed with both labeled cutin and *p*-nitrophenylbutyrate as substrates and both assays gave similar patterns of induction (Fig. 5). To determine whether cutin hydrolysate would induce cutinase in the bacterium as previously observed for fungi, cutin hydrolysate was included in the bacterial medium at 0.2 or 1.0 mg/ml. Under these conditions cutin hydrolysate failed to induce cutinase synthesis in *P. putida* (Fig. 5). The mechanism of induction of cutinase in the bacterium might be different from that operating in fungi.

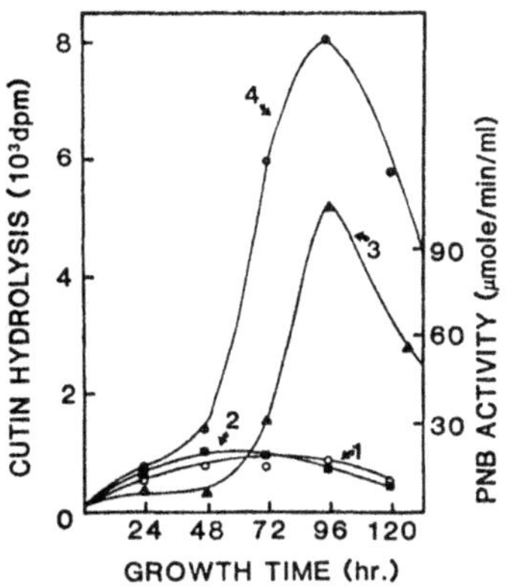

FIGURE 5. Induction of cutinase activity in *P. putida* cultures in nutrient broth-yeast extract media alone (1), or supplemented with cutin hydrolysate (2), or cutin (3); the activity was measured spectophotometrically. The induction of activity in cutin sup- plemented media was also measured using tritiated apple cutin (4).

4.3 Purification of Cutinase Produced by *P. putida*. Cutinase from the bacterium was purified to homogeneity using acetone precipitation, followed by fractionation with DEAE 52, QAE-Sephadex, 6B-Sepharose and Sephadex G-100 (Joseph Sebastian and P. E. Kolattukudy, manuscript in preparation). When passed through the ion exchange column the colored materials were retained on the column while cutinase was not. Combination of the above purification steps gave a 200-fold purification.

4.4 Properties of Bacterial Cutinase.

Cutinase from *P. putida* consisted of a single peptide of 30 kDa. The amino acid composition of the bacterial cutinase was different from that of fungal enzymes. The isoelectric point of the enzyme was 8.8. Rabbit antibodies prepared against fungal cutinase neither cross-reacted with the bacterial enzyme nor inhibited it, whereas antibodies prepared against bacterial cutinase completely inhibited the enzyme. Thus, the bacterial and fungal cutinases are immunologically quite dissimilar.

Catalytic properties of bacterial cutinase resemble those of fungal enzymes. The optimum pH of hydrolysis of cutin by this enzyme was between 8-10.5. This enzyme was stable at either acidic or basic conditions. The bacterial enzyme was extremely sensitive to diisopropylfluorophosphate and to alkyl boronic acids; it was not affected by reagents such as N-ethyl maleimide or *p*-hydroxymercuribenzoate suggesting that thiol groups are not essential for catalysis. Thus, it appears that the bacterial enzyme is also a serine hydrolase like the fungal enzymes. Substrate specificity of the bacterial enzyme was similar to the fungal enzymes. With *p*-nitrophenyl esters, V_{max} and Km showed only small changes when the length of the acyl chain was varied from 4 to 16.

4.5 Temperature Stability of Bacterial Cutinase.

Cutinase produced by the bacterium was completely stable at 60° for 1 h and retained 85% of the activity after 1 h at 70° (Joseph Sebastian, A. K. Chandra, and P. E. Kolattukudy, manuscript submitted). On the other hand, the four fungal cutinases tested were unstable above 45° and lost more than 80% activity in 1 h at 60°C. This stability might be advantageous or even necessary for the function of the bacterial enzyme on the plant surface.

4.6 Cloning and Identification of Bacterial Cutinase gene.

A genomic library of the cutinase-producing *P. putida* was constructed in λgt11 vector (Joseph Sebastian and P. E. Kolattukudy, unpublished). Bacterial DNA was digested with Hae III and 1-5 Kb fragments were isolated from an electrophoretic gel. After adding EcoRI linkers the fragments were ligated into the EcoRI site of λgt11. A library containing 120,000 plaques was screened with the bacterial cutinase antibody. The cutinase encoding clone was isolated and is being sequenced.

5.0 CONCLUSION

The molecular basis of interaction between the plant surface and microorganisms is only beginning to be elucidated. A better understanding of the beneficial interaction with microbes such as the example noted above and the detrimental interaction such as that involved in pathogenesis could allow us to modify such interactions for the benefit of man.

REFERENCES

1. Kolattukudy PE: Structure, Biosynthesis, and Biodegradation of Cutin and Suberin. Ann Rev Plant Physiol 32: 539-567, 1981.
2. Kolattukudy PE: PE Stumpf(ed): The Biochemistry of Plants (vol. 4.): Cutin, Suberin, and Waxes. New York: Academic Press, 1980.
3. Kolattukudy PE: Enzymatic Penetration of the Plant Cuticle by Fungal Pathogens. Ann Rev Phytopathol 23: 223-250, 1985.
4. Kolattukudy PE: Borgstrom B., Brockman H.(eds.): Lipases: Cutinases from Fungi and Pollen. Amsterdam: Elsevier/North Holland, 1982.
5. Soliday CL, WH Flurkey, TW Okita, PE Kolattukudy: Cloning and Structure Determination of cDNA for Cutinase, An Enzyme Involved in Fungal Penetration of Plants. Proc Natl Acad Sci USA 81: 3939-3943, 1984.
6. Koller W, CR Allan, and PE Kolattukudy: Role of Cutinase and Cell wall Degrading Enzymes in Infection of *Pisum sativum* by *Fusarium solani* f. sp. *pisi*. Phys Plant Pathol 20: 47-60, 1982.
7. Woloshuk CP and PE Kolattukudy: Mechanism by Which Contact with Plant Cuticle Triggers Cutinase Gene Expression in Spores of *Fusarium solani* f. sp. *pisi*. Proc Natl Acad Sci USA 83: 1704-1708, 1986.
8. Ruinen J.: A Quispel(ed.): The Biology of Nitrogen Fixation: Nitrogen Fixation in the Phyllosphere. Amsterdam: Elsevier/North Holland, 1974.
9. Vasantharajan VN, JV Bhat: Interrelations of Microorganisms and Mulberry. II. Phyllosphere Microflora and Nitrogen Fixation in Leaf and Root Surfaces. Plant and Soil 28: 258-267, 1968.
10. Pati BR., AK Chandra: Effect of Spraying Nitrogen-fixing Phyllospheric Bacterial Isolates on Wheat Plants. Plant and Soil 61: 419-427, 1981.

SIDEROPHORE BIOSYNTHESIS, UPTAKE AND EFFECT ON POTATO GROWTH OF RHIZOSPHERE STRAINS

Peter Weisbeek, Joey Marugg, Gerard van der Hofstad, Peter Bakker and Bob Schippers.
Department of Molecular Cell Biology, Institute of Molecular Biology and Department of Phytopathology. University of Utrecht, Utrecht, The Netherlands.

INTRODUCTION

Treatment of seed potatoes with certain root-colonizing Pseudomonas putida and fluorescens strains has resulted in protection of the potato tuber yield against the effects of narrow rotation cropping (1:3) and against the effects of certain microbial pathogens. This protective activity of the bacteria is thought to be caused by the production and excretion of large quatities of siderophores with high affinity for binding of iron(III) and uptake of the siderophore-iron(III) complex. The subsequent decrease in iron(III) around the root-surface prevents or delays the growth of other (pathogenic) micro-organisms.
Our research is focused on the analysis of the siderophore-biosynthesis and on its relationship with the effect on the potato-growth.

Mutants and genes

Pseudomonas putida WCS358, isolated from potato-roots, was mutagenized by random Tn5 integration using a RP4-dependent mobilization system. Mutants were Km- and str-resistant. A selection was made for flu$^-$ (non-fluorescing) and sid$^-$ (no high-affinity iron-uptake system) mutants. These mutants were used in a complementation test with a genome bank of the wild-type strain to isolate the corresponding genes. The purified siderophore from the wild-type and some of the mutants was used for the determination of the chemical structure. Outermembrane protein profiles of WCS358 grown in medium with and without added iron(III) showed one major protein band (ca. 70 kD) that is induced by limiting iron concentrations.

Effect of mutants on growth

The mutants were tested in pot- and limited fieldexperiments for their effect on potato-growth in narrow rotation soil and their interaction with each other on the root surface. Both the pot- and the fieldtests gave significant increase in rootdevelopment and tuberformation for mutants that still produce and excrete a normal siderophore but no effect with sid$^-$ mutants (Table I).

The sid$^-$ Km-resistant mutants were used to study the interaction of bacteria on the rootsurface. They were applied to the root either alone or in combination with their wildtype parents. Figure 1 shows that the parent Pseudomonas WCS358 can help its sid$^-$ mutant but also that it interferes strongly with the growth of the WCS374 sid$^-$ mutant. WCS374 can not use the siderophore of WCS358 allthough WCS358 is well able to take up the siderophore of WCS374. In the soil used there is apparently siderophore-production and interaction between the mutant and wildtype cells.

1:3 rotation	tuber fresh weight (kg)	number of tubers
untreated	4.76	90.2
WCS358 sid+	5.31	96.3
JM217 sid-	4.83	86.0
1:6 rotation		
untreated	5.36	99.1
WCS358 sid+	5.38	96.5
JM217 sid-	5.32	100.8

Table I Results of field tests with bacteria-treated potato-tubers.

Pseudomonas WCS358 sid- (kan^R)
+ Ps. WCS358 (kan^S)
+ Ps. WCS374 (kan^S)
+ Ps. WCS358 sid- (kan^R)

Pseudomonas WCS374 sid- (kan^R)
+ Ps. WCS358 (kan^S)
+ Ps. WCS374 (kan^S)
+ Ps. WCS374 sid- (kan^R)

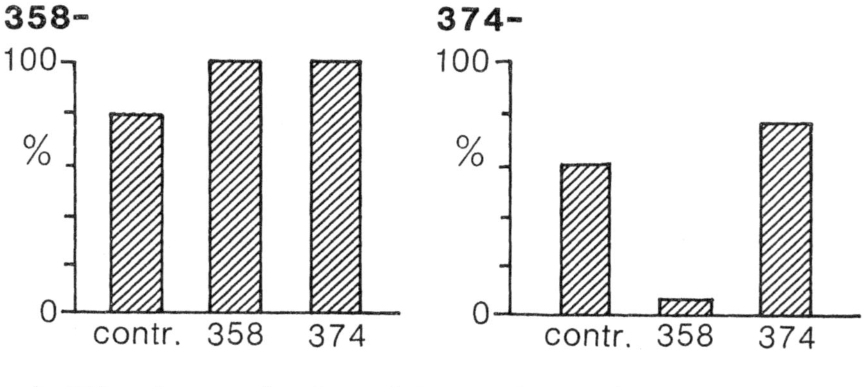

Figure 1: Siderophore-production and interaction on the rootsurface

Genetics analysis
Complementation and DNA hybridization was used to align the different genomic clones containing information for the siderophore system. This resulted in five separate gene clusters of which the largest one was analysed further. In vitro synthesized RNA molecules from selected regions of this cluster were used to probe the transcriptional activity and direction of this region. Fusions to non-expressed coding regions were used to identify iron-dependent controlling regions and minicell analyses gave information on the minimal number and size of the genes in this cluster. This information is compiled in Fig. 2.

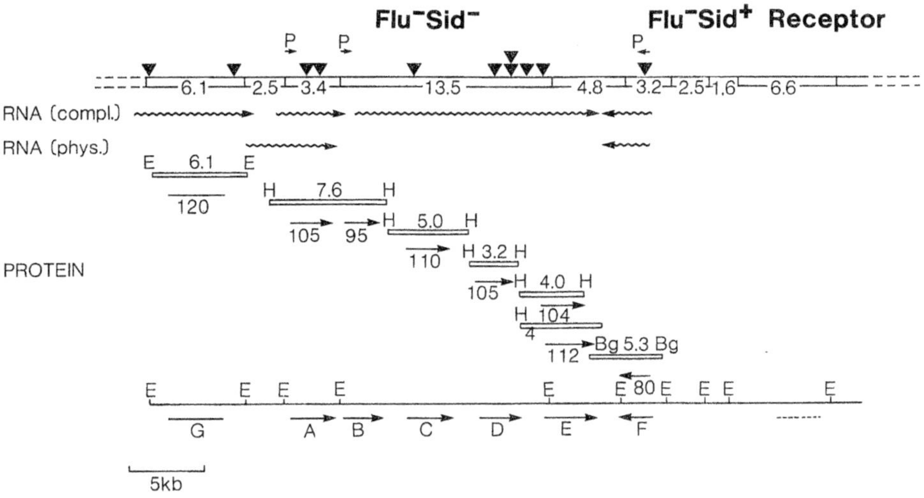

Figure 2. Genetic map of the major biosynthetic gene-cluster.

Siderophore receptor
By using a Pseudomonas strain that can not take up the siderophore of WCS358, a cosmid clone was identified that contains information for the WCS358 siderophore uptake, possibly the receptor. This cosmid is linked to the major gene cluster (see Fig. 2).

Conclusion
The use of the sid^- and sid^+ mutants on potato-tubers and roots confirmed the correlation between growth-stimulation and siderophore-biosynthesis. A large part of the genes involved in biosynthesis of the siderophore are clustered on the chromosome and some of the genes are organized in an operon. Iron-dependent promotors are now being analysed and used to identify the regulation gene.

A GENE CLUSTER IN XANTHOMONAS CAMPESTRIS PV CAMPESTRIS REQUIRED FOR PATHOGENICITY CONTROLS THE EXCRETION OF ENZYMES

J.M. DOW AND G. SCOFIELD

1. INTRODUCTION

Xanthomonas campestris pv. *campestris* Pammel (Dowson) is the causal agent of black rot of crucifers(6). As an approach to studying the molecular biology of *Xanthomonas* pathogenicity we have isolated mutants, derived by NTG mutagenesis, with altered pathogenicity on host plants but unimpaired growth *ex planta*(2). Complementation of the lesions in these mutants with recombinant plasmids containing wild-type DNA has allowed us to identify and clone genes involved in pathogenicity(3). Physiological and genetic studies of the mutants and the cloned genes should increase our understanding of the mechanism of bacterial pathogenicity in plants.

Mutant 8237 was defective in the production of a number of extracellular enzymes including polygalacturonate lyase (PGL)(3). Plasmid pIJ3020 concomitantly restored pathogenicity and PGL production to this mutant on complementation. Mutant 8288 had a lesion of unknown nature and was selected for further genetic study. Plasmid pIJ3000, which is different from pIJ3020, restores pathogenicity to mutant 8288 on complementation. Transposon mutagenesis of pIJ3000 with Tn*5* followed by marker exchange recombination into the chromosome of the wild-type has generated a series of mutants with Tn*5* inserted at different sites within a 25.8 kilobase region of the genome(5). With one exception, all mutants with insertions in a particular 10 kilobase region were non-pathogenic whereas insertions outside this region had no detectable phenotypic effects. This region was considered to be a cluster of pathogenicity genes split by a single insertion into two sub-regions. Here we summarise recent work(4) on the phenotypic characterisation of mutants with lesions in this gene cluster.

2 RESULTS

The mutants were screened for their ability to produce PGL both intra- and extracellularly. Intracellular activity was released by lysozyme/EDTA treatment. Total enzyme activities were comparable in mutants and wild type but the non-pathogenic mutants retained a greater proportion of the total activity within the cells compared to the mutants which retained pathogenicity or the wild-type. (The enzyme activity is predominantly extracellular in the wild-type.) The rates of enzyme induction on adding polygalacturonate to cultures were however similar.

These effects were investigated at a more detailed level by a study of individual PGL isozymes rather than the total activity. The extracellular PGL activity of the wild type could be resolved into 3 major isozymes by fast protein liquid chromatography on mono S, a cation exchange resin. Isozymes of indistinguishable molecular weight (as judged by SDS-PAGE) and ion-exchange characteristics were also present within the cells of the wild type. The patterns of isozymes in mutant 8288 and selected transposon-generated mutants were similar to that seen in the wild type. In addition equivalent isozymes had indistinguishable molecular weights indistinguishable from the wild type. Thus all isozymes appear to be expressed in the mutants and the distribution of each between the cells and the medium was affected.

The distribution of a number of other enzyme activities (carboxymethylcellulase, amylase and protease) was similarly affected by lesions in the pathogenicity gene cluster. However, SDS-PAGE of culture filtrates of the various strains suggested that the excretion of a considerable number of proteins to the medium was not impaired. Preliminary evidence that the non-pathogenic mutants retain the enzyme activities largely in the periplasm has been obtained; treatment of the cells with EDTA alone released the majority of the intracellular PGL and carboxymethyl cellulase activity from all strains tested, although less than 6% of the cytoplasmic marker, malate dehydrogenase, was released under the same conditions. The position of the transposon insertion within the gene cluster was unimportant for the excretion-defective phenotype.

3. DISCUSSION

Our results suggest that the genes in the pathogenicity cluster are not structural or regulatory genes for the enzymes studied but are rather genes involved in their excretion from the cell, perhaps through the outer membrane. It is perhaps not surprising that mutants defective in the excretion of a range of plant tissue degrading enzymes are non-pathogenic although it must be emphasised that the excretion of a number of other proteins is unimpaired. In addition transposon insertion within the gene cluster does not have any major effect on the polypeptide profile of the membranes of *Xanthomonas* (D. Collinge, unpublished). Analogous mutants of *Erwinia* with defects in the excretion of pectinase and cellulase have been described(1). These mutants retain active enzymes within the cell, also probably in the periplasmic space and are non-pathogenic.

The factors determining the distribution of PGL activity between the cells and the medium in the wild type of *Xanthomonas campestris* are unknown; the excretion gene products may be rate-limiting for excretion in simple liquid media. It will be of interest to see if the expression of these genes is modulated *in planta* where the expression of the structural genes for PGL undoubtedly is.

4. REFERENCES

1. Andro T, Chambost J-P, Kotoujansky A, Cattaneo J, Bertheau Y, Barras F, Van Gijsegem F, Coleno A: Mutants of *Erwinia chrysanthemi* defective in secretion of pectinase and cellulase: J Bacteriol 160:1199-1203 (1984).
2. Daniels MJ, Barber CE, Turner PC, Cleary WG, Sawczyc MK: Isolation of mutants of *Xanthomonas campestris* pv *campestris* showing altered pathogenicity: J Gen Microbiol 130:2447-2455 (1984).
3. Daniels MJ, Barber CE, Turner PC, Sawczyc MK, Byrde RJW, Fielding AH: Cloning of genes involved in pathogenicity of *Xanthomonas campestris* pv. *campestris* using the broad host range cosmid pLAFR1: EMBO Journal 3:3323-3328 (1984).
4. Dow JM, Scofield G, Turner PC, Daniels MJ: A gene cluster in *Xanthomonas campestris* pv *campestris* required for pathogenicity controls the excretion of polygalacturonate lyase and other enzymes: submitted for publication.

5. Turner P, Barber C, Daniels M: Evidence for clustered pathogenicity genes in Xanthomonas campestris pv campestris: Mol Gen Genet 199:338-343 (1985).
6. Williams PH: Black rot: a continuing threat to world crucifers: Plant Disease 64:736-742 (1980).

5. ACKNOWLEDGEMENTS

This work was supported by the Agricultural and Food Research Council and by the Gatsby Foundation.

Authors' address: John Innes Institute, Colney Lane, Norwich, U.K.

DIRECT ANALYSIS OF THE INVASIVENESS OF XANTHOMONAS CAMPESTRIS MUTANTS GENERATED BY Tn4431, A TRANSPOSON CONTAINING A PROMOTERLESS LUCIFERASE CASSETTE FOR MONITORING GENE EXPRESSION

JOE J. SHAW AND CLARENCE I. KADO
Department of Plant Pathology, University of California, Davis, California 95616

1. INTRODUCTION

Difficulties in the detection and quantification of small numbers of bacteria have long been apparent. In most cases, detection of invading bacteria is not possible prior to the onset of visible symptoms. Even in the presence of symptoms, sampling is done by extracting bacteria from the invaded tissues. The procedure is always disruptive of the ongoing disease process and the information obtained is dated as it often depends on the formation of colonies. The same difficulties apply in studying bacterial genetics in planta.

We have alleviated these problems by the use of a novel marker in the form of bacterial bioluminescence. Bacteria harboring bioluminescence genes can be detected and quantified simply by detecting and measuring light, without ever touching or physically manipulating the bacteria or their host plant. Likewise, gene expression can be monitored if promoters of interest are fused to the bioluminescence genes, coupling light production to gene activity. Thus, bacteria invading a plant can be detected and localized in the plants they are invading by monitoring the light which they emit.

Bacterial bioluminescence (lux) occurs primarily in marine bacteria and is widespread (1). Recently the genes which are required for sustained light production were cloned from *Vibrio fischeri* and expressed in *E. coli* (2). Seven genes were cloned but only five are necessary whereas two encode regulatory functions. In several steps the regulatory genes were removed and a promoterless lux cassette was constructed (3 and Tim Close unpublished results). These genes are expressed under the control of heterologous promoters via transcriptional fusions. Thus inducible promoters allow inducible bioluminescence and constitutive promoters direct constitutive light production.

The biochemistry of light production in bacteria has been widely studied (1). Basically, an aldehyde and reduced flavin mononucleotide are oxidized in a reaction catalyzed by luciferase. The products of this reaction include a fatty acid and photons. In a follow-up reaction the fatty acid is reduced to the aldehyde form by fatty acid reductase (4). These two enzymes are encoded by the five genes of the lux cassette. It is possible to omit the three genes that encode the fatty acid reductase subunits and supply aldehyde exogenously to generate light. Aldehyde can also be supplied to bacteria carrying the complete set of lux genes to boost light production. However, we do not recommend this process for either system because of the high degree of variation in light production (see below).

2. PROCEDURE

Media and Cultural Conditions

The bacteria were maintained and grown as previously described (5). Bacteria were grown at 24°C where light was monitored or quantified. This was to avoid variability in bioluminescence due to temperature.

Light Measurements and Detection

The luminometer is described elsewhere (5) and is a photomultiplier adapted for our use in collaboration with Leo Clougherty and his associates at Beckman Instruments Inc. (Fullerton CA).

DNA Manipulations

Competent *E. coli* cells were stored at -65°C and transformed with plasmid DNA as described by Maniatis et al. (6). All enzymatic reactions and gel electrophoresis was also done as described by those authors. Restriction endonucleases were purchased and used according to the supplier (New England Biolabs, Beverly, MA).

3. RESULTS

Lux genes have been placed in a variety of bacteria and they all seem to emit the same kind of blue green light. The amount of light emitted depends on the strain of bacteria used, even within a pathovar. Also, copy number and promoter strength will affect bioluminescence levels. The light may be visible in a darkened room or may be dimmer and detected with photomultipliers or photographic or x-ray film.

We have constructed many vectors to introduce the lux genes into plant pathogenic bacteria. pUCD607 is a broad-host-range plasmid with a variety of selectable markers (5). With pUCD607 the lux cassette is constitutively expressed in bacteria including *Agrobacterium*, *Rhizobium*, *Erwinia*, *Pseudomonas*, and *Xanthomonas*. Bioluminescence does not appear to interfere with the pathogenic abilities of these bacteria and their light can be detected in planta during an ongoing, uninterrupted infection. pUCD607 is mobilizable by the helper plasmid pRK2013 (7) and is amplifiable in *E. coli*.

Unique cloning sites are available in pUCD607 including some in the lux genes where insertions will cause the dark phenotype which can be screened on film. Colonies on filter paper or agar will make dark impressions on x-ray film whereas those colonies with inserts will make no mark as the lux genes are interrupted and non functional.

It is not necessary to add aldehyde to bacteria carrying pUCD607 because all five genes are contained in the lux cassette. We find this beneficial as n-decyl aldehyde (the usual added substrate) has a noxious odor. It is applied in vapor form and so is hard to quantitatively apply, this in turn affects bioluminescence levels. The net result is that attempts to quantify bioluminescence are virtually unrepeatable from day to day although relative intensities can be ascertained and reproducibly obtained. With the five gene lux cassette absolute levels of bioluminescence can be reproducibly determined. Another problem we have found with n-decyl aldehyde is that it is toxic to both plants and bacteria in pure form, actually dissolving holes in leaves in a short time. We have found that it is usually preferable to work with a naturally efficiently bioluminescing bacterial isolates or to use more sensitive detection equipment rather than to use exogenously applied substrate. Thus, while aldehyde can be used in conjunction with pUCD607 to boost light production the problems of reproducibility and toxicity do not weigh in its favor.

For studying the regulation of genes involved in pathogenesis, we have constructed the transposon Tn4431 (Fig. 1). We have used it to mutagenize

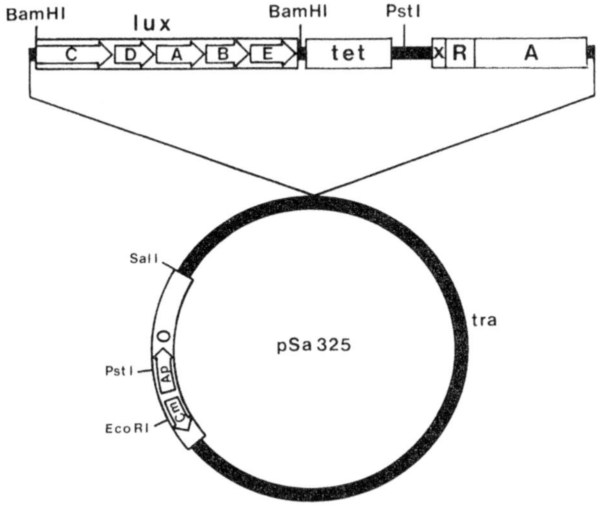

Agrobacterium, Pseudomonas, Xanthomonas, and E. coli. The lux genes were inserted into an unpublished derivative of Tn1721 (8). The resultant 15 kb transposon seems to have the same general characteristics as its original progenitor, Tn1721. It inserts relatively randomly into DNA and once inserted prevents second insertions into the same replicon. We have used Tn4431 to generate auxotrophs, carbon utilization mutants, pigment mutants and mutants in pathogenicity in Xanthomonas campestris pv. campestris (XCC). It is delivered to bacteria on the suicide vector pSa325 (9). The entire replicon pSa325::Tn4431 is called pUCD623.

On studies on XCC, we concur with Mike Daniels that a variety of genes are involved in pathogenicity and that many will be expressed only during the disease and not in culture. One way to find these genes is to screen Tn4431 mutants for differential light producing abilities on agar and in the plant. Work is in progress to identify genes involved in pathogenicity in XCC and to study their expression in planta by use of lux::promoter fusions.

4. DISCUSSION

The application of molecular genetic techniques to Plant Pathology is relatively new. Expression of symptoms in the host is an easily assayed function of pathogenicity and an obvious place to apply molecular biological methods. However, other attributes of the pathogen are equally important but more subtle and thus difficult to assay; including vector acquisition, overwintering abilities and epiphyte capabilities. Bioluminescence may provide a sensitive means to locate and quantify bacterial populations, simplifying such studies. We also think that bioluminescence will be useful to study bacterial gene expression in planta. Finally, we propose the bioluminescent phenotype for use in tracking genetically engineered organisms which have been released into the environment.

5. ACKNOWLEDGEMENTS

This work was supported in part by the McKnight Foundation graduate fellowship in plant science, the Jesse D. Carr graduate fellowship in agriculture and the Jastro Shields graduate research fellowship. We would like to thank Lynette Settles for invaluable technical assistance and R. Schmitt for the generous gift of the Tn1731-Ba152.

6. REFERENCES

1. Hastings, J. W. (1977) Bacterial Bioluminescence. Ann Rev. Microbiol. 31:549-595.
2. Engebrecht, J., Nealson, K. and Silverman, M. (1983) Bacterial bioluminescence: isolation and genetic analysis of functions from Vibrio fischeri, Cell, 32:773-781.
3. Engebrecht, J., Simon, M. and Silverman, M. (1985) Measuring gene expression with light, Science, 227:1345-1347.
4. Wall, L., Byers, D. and Meighen, E. (1984) In vivo and in vitro acylation of polypeptides in Vibrio harveyi: identification of proteins involved in aldehyde production for bioluminescence, J. Bacteriol., 159:720-724.
5. Shaw, J. and Kado, C. (1986) Development of a Vibrio Bioluminescence gene-set to monitor phytopathogenic bacteria during the ongoing disease process in a non-disruptive manner, Bio/technology, 4:560-564.
6. Maniatis, T., Fritsch, E. and Sambrook, J. (1982) Molecular Cloning: A Laboratory Manual, Cold Spring Harbor Laboratory, Cold Spring Harbor, N.Y.
7. Figurski, D., and Helinski, D. (1979) Replication of an origin containing derivative of plasmid RK2 dependent on a plasmid function provided in trans. Proc. Natl. Acad. Sci., USA, 76:1648-1652.
8. Schmitt, R., Bernhard, E. and Mattes, R. (1979) Characterization of Tn1721, a new transposon containing tetracycline resistance genes capable of amplification, Mol. Gen. Genet. 172:53-65.
9. Zaitlin, D. (1984) Ph. D. dissertation, Department of Biochemistry and Biophysics, University of California at Davis, Davis, CA, USA.

ANALYSIS OF THE SPONTANEOUS MUTATION TO AVIRULENCE BY PSEUDOMONAS SOLANACEARUM

MARK A. SCHELL,[+*] DANIEL P. ROBERTS,[+*] AND TIMOTHY P. DENNY[*]

Department of Microbiology[+] and Department of Plant Pathology[*], University of Georgia, Athens, GA 30602

ABSTRACT

A gene from P. solanacearum encoding an endoglucanase (EGase) enzyme was cloned on a 30-kb DNA fragment in the cosmid pLAFR3 by virtue of its expression in E. coli. Activity inhibition analysis with antiserum against a purified EGase of P. solanacearum showed that this gene (egl) encodes the major endoglucanase excreted by virulent P. solanacearum. This 38-kDA EGase enzyme is produced at 25-fold lower levels in spontaneous, non-mucoid, avirulent mutants. After mapping and subcloning, Southern hybridization analysis showed that there are no major structural differences between the egl gene in the wild type virulent parent and in avirulent mutants. However, RNA measurements showed that the egl gene was transcribed nearly 10-fold less in avirulent mutants. Transfer of the wild type egl gene into avirulent mutants resulted in non-mucoid strains which produce nearly normal levels of egl gene product. Attempts to spontaneously mutate the cloned egl gene in merodiploids were unsuccessful. The available evidence suggests that a small cis-acting promoter mutation results in the reduced egl expression in the spontaneous, pleiotrophic mutants.

INTRODUCTION

Pseudomonas solanacearum is an important bacterial phytopathogen which produces a bacterial wilt disease of many solanaceous plants all over the world (1). The virulent phytopathogen produces large quantities of an extracellular polysaccharide slime which is believed to be involved in wilting and killing susceptible plants (2). The ability of virulent, mucoid P. solanacearum strains to spontaneously mutate at high frequency to a non-mucoid, avirulent phenotype was first reported by Kelman (3). The transition from virulent, mucoid to the avirulent, non-mucoid phenotype is accompanied by many other biochemical and morphological changes (e.g. reduced cellulase production, increased pilus formation, increased motility, alteration in LPS structure) (4,5).

Virulent P. solanacearum strains excrete a 38-kDa CMCase enzyme which represents at least 30% of the total protein excreted by P. solanacearum (4,6). This enzyme has been purified, characterized, and shown to probably be a β-1,4 endoglucanase (EGase) (6). Activity inhibition experiments with antiserum against the purified protein showed that an antigenically similar enzyme is produced by many virulent strains of races 1 and 2, whereas all spontaneous, non-mucoid avirulent mutants produce at least 25-fold less EGase activity (6). However, the role of the EGase in disease is unknown as is the genetic mechanism underlying the spontaneous pleiotrophic mutation to avirulence which affects its expression. Many hypotheses have been proposed to explain this apparently irreversible, pleiotrophic transition (mutation) such as loss of plasmids, DNA rearrangements, and partial

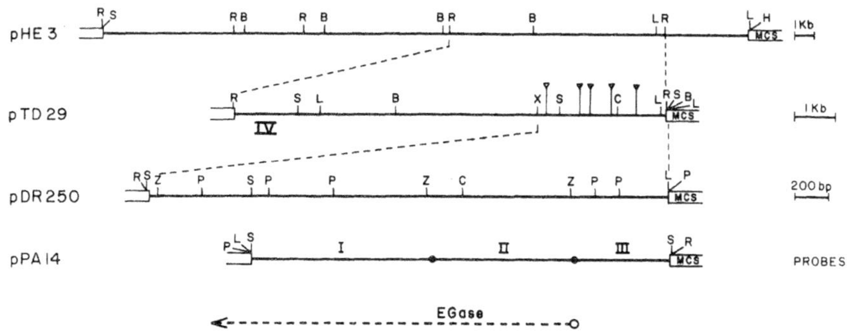

Fig. 1. Physical map of plasmid inserts containing cloned egl gene. Restriction endonuclease cleavage sites are designated as follows: BamHI (B), ClaI (C), EcoRI (R), PstI (P), SalI (L), SmaI (S), SphI (Z), XhoI (X). Position of Tn5 inserts in pTD29 which inactivate endoglucanase activity are represented by (▼); position of Tn5 insertions which do not affect endoglucanase activity are indicated by (▽). The approximate extent and location of egl and probable direction of transcription is indicated by the arrow at the bottom; MCS- multiple cloning site of vectors: pUC9 for pTD29 and pDR250; pUC8 for pPA14; pLAFR3 for pHE3. Location of probes derived from pPA14 used in Southern hybridization analysis and RNA measurements are designated I-IV.

genomic deletions. However none of these hypotheses have been confirmed by subsequent experiments. We have attempted to probe the genetic mechanism underlying this pleiotrophic mutation by utilizing a cloned gene (egl) whose expression is affected by the transition to the non-mucoid, avirulent state (4). Our evidence suggests, that for the egl gene encoding the major endoglucanase of *P. solanacearum*, the mutation producing lowered expression does not involve major DNA rearrangements, insertions, or deletions. Lowered expression appears to result from a reduction of transcription of this gene due to an unknown type of *cis*-acting mutation.

RESULTS AND DISCUSSION

Cloning of the endoglucanase gene.

A genomic library of *P. solanacearum* DNA was constructed in the broad host range cosmid pLAFR3, transformed into *E. coli*, and the library screened *en masse* for production of EGase activity utilizing CMC-agar plates and Congo Red staining. Two *E. coli* clones of 1000 tested produced EGase activity and were found to contain identical cosmids. The restriction map of the 30-kb insert on one of these cosmids (pHE3) is shown in Fig. 1. The EGase gene was mapped by transposon mutagenesis with Tn5 and the active gene subcloned utilizing expression in *E. coli* (See Fig. 1). Plasmids pTD29 and pDR250 contain the entire, active egl gene while pPA14 contains only a portion of the gene.

TABLE 1. Analysis of the product of cloned *egl* gene

Bacterial Strain	Endoglucanase Activity+	
	Intracellular	Extracellular
E. coli JM83 (pDR250)	0.080	<0.001
P. solanacearum AW	<0.001	1.30
E. coli JM83 (pDR250) + 5 µl Anti-EGase	<0.001	-
E. coli JM83 (pDR250) + 5 µl preimmune	0.050	-
P. solanacearum AW + 5 µl Anti-EGase	-	0.008
P. solanacearum AW + 5 µl preimmune		0.900

+ µmole cellobiose produced from carboxymethyl cellulose at 50°C per min per ml of culture equivalent; intracellular = whole cell extract prepared by sonication; extracellular = culture supernatant.

E. coli cells containing the cloned *egl* gene on pDR250 were analyzed for the levels of expression (TABLE 1). All detectable EGase activity was found in extracts of the *E. coli* cells, in contrast to *P. solanacearum* which excretes EGase into the culture supernatant. To show that these two enzymes were identical, aliquots of each enzyme fraction were incubated with antiserum against the 38-kDa excreted EGase of *P. solanacearum*. The activity of both the EGase encoded on pDR250 in *E. coli* and that excreted by *P. solanacearum* were completely inhibited by the antiserum (TABLE 1). We conclude from these data: 1) the cloned *egl* gene on pDR250 encodes the major 38-kDa EGase enzyme excreted by *P. solanacearum* and is also the one affected by the spontaneous, avirulence mutation; 2) *E. coli* is incapable of excreting the *egl* gene product.

Structure of egl gene in virulent strains and avirulent mutants.

Labelled DNA fragments from pTD29 and pPA14 (Fig. 1) were hybridized to Southern blots of restriction endonuclease digested DNA from virulent and avirulent mutants to detect insertions, deletions, and/or DNA rearrangements in or near the *egl* gene which could explain lowered expression in the mutants. An example of the results is shown in Fig. 2. No major structural differences in or near the *egl* gene were detected in DNA isolated from avirulent mutants probed with a 950-bp *SmaI*-*SphI* fragment of pPA14 (I). This is an internal fragment from the *egl* gene as determined by transposon mutagenesis and subcloning. Other probes hybridized to identical filters were a 750-bp *SphI*-*SphI* fragment of pPA14 (II), a 500-bp *SphI*-*EcoRI* fragment from pPA14 (III) and a 710-bp *SmaI*-*XhoI* fragment from pTD29. No major structural differences between *egl* in wild type, virulent *P. solanacearum* and spontaneous avirulent mutants were observed with these probes which cover the entire *egl* gene. We conclude from these results that reduction of *egl* expression in spontaneous, non-mucoid mutants is not due to: 1) deletion of any major portion of the *egl* gene (> 50 bp); 2) insertions of DNA (>50 bp) into the *egl* gene; 3) major rearrangements of DNA in or near the *egl* gene. However, minor DNA rearrangements (involving <200bp) would not be detected without more extensive analysis.

Analysis of egl expression in avirulent mutants.

The Southern hybridization experiments above suggested that the

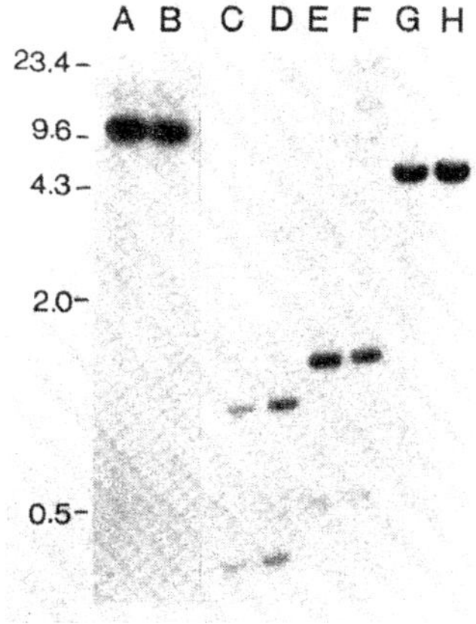

Fig. 2. Southern hybridization analysis of the structure of egl gene in wild type and avirulent mutants. Autoradiograph of nitrocellulose filter containing fractionated chromosomal DNA from wild type (lanes A, C, E and G) and spontaneous, non-mucoid mutants (lanes B, D, F, and H) digested with EcoRI (lanes A and B), PstI (lanes C and D), SphI (lanes E and F), and SmaI (lanes G and H), and hybridized with ^{32}P-labelled 950-bp SmaI - SphI fragment of pPA14. Migration of molecular weight markers (in Kb) is shown at left.

expression of egl in avirulent mutants may be reduced because of an effect on transcription or translation of the gene rather than resulting from insertions or deletions. To test this hypothesis, labelled DNA probes from egl were hybridized to RNA isolated from virulent and avirulent mutant strains to measure transcription of egl in each strain (TABLE 2). Four probes were used; I-III described previously and IV a 1500-bp EcoRI - SmaI fragment of pTD29 that is totally outside egl. The 950-bp SmaI-SphI fragment of pPA14 (I), located totally within the egl gene, strongly hybridized to RNA from wild type virulent strains but hybridized nearly 10-fold less to RNA isolated from spontaneous avirulent mutants as did fragment II (TABLE 2). Fragment III, of which at least part lies outside egl, hybridized weakly to both virulent and avirulent RNA as did the 1500-bp EcoRI-SmaI fragment of pTD29 (IV) which is totally outside the egl gene. These results suggest that the egl gene is transcribed nearly 10-fold less in the avirulent mutants. Thus it seems that lower egl expression in the spontaneous avirulent mutant results from a decreased level of transcription and not from a major gene alteration that inactivates its product.

TABLE 2. Comparison of levels of egl RNA in wild type and avirulent mutants of P. solanacearum

Probe	Location	Relative Hybridization^{+} to Immobilized RNA from:	
		Virulent	Avirulent
I	egl internal	9.2	1.0
II	egl internal	8.6	1.0
III	flanking	1.2	1.0
IV	flanking	0.8	1.0

$^{+}$Nitrocellulose filters containing spots of between 1 and 10 μg of immobilized RNA from each strain were hybridized with ^{32}P-labelled probes, autoradiographed, and spot intensities determined by densitometry.

Expression of the cloned *egl* gene in avirulent mutants.

To determine if avirulent mutants contain a *trans*-dominant regulatory mutation which lowers *egl* transcription, pHE3 was mobilized into an avirulent mutant (*P. solanacearum* AR) and EGase production measured (TABLE 3). Avirulent mutants containing pHE3 produce 15-fold more EGase activity than the avirulent mutant without pHE3 but remained morphologically identical. Thus it appears that the spontaneous mutants containing pHE3 can express the cloned *egl* gene at 50% of the level observed in wild type cells. This result appears to exclude a *trans*-dominant mutation in a pleiotrophic regulatory gene as the cause of reduced *egl* expression.

TABLE 3. EGase activity in strains carrying cloned *egl* gene.

Strain	Endoglucanase Activity+
P. solanacearum AW; wild type, virulent	0.90
P. solanacearum AR; non-mucoid, avirulent	0.03
P. solanacearum AR (pHE3)	0.45
P. solanacearum AW (pHE3)	2.00
P. solanacearum AWR (pHE3)	0.30

+μmole cellobiose produced per min per ml culture supernatant at 50°C.

The pHE3 cosmid was mobilized into a wild type virulent strain producing *P. solanacearum* AW (pHE3). Spontaneous, non-mucoid mutants were then selected from the merodiploid (producing *P. solanacearum* AWR (pHE3)) and then analyzed for EGase production (TABLE 3). Apparently, the *egl* gene on the pHE3 cosmid is "immune" to the spontaneous, pleiotrophic mutation, since all resultant merodiploid mutants produced EGase levels 10-fold higher than non-mucoid mutants lacking pHE3. The levels of EGase produced by *P. solanacearum* AWR (pHE3) are 66% of the levels produced by non-mucoid mutants containing the wild type pHE3 cosmid. In addition, transfer of cosmids from *P. solanacearum* AWR (pHE3) isolates into *P. solanacearum* AR, produced strains with EGase levels identical to those of *P. solanacearum* AR (pHE3). In summary, these results suggest that the *egl* gene on pHE3 was not affected by the transition to the non-mucoid, avirulent state.

Although we have not determined the exact molecular basis of the spontaneous, pleiotrophic avirulence mutation, the possiblities have been narrowed. Most available evidence favors the conclusion that a *cis*-dominant mutation occurs near the *egl* gene promoter as a result of other unknown events occurring during the spontaneous mutation to the non-mucoid state. Perhaps there is a genetic switch contained on a small DNA fragment in the *egl* promoter region which was not detected in Southern hybridization experiments. Comparison of the DNA sequence of the *egl* promoter from the wild type and spontaneous avirulent mutants of *P. solanacearum* may provide a more definitive answer.

REFERENCES

1. Buddenhagen, I. W. and A. Kelman. 1964. Ann. Rev. Phytopathol. 2:203-230.
2. Husain, A. and A. Kelman. 1958. Phytopathol. 48:155-165.

3. Kelman, A. 1954. Phytopathol. 44:693-695.
4. Schell, M. A. and J. D. Vinson. 1985. In: Advances in Molecular Genetics of the Plant-Bacterial Interaction (A.A. Szalay and R.P. Legocki, eds.), Media Services Cornell University, Ithaca, N.Y., pp. 182-187.
5. Morales, V. M., Stemmer, W. P. C. and L. Sequeira. 1985. In: Current Communications in Molecular Biology: Plant Cell/Cell Interactions, Cold Spring Harbor Lab, Cold Spring Harbor, N.Y., pp. 89-91.
6. Schell, M. A. 1986. Appl. Env. Micro., Submitted.

This research was supported by grants from the University of Georgia Research Foundation and University of Georgia Biotechnology Program.

CHARACTERIZATION OF PATHOGENICITY GENES OF *ERWINIA CAROTOVORA*

A.K. Handa[1], R.A. Bressan[1], L. Lee[2], D.J. Charles[1], R.K. Jayaswal[1], J. Chiu[1] and J.L. Bennetzen[2], Departments of Horticulture[1] and Biological Sciences[2], Purdue University, W. Lafayette IN 47907, USA

INTRODUCTION

The plant pathogenic bacterium, *Erwinia carotovora* subsp. *carotovora* (*Ecc*), causes soft rot on a number of plant species. The most striking characteristic of soft-rot *Erwinias* is the production of large amounts of extracellular pectolytic enzymes (2). *Ecc* produces several pectate lyases and polygalacturonases (7,8,10). In addition to pectic enzymes, several environmental components, especially anaerobic conditions, have been implicated in the development of severity of soft rot disease (3). During the last five years, several laboratories have cloned DNA sequences encoding pectolytic enzymes in order to clarify their role in the disease development (7,8,10, Handa *et al*. unpublished results). Isolated PL and PG can cause soft rot symptoms on potato tuber (7,8,10). However, an *E. coli* strain expressing both PL and PG from cloned genes caused only a limited tissue maceration of potato tuber (8), thus suggesting the involvement of other gene product(s) in the development of disease. At present, little is known about the nature of these gene products. To investigate the genetic factors responsible for the virulence of *Ecc* we have isolated and characterized a large number of nonpathogenic mutants by transposon mutagenesis. Partial characterization of some of these mutants is presented in this paper.

MATERIALS AND METHODS

Mutagenesis of *Erwinia carotovora* subsp. *carotovora* with Mudl was performed as described previously (5). Following purification on selective medium the transductants were grown overnight on LB plates containing 30 $\mu g\ ml^{-1}$ ampicillin and were tested for auxotrophy and virulence. Putative nonpathogenic mutants were selected based on their inability to cause soft rot on the cut surface of potato tuber 72 h following stab inoculations. To quantitate pathogenicity, cultures (10 μl) of each mutant grown overnight in LB containing ampicillin were injected into potato tubers according to the method described by Debore and Kelman (3) and the diameters of rot were determined at different levels of O_2 after 48 h. Pectate lyase and polygalacturonase were assayed by using thiobarbituric acid and arsenomolybdate methods, respectively. The genomic libraries of *Ecc* AH2 were constructed in cloning vectors lambda EMBL-4 and pLAFR-3 by ligating *Sau*3A partial digested DNA, ca 15-30 Kb, into the respective vectors. Among 1100 genomic clones of Ecc AH2 in pLAFR-3 four clones produced pectic enzymes, 2 were resistant to nalidixic acid and 1 showed streptomycin resistance. A DNA sequence representing the right end of Mud1 was isolated from pPO1681 (1) after digesting with BamH1 and used as a probe.

RESULTS AND DISCUSSION

Classes of Mud1-induced nonpathogenic mutants of Ecc: Mutagenesis of Ecc AH2 with bacteriophage Mud1 (Ap^r lac cts62) resulted in several classes of apparently nonpathogenic mutants. Class 1 mutants (42 isolates) were impaired in secretion of pectolytic enzymes. Class 2 mutants (3 isolates) were not impaired in secretion or synthesis of pectolytic enzymes. Class 3 mutants (16 isolates) were blocked in biosynthesis of pyrimidines or purines. Class 4 mutant (1 isolate), which was unable to utilize various carbon sources, was characterized to be defective in synthesis of UDP-glucose pyrophosphorylase (6). The nonpathogenic phenotype of this mutant was caused by its sensitivity to galactose.

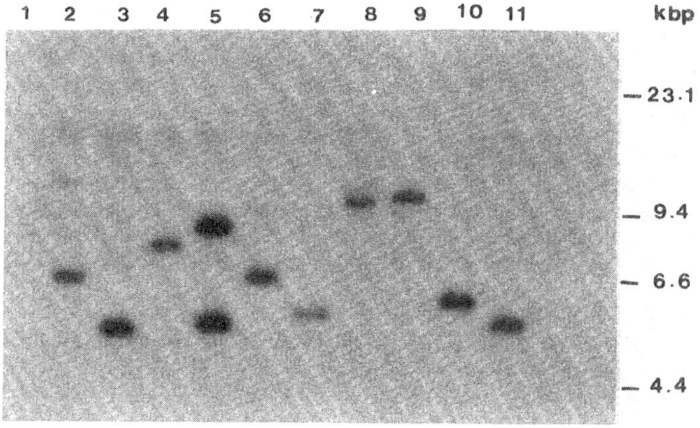

Figure 1. Hybridization of total DNA from nonpathogenic AH2::Mud1 mutants with ^{32}P-labelled probe representing the right end of Mud1. Total DNA isolated from 10 mutants was digested with EcoR1 and hybridized to a ^{32}P-labelled probe from pPO1681 (1). Plasmid pPO1681 was extracted and purified over CsCl gradients to eliminate chromosomal DNA contamination. Purified plasmid was digested with Bam H1 and a DNA probe representing the right end of Mud1 was purified from agarose gels. Lane 1, wild type AH 2; Lane 2 AH 174; Lane 3, AH 453; Lane 4, AH 1028; Lane 5, AH 2024; Lane 6, AH 2552; Lane 7, AH 2819; Lane 8, AH 4334; Lane 9, AH 4612; Lanes 10 and 11 are AH 4688 and AH 4690, respectively, which are pathogenic strains obtained after Mud1 mutagenesis.

Southern blot hybridizations were performed to determine whether the observed nonpathogenic phenotypes were due to single Mudl insertions. Figure 1 shows the hybridization patterns of *EcoRl* digested DNAs from several nonpathogenic mutants with a ^{32}p-labeled right end-Mudl probe. Except for mutant AH2024, all other mutants showed single bands of varying sizes. Mutant AH2024 has two separate insertions of Mudl. No hybridization was observed when Ecc AH2 DNA was used. This data shows that most mutants arose due to single insertions of Mudl at different sites into the genome of *Ecc*.

Pathogenicity of auxotrophic mutants: Among auxotrophic mutants, only those requiring either purine or pyrimidine or pyrimidine plus arginine were not able to cause maceration of potato tuber tissue (Table 1). Virulence of these mutants was restored partially by supplying the required nutrient during the tuber maceration assay. Under low O_2 tension (2% O_2) these mutants become partially virulent, suggesting that at reduced O_2 tension the integrity of the plant cell membranes is altered causing leakage of solutes from potato cells and enhancing growth of bacteria. However, further experiments are needed to establish the molecular basis of this observation.

Characterization of mutants defective in secretion of pectolytic enzymes: Based on intracellular and extracellular activities of polygalacturonase (PG) and pectate lyase (PL),

Table 1. Pathogenicity of Auxotrophic Mutants of *Erwinia carotovora*: Effects of supplements and O_2 tension.

Phenotype	Number of Mutants	Tissue Maceration, % AH2::Mudl		
		21% O_2		2% O_2
		Suplement[1]		
		-	+	-
AH2::Mudl	-	100	100	100
Pur	5	0	50 ± 10	22 ± 6
Pyr	6	0	66 ± 9	50 ± 4
Pyr A	5	0	87 ± 4	36 ± 10

Auxotrophic mutants requiring Arg, Cys, His, Ile, Ile + Val, Leu, Lys, Met, Thr and Trp, respectively showed pathogenicity similar to that of EccAH2::Mudl. + and - indicate presence or absence of supplement, respectively.

Pyr A: mutation in carbamoyl phosphate synthetase gene, auxotroph require Arg + Pyr for growth.

[1]Pur, Pyr, PyrA mutants were tested in the presence of 10 mM adenosine, 10 mM uridine or 10 mM uridine + 60 mM arginine, respectively.

Table 2. Pathogenicity and levels of different pectic enzymes in representative isolates of different classes of nonpathogenic mutants of *E. carotovora*.

Isolate	Pathogenicity Tuber Decay % AH2::Mudl		Pectolytic Activity % AH2::Mudl			
			Polygalacturonase		Pectate lyase	
	$21\% \ O_2$	$2\% \ O_2$	Extra-cellular	Intra-cellular	Extra-cellular	Intra-cellular
Mutants impaired in secretion of PG and PL (34 isolates)						
AH 453	0	10	17	80	13	143
AH 114	14	87	7	58	18	58
AH 2728	51	68	1	25	1	50
Mutants impaired in secretion of PG (8 isolates)						
AH 755	33	50	3	92	73	56
AH 4622	16	52	41	107	117	78
Mutant impaired in secretion of PL (1 isolate)						
AH 2820	15	76	104	75	22	94
Mutants not impaired in secretion of PG or PL (3 isolates)						
AH 2024	10	79	79	88	78	117
AH 2552	11	68	76	86	77	130
AH 4334	15	82	84	74	98	78
Mutant impaired in utilization of various carbon sources						
AH 1028	0	4	25	120	20	25

Pectic enzymes were assayed after 24 h growth in a medium containing 1% yeast extract and 0.5% polygalacturonate.

these mutants can be divided into three categories: i) mutants defective in secretion of PG; ii) mutants defective in secretion of PL; and iii) mutants defective in secretion of both PG and PL (Table 2). The degree of secretion of PG or PL in these mutants varied from 0 to 70% of that of the wild type strain. Several interesting conclusions can be drawn from this data. First, the high frequency of mutation resulting in impaired secretion of PG, PL or both enzymes compared to auxotrophic mutations indicates that well over 50 genes may be involved in regulating secretion of these enzymes. Furthermore, most mutants (34 isolates) were defective in secretion of both enzymes, suggesting that there are several single genes, each of which is capable of controlling secretion of both of these enzymes. Single site Tn5 insertion mutants affecting the export of both pectolytic and cellulytic enzymes in *E. chrysanthemi* have been reported (9). Finally, this data suggests that *Ecc* has mechanisms which can influence selectively the export of either PL or PG. These types of mutants may provide a model system to investigate specific protein secretion mechanisms in gram-negative bacteria.

Isolation of nonpathogenic mutants of Ecc not defective in secretion or synthesis of pectolytic enzymes: During the present investigation we have isolated 3 mutants which were not significantly impaired in secretion of pectolytic enzymes (Table 2). Apart from the fact that these mutants secrete normal levels of both pectate lyase and polygalacturonase, little is known about the biochemical lesion. However, some of these mutants complemented the defect of other nonpathogenic mutants and cause tissue maceration when inoculated on potato tuber together, thus suggesting production of a factor(s) involved in pathogenicity (4).

Role of O_2 in development of soft-rot disease: It has been shown that reduction in O_2 tension greatly enhances the ability of soft rot *Erwinias* to cause tuber decay (3). Among the prototrophic mutants impaired in pathogenicity, 44 became more pathogenic at 2% O_2, while 3 mutants remained nonpathogenic at 2% O_2. Pathogenic behavior of some of these mutants at two different levels of O_2 are shown in Table 2. We have cloned mutated genes from two of these mutants. One of these mutants becomes pathogenic at low O_2 tension while the other remains nonpathogenic. It is our hope that understanding the function of these cloned genes will provide useful information concerning the role of O_2 tension in the development of soft rot disease.

Molecular cloning of pathogenicity genes of *Ecc*: Using the right end of Mudl as probe we have now cloned DNA sequences flanking the right end of Mudl insertions from mutants AH453, AH2024, AH2552, AH4334 and AH4612. Using these cloned partial genes as probes several putative complete pathogenicity genes have been isolated for each of these mutants from a genomic library of *Ecc* AH2 prepared in the bacteriophage lambda. Presently the characterization of these genes is being performed by complementation of respective nonpathogenic mutants. After subcloning of these genes, an attempt will be made to characterize the protein product encoded by the cloned DNA sequences. Also experiments are in progress to establish the role of these genes in pathogenicity of other soft rot *Erwinias*.

SUMMARY

Development of pathogenesis undoubtedly results from a series of complex interactions between the host and pathogen. At present, the biology of interaction of soft rot bacteria with their host is poorly understood. In addition, little is known about the mechanisms of resistance of plant species to soft rot *Erwinias*. While many species of plants are resistant to *E. carotovora*, different cultivars of susceptible species are reported to vary only slightly in their resistance properties. Although, it is clear that pectic enzymes play a major role in the development of soft-rot disease, it is not clear why genomes of soft rot *Erwinias* carry multiple genes for isofunctional pectolytic enzymes. Results presented here for the first time indicate the role of other genetic determinants of *E. carotovora* in the disease development. Molecular cloning and characterization of these genes will provide some insight towards understanding the interactions between soft rot pathogens and their host. Finally understanding the regulation and expression of these pathogenicity genes may allow development of specific controls for soft rot causing bacterial pathogens.

Acknowledgement: We thank D. Tieman and A. Korty for excellent technical assistance. This research was supported by a USDA CRGO grant (85-CRCR-1-1595) and a grant from Indiana Corporation of Science and Technology.

REFERENCES

1. Castilho, BA, Olfson P, Casadaban MJ (1984) Plasmid insertion mutagenesis and lac gene fusion with mini-Mu bacteriophage transposon. J. Bacteriol 158:488-495.
2. Chatterjee AK, Starr MP (1980) Genetics of Erwinia species. Ann. Rev. Microbiol. 34:645-676.
3. DeBoer SH, Kelman A (1978) Influence of oxygen concentrationand storage factors on susceptibility of potato tuber to bacterial soft rot (Erwinia carotovora). Potato Res 21:65-80.
4. Handa AK, Bressan RA, Korty AG, Jayaswal RK, Charles DJ. Isolation and characterization of pectolytic nonpathogenic mutants of Erwinia carotovora subsp. carotovora. Proc. VIth Int. Conf. Plant Pathogenic Bacteria (In Press).
5. Jayaswal RK, Bressan RA, Handa AK (1984) Mutagenesis of Erwinia carotovora subsp. carotovora with bacteriophage Mud1 (Ap^r lac cts62): Construction of his-lac gene fusions. J. Bacteriol. 158:764-766.
6. Jayaswal RK, Bressan RA, Handa AK (1985) Effects of a mutation that eliminates UDP-glucose-pyrophosphorylase on the pathogenicity of Erwinia carotovora subsp. carotovora. J. Bacteriol. 164:473-476.
7. Lei S-P, Lin H-C, Hefferman L, Wilcox G (1985) Evidence that polygalacturonase is a virulence determinant in Erwinia carotovora. J. Bacteriol. 164:831-835.
8. Roberts DP, Berman PM, Allen C, Stromberg VK, Lacy GH, Mount MS (1986) Requirement for two or more Erwinia carotovora subsp. carotovora pectolytic gene products for maceration of potato tuber tissue by Escherichia coli. J. Bacteriol. 167:279-284.
9. Thurn KK, Chatterjee AK (1985) Single-site chromosomal Tn5 insertions affect the export of pectolytic and cellulolytic enzymes in Erwinia chrysanthemi EC16. Appl. Environ. Microbiol. 50:894-898.
10. Zink RT, Chatterjee AK (1985) Cloning and expression in Escherichia coli of pectinase genes of Erwinia carotovora subsp. carotovora. Appl. Environ. Microbiol 49:714-717.

CHARACTERIZATION OF A NOVEL ESTERASE PRODUCED BY PLANT PATHOGENIC STREPTOMYCES

McQUEEN, D.A.R. AND J.L. SCHOTTEL
Departments of Plant Pathology and Biochemistry, University of Minnesota, St. Paul, MN 55108 U.S.A.

1. INTRODUCTION

A group of Streptomyces, termed S. scabies, are pathogenic on a variety of underground vegetables including potato (3,4). Infection results in the formation of erumpant as well as deep pitted lesions on the surface of the potato tubers (1) (Figure 1). Little, however, is known about the mechanism of pathogenicity. We have been interested in identifying common properties of pathogenic isolates that are involved in causing scab disease. One particular feature of the disease process may be the production of specific extracellular proteins by the pathogen that are involved either in the degradation of suberin, a protective coating on the surface of the tuber (6), or in plant cell recognition. An esterase is one type of enzyme that may be important for both of these processes. Our current studies are focused on the purification and characterization of a novel esterase that is uniquely produced by plant pathogenic Streptomyces.

2. SURVEY OF ESTERASE PRODUCTION IN STREPTOMYCES

Nine strains of Streptomyces sp. isolated from potato scab lesions (9), the ATCC S. scabies strain 10246, S. lividans strain 1326 (2) and S. coelicolor strain M124 (2) were tested for the production of an extracellular esterase that could utilize either p-nitrophenyl butyrate (PNB) or p-nitrophenyl palmitate (PNP) as a substrate (Table 1). In terms of esterase production there were three different classes of Streptomyces strains. The first class of strains (FL1, CBL1, CBL2 and ME1) produced an esterase that preferred PNB as a substrate. ME2, RB1 and 10246 comprised the second class of isolates that produced one or more esterases which utilized both PNP and PNB as substrates. The remaining isolates produced either a very low level of esterase activity (CRYS1) or no detectable esterase (WRB1, PNT1, 1326 and M124) when grown on cutin (a polyester compound similar to suberin (6)) as a carbon source.

Four isolates were chosen for more extensive analysis. One isolate was taken from each of the three categories of Table 1; additionally, PNT1 was included because it is pathogenic (9) but was showing only background PNB esterase activity when grown on cutin. Minimal medium (5) containing suberin or cutin as the sole carbon source was inoculated with the four strains. After three days of growth, the culture filtrates were assayed for PNB activity (data not shown). FL1 and ME2 produced a PNB esterase activity when grown on cutin or suberin. Surprisingly, the PNT1 strain produced a PNB esterase activity, but only when grown on suberin. S. lividans showed no evidence of PNB esterase activity when grown on either carbon source. These results suggested that a novel esterase may be produced by plant pathogenic Streptomyces when suberin is used as the carbon source for growth.

TABLE 1. Esterase activity produced by Streptomyces isolates. All strains were grown in minimal medium (5) with cutin (2.5 mg/ml) as carbon source. The culture filtrates were assayed with PNP or PNB as substrate. The results are the average of three experiments. Protein was measured by the method of Lowry et al. (8).

Strains	Esterase activity (nmoles/min/mg)	
	PNP	PNB
FL1	15	1900
CBL1	30	700
CBL2	<3	230
ME1	<3	60
ME2	1100	280
RB1	330	130
10246	160	56
CRYS1	42	56
WRB1	<3	<3
PNT1	<3	<3
1326	<3	<3
M124	<3	<3

FIGURE 1. Scab lesions on the cultivar Pontiac resulting from infection by S. scabies.

We have carried out additional growth experiments to determine whether utilization of other carbon sources would result in esterase production by the FL1 isolate (data not shown). Growth on the esters palmitic acid methyl ester, oleic acid palmityl ester, or lauric acid myristyl ester resulted in production of esterase activity, but at levels three to five times less than that seen in cutin grown cells. Intermediate amounts of esterase were found when chitin, n-nonocosane, docosane, α-cellulose, or octadecane were utilized as the carbon sources. Barely detectable amounts of esterase were produced when FL1 was grown on either glucose or glycerol as the carbon source. We have found, however, that if the glucose minimal medium is supplemented with 2 μM zinc, esterase production is comparable to cutin grown cultures. A further 2 to 3 fold increase in activity is achieved when 2 μM iron is added to the medium in addition to the 2 μM zinc. It is not yet understood as to how these trace elements stimulate esterase production.

3. POLYACRYLAMIDE GEL ANALYSIS OF ESTERASE ACTIVITY

Native polyacrylamide gels of extracellular proteins were stained in situ with α-napthyl butyrate to visually detect esterase activities (10). The results of the activity stain indicated that FL1 and PNT1 produced one major species of α-NB esterase activity with similar migration rates when grown on suberin. ME2 and S. lividans 1326 showed minor α-NB esterase activities, but the relative migration rates of these proteins were different from the major FL1 and PNT1 esterase activities (data not shown). To demonstrate the relationship between the PNB activity of the culture filtrate and the α-NB activity of the in situ stained gels, the region of FL1 α-NB activity was cut from a non-denaturing polyacrylamide gel. The protein was electroeluted from the gel slice and was found to contain PNB activity (data not shown).

4. PURIFICATION AND CHARACTERIZATION OF THE ESTERASE FROM FL1

The major esterase activity from the FL1 isolate was purified by ion exchange chromatography. Concentrated culture filtrate was loaded onto a DEAE Sephadex column equilibrated with 50 mM Tris pH 7.5, 0.1 M NaCl. The esterase was eluted from the column with a salt gradient at 0.6 M NaCl.

The molecular weight of the esterase was estimated to be 36,000 by Sephadex G-100 gel filtration and by denaturing polyacrylamide gel electrophoresis. This enzyme is quite heat stable, losing only 10% of its activity when treated at 80°C for 10 minutes. The apparent Km of the esterase is 125 μM PNB and does not show Michaelis-Menten kinetics. These properties of the esterase have indicated that this enzyme is different from the esterase characterized by Lin and Kolattukudy (7).

In addition to the esterase activity of this protein, we have also detected a haemagglutination activity. This ability to agglutinate red blood cells is inhibited by D-glucosamine. The agglutination activity appears to be somewhat unstable, and the conditions necessary for its stabilization are presently being determined.

5. FUTURE PROSPECTS

Esterase minus mutants of the FL1 strain are currently being isolated to determine the role that this protein plays in pathogenicity. Additional areas of interest are to investigate the effect of trace elements on the production of the esterase by these organisms and to determine whether the agglutination activity of this protein is a characteristic important to the disease causing process. Because of its extracellular location, this enzyme will also be used to study the mechanism of protein secretion in Streptomyces.

REFERENCES

1. Archuleta, J.G. and G.D. Easton. 1981. Amer. Potato J. 58:385-392.
2. Bibb, M.J. and D.A. Hopwood. 1981. J. Gen. Microbiol. 126:427-442.
3. Davis, J.R. and J. Garner. 1978. University of Idaho Agricultural Experiment Station. Current Information Series No. 386.
4. Harrison, M.D. 1962. Amer. Potato J. 39:368-387.
5. Hopwood, D.A. 1967. Bacteriol. Rev. 31:373-403.
6. Kolattukudy, P.E. 1980. Science 208:990-999.
7. Lin, T.S. and P.E. Kolattukudy. 1980. Physiol. Plant Path. 17:1-15.
8. Lowry, O.H., N.J. Rosebrough, A.L. Farr and R.J. Randall. 1951. J. Biol. Chem. 193:265-275.
9. McQueen, D.A.R., N.A. Anderson and J.L. Schottel. 1985. J. Gen. Microbiol. 131:1149-1155.
10. Rosenburg, M., V. Roegner and F.F. Becker. 1975. Analytical Biochem. 66:206-212.

Section III

MOLECULAR GENETICS OF THE HOST (SYMBIOSIS/PATHOGENICITY)

INDUCED SYMBIOSIS MUTANTS OF PISUM SATIVUM

B.E. KNEEN, D. VAM VIKITES AND T.A. LARUE
Boyce Thompson Institue, Ithaca, NY 14853 USA

INTRODUCTION

The availability of mutants is of fundamental importance to biological research. Complex metabolic pathways and developmental processes can be analyzed using a collection of mutants which are defective at various stages. Mutations in the plant genes are needed to study symbiotic nitrogen fixation, for the use of rhizobial mutants alone will not indicate the developmental stages controlled by plant genes. We plan to identify host factors involved in recognition, infection and control of nodule number through the use of mutants of *Pisum sativum*. After a transformation and regeneration system is established for *Pisum*, the function of cloned symbiotic genes may be discerned via complementation of host mutants.

Lie discovered several naturally occurring nodulation resistant varieties of pea including temperature-sensitive 'Iran' (*sym* 1) (11), strain-dependent 'Afghanistan' (*sym* 2, *sym* 3, *sym* 6) (5,8,12,14,18), and another wild pea with strain-specific nodulation (*sym* 4) (13). The best studied natural variant is the primitive cultivar 'Afghanistan'. It forms few or no nodules with strains of *R. leguminosarum* from temperate soils, but nodulates with some strains, typified by TOM, from Middle Eastern soils. The gene conditioning strain specific nodulation, *sym* 2, has not been easily characterized because 'Afghanistan' also has two semi-dominant genes modifying nodule number (16).

Because naturally occurring variants are rare and genetically complex, researchers have turned to induced mutagenesis to obtain near-isogenic lines of pea altered in the symbiosis (3,4,9,15).

MUTATION, SELECTION AND CHARACTERIZATION

'Sparkle' is an early flowering small-statured garden pea (Rogers Bros. Seed Co.) which nodulates well (~100) with every strain of *R. leguminosarum* we have tested. Ethyl methanesulfonic acid (0.5-1.0%) for 0.5-2.0 hours, nitroso urea (0.02-0.2%) for 3-6 hours, 15 or 25 kR gamma radiation or 0.5-1.5 kR neutron radiation (1,2) all produced mutations. M_1 seedlings were grown in the greenhouse or field.

M_2 seeds were planted in Conetainers filled with coarse vermiculite, inoculated with *R. leguminosarum* 128C53 (Nitragin Co.). and subirrigated with N-free nutrient. Roots were examined three weeks after planting, and putative mutants transplanted. If the phenotype was stably inherited in the M_3 generation, those selections were grown for seed increases and crosses to 'Sparkle'. F_1 data provides information about the dominance or recessiveness of the mutant allele at the *sym* locus. F_2 segregants with the mutant phenotype are back-crossed again to the parent to assure that undetected mutations at other loci are returned to parental type. Crosses are made among mutants to determine allelism. After we consult with other workers, gene designations are published in the "Pisum Newsletter."

To determine if the phenotype is shoot or root-controlled, control and reciprocal shoot grafts are made between the mutants and the parent, inoculated with R. leguminosarum 128C53, kept in dim light 3 days, and grown in the lightroom for 14 days before nodulation is scored. To determine if the mutation involves a temperature sensitive gene, some plants are grown in a chamber in which the roots are kept cool at 9-12C while the shoots are in the regular 20/15C day/night cycle. Plants are harvested 28 days after planting. Sibling plants are grown in the 20/15C and in a warm (26/23C) day/night cycle, and harvested at 21 days. To determine if the non-nodulation is strain dependent, seeds are surface-sterilized and planted in sterile 180 ml Dispo bottles and inoculated with one of nine test strains. Five strains are from temperate soils and four (including TOM) are strains from Middle Eastern soils.

We obtained over 30 lines with aberrant nodulation. These include mutants which do not nodulate, or which have lower than normal nodule numbers, or which have nodules in which nitrogenase (C_2H_2) activity is low or absent. The reciprocal diallel crossing program is not complete. The first mutant (E2) obtained through EMS treatment is controlled by a single gene, sym 5, which is nonallelic to Afghanistan's sym 2 (9). The selection has infection threads and occasionally a few nodules are observed. This line does not display the strain specificity of 'Afghanistan' with Middle Eastern strains of Rhizobium leguminosarum. Grafting experiments showed that the non-nod factor is associated with the roots, not shoots. Sym 5 is a mutational hotspot; we obtained 6 mutants from EMS and one from γ-radiation with alleles at this locus, as evidenced by lack of nodulating recombinants in the F_1 and F_2 generations of crosses among the mutants. This shows that different mutagens may act at the same gene locus.

Although E2 was non-nodulated in vermiculite or soil at a 20/15C day/night cycle, we found in field plots that E2 planted in early Spring nodulated. This effect was traced to the soil temperature. In our lightrooms, E2 grown in vermiculite forms nodules when the roots are maintained day/night at 9-12 C. Thus, we have a temperature sensitive mutation.

Other non-nodulating mutants were characterized as non-allelic monogenic recessive mutations: E69 (sym 7), R25 (sym 8), R72 (sym 9), N15 (sym 10) and N24 (sym 11). These selections have apparently normal appearance and have normal seed yield when grown in Redi-earth. They are not nodulated by any rhizobial strain we have tested.

Jacobsen and Feenstra (7) obtained an EMS-induced mutant of 'Rondo' with higher than normal nodule numbers even in the presence of 15 mM KNO_3. The mutant gene was tentatively designated nod 3 nod 3. Jacobsen (4,6) obtained two non-nodulating pea mutants during his search for nitrate resistant mutants. One of these (K5) is at a novel locus (sym 12), while K24 is allelic with two of our mutants and 'Afghanistan'.

We selected pleiotropic mutants in which the nodulation defect segregates together with a change in plant morphology. This suggests that these genes have another function in the plant in addition to being involved in the symbiosis. An EMS-derived mutant, E140 is allelic with the sym 8 gene found in the γ-radiation induced mutant R25. E140 has shorter internodes than the parent 'Sparkle' and R25. This indicates that phenotypically different symbiotic mutant alleles may arise at the same locus.

An EMS-derived mutant, E135, forms a normal number of white nodules lacking nitrogenase (C_2H_2) activity. This is conditioned by a monogenic recessive allele at the sym 13 locus. Non-nodulating plants appeared in the F_2 progeny from a backcross to parent, and genetic analysis indicates that isolate E135 also carried an independent heterozygous mutant sym gene conditioning non-nodulation when homozygous. Postma *et al.* (17) also found an EMS-derived nodulated mutant (85.706-1) lacking nitrogenase activity. This mutant line also segregated for a non-nodulated phenotype, indicating a mutation at another sym locus. These observations show that EMS treatment may mutate more than one sym gene. Researchers should adequately characterize their mutant lines by two series of back-crosses before starting physiological experiments, to insure they are studying the effects of a single gene.

Strain-specificity is not an all or nothing phenomenon; rather it is a host-strain-temperature interaction that results in differing numbers of nodules. There appear to be genes in both the bacteria and the plant limiting infections and/or nodule number subject to environmental variation or influenced by environmental conditions.

Three selections, E54, N25 and N27 exhibit strain specific nodulation similar to 'Afghanistan' in being well nodulated by Middle Eastern strains of *R. leguminosarum*. However, these mutants differ in their capacity to nodulate with various temperate strains (Table 1). Whereas mutant line E54 nodulates best with temperate strains R1300 and PF2, lines N25 and N27 interact most favorably with strains 128C53 and ATCC 10004. The following genetic analyses were performed at 20C with strain 128C53. The F_1 progeny of crosses between wildtype 'Sparkle' and mutant E54 provide evidence for a dominant allele for intermediate numbers of nodules segregating 3 intermediate:1 high in the F_2. In contrast, the intermediate nodulation phenotypes of lines N25 and N27 are recessive to high nodulation. Testing for allelism, crosses were made among the mutant lines and 'Afghanistan' derived lines ('Afgh' backcross to 'Spkl', carrying the sym 2 gene for strain specificity). The F_1 progeny of the ('Afgh' x 'Spkl') backcross x N25 or N27 were all parental type (non-nod, <20) indicating allelism. F_1 data from crosses of E54 to other strain specific lines are difficult to interpret because E54's intermediate nodulation phenotype is dominant over high nodule number.

Although crosses between lines N25 and N27 indicate allelism, the morphology of these plants are different. N25 is like 'Sparkle' in stature and days to flower, while N27 is taller and flowers several days later. This suggests that these lines are controlled by different alleles at the same locus, one having a pleiotropic effect.

Other mutant selections, NEU5, NMU1, and K24 (4) are not nodulated by any tested strain, but are allelic with a gene in 'Afghanistan' controlling nodule number. NEU5 x 'Sparkle' crosses show that the non-nod phenotype is recessive to wildtype and the F_2 data fit the expected 3 nod:1 non-nod ratio for monogenic recessive control. Using the TOM strain, close to 25% of the F_2 population was non-nodulated . Whereas F_1 progeny of crosses of either NEU5 or 'Afghanistan' derived lines with 'Sparkle' are nodulated, F_1 progeny of crosses between NEU5 and 'Afghanistan' have no or very few nodules, indicating allelism. Morphological markers, such as plant height and seed coat mottling, were used to insure that the F_1's were not accidental selfs. Approximately 86% of the NEU5 x 'Afgh' F_2 population had no or < 20 nodules, while the remaining 14% had > 60 nodules. The lack of intermediates (20-60 nodules, as found in 'Afgh' x 'Spkl' F_2 populations), suggests that the gene controlling non-nodulation in NEU5 is allelic with a gene modifying nodule number in 'Afghanistan'. The 14% high-nod recombinants in the F_2 fits a hypothesis that 3 genes are segregating in this cross. The NEU5 non-nod allele is recessive to the wildtype nodulation allele, but dominant to the allele for intermediate nodule number and epistatic to other nodulation loci in 'Afghanistan'. Using strain TOM, the nodulation-resistance genes of 'Afghanistan' are not expressed and F_2 progeny of the NEU5 x 'Afgh' cross fit the expected 3 nod:1 non-nod segregation ratio for a non-strain specific allele.

In summary (Table 2), there are non-nodulating mutations at least seven different loci (<u>sym</u> 5,7,8,9,10,11,12). There are several mutants with pleiotropic effects on nodulation and plant morphology, and an ineffective mutant designated <u>sym</u> 13. The remaining two groups of mutants are allelic with genes controlling nodulation in the primitive pea 'Afghanistan'. The first group exhibits strain-specificity, while the second group does not nodulate with any strain tested. These induced

TABLE 1. Interaction of pea mutants and rhizobial strain on nodule number.

Pea Line	Strain: 128C53	R1300	ATCC 10004	PRE	PF2	TOM
	Average Nodule Number					
'Sparkle'	119	105	138	122	104	110
'Afghanistan'	0	1	0	0	1	46
Spkl x Afgh-BC_5	1	6	1	2	2	110
Mutants						
E54 Sparkle	23	43	21	10	84	75
N25 Sparkle	37	7	61	1	24	114
N27 Sparkle	22	6	32	1	8	70

The number of nodules is the average ($n \geq 4$) on 3 week old seedlings inoculated with a single strain of *Rhizobium leguminosarum* at 20C.

Source of strains:

128C53	Nitragin Co., Milwaukee, WI, USA.
R1300	John Innes Institute, Norwich, UK.
ATCC 10004	American Type Culture Collection, Beltsville, MD, USA.
PRE, PF2, TOM	T. Lie, Wageningen, The Netherlands.
BB54b	Cent. for Agr. Res. Alleppo, Syria } (Similar to TOM;
510P, 511P	Volcani Inst., Bet Dagen, Israel } not incl. in table)

single-gene mutations will eventually provide valuable information about host factors involved in specificity and control of nodule number.

TABLE 2. Induced nodulation mutants of *Pisum sativum* used at our lab.

Line	Gene Designation	Comments
E2, E77, E11, E143, E166, E169, R88	*sym*-5	Temperature sensitive. Nodulates if roots are maintained at 9-12C. Few or no nodules when grown at 20/15C day/night.
E69, N12	*sym*-7	Do not nodulate with any tested strain of *R. leguminosarum*, or at low root temperatures. All lines appear normal except for E140 (*sym*-8) which has shorter internodes than the parent.
R25, E140, R19	*sym*-8	
R72	*sym*-9	
N15	*sym*-10	
N24	*sym*-11	
K5	*sym*-12	
E135	*sym*-13	Small white nodules lacking nitrogenase (C_2H_2) activity.
E107 (bronze leaf spot)	-	Tests for allelism still in progress. In these lines, low or non-nodulation segregates together with a morphological character.
E132 (short root laterals)	-	
E134 (abnormal flower)	-	
E147 (short root laterals)	-	
E151 (short root laterals)	-	
R50 (short internodes)	-	
R82 (stubby root)	-	
R96 (short root laterals)	-	
E54 (late flowering)	-	Allelic with strain-specificity gene in 'Afghanistan'. Well nodulated by strain TOM; with other strains nodules are low or intermediate in number.
N25	-	
N27 (late flowering)	-	
NEU5, NMU1 (low fertility)	-	Allelic with nodule number gene in 'Afghanistan'.
K24		Do not nodulate with any tested strain of *R. leguminosarum*.

REFERENCES

1. Arenaz P and BK Vig: Somatic crossing-over in *Glycine max* (L.) Merrill: activation of dimethyl nitrosamine by plant seed and comparison with methyl nitrosourea in inducing somatic mosaicism. Mutat. Res. 52: 367-380, 1978.

2. Constantin MJ, WD Klobe and LN Skold: Effects of physical and chemical mutagens on survival, growth, and seed yield of soybeans. Crop Sci. 16:49-52.
3. Engvild KC: Nitrogen fixation mutants of pea *in* Analysis of the Plant Genes Involved in the Legume-*Rhizobium* Symbiosis. R. Marcellin, ed. OECD, Paris p49, 1985.
4. Feenstra WJ and E Jacobsen: Pea mutants with an altered response to *Rhizobium leguminosarum*. *in* Analysis of the Plant Genes Involved in the Legume-*Rhizobium* Symbiosis, R. Marcellin, ed. OECD, Paris. pp. 50-51, 1985.
5. Holl FB: Host plant control of the inheritance of dinitrogen fixation in the *Pisum* - *Rhizobium* symbiosis. Euphytica 24:767-770, 1975.
6. Jacobsen E: Modification of symbiotic interaction of pea (*Pisum sativum* L.) and *Rhizobium leguminosarum* by induced mutations. Plant and Soil 82:427-438, 1984.
7. Jacobsen E and WJ Feenstra: A new pea mutant with efficient nodulation in the presence of nitrate. Plant Sci. Letters 33:337-344, 1984.
8. Kneen BE and TA LaRue: Peas (*Pisum sativum*) with strain specificity for *Rhizobium leguminosarum*. Heredity 52:383-389, 1984.
9. Kneen BE and TA LaRue: Nodulation resistant mutant of *Pisum sativum* L. J. Heredity 75:238-240, 1984.
10. LaRue TA, BE Kneen and E Gartside: Plant mutants defective in symbiotic nitrogen fixation. *in* Analysis of the Plant Genes Involved in the Legume-Rhizobium Symbiosis, R. Marcellin, ed. OECD, Paris. pp. 39-48, 1985.
11. Lie TA: Symbiotic nitrogen fixation under stress conditions. Plant and Soil, Spec. Vol.:117-127, 1971.
12. Lie TA: Symbiotic specialization in pea plants: the requirement of specific *Rhizobium* strains for pea from Afghanistan. Ann. Appl. Biol. 88:445-487, 1978.
13. Lie TA: Host genes in *Pisum sativum* conferring resistance to European *Rhizobium leguminosarum*. Plant and Soil 82:415-425, 1984.
14. Lie TA and PCJM Timmermans: Host genetic control of nitrogen fixation in the legume - *Rhizobium* symbiosis: complication in the genetic analysis due to maternal effects. Plant and Soil 75:449-453, 1983.
15. Messager A: Selection of pea mutants for nodulation and nitrogen fixation. *in* Analysis of the Plant Genes Involved in the Legume-Rhizobium Symbiosis, R. Marcellin, ed. OECD, Paris. pp. 52-60, 1985.
16. Ohlendorf H: Genetic studies of resistance to strain 311d in *Pisum sativum*. Z. Pflanzenzuchtg 91:13-24, 1983.
17. Postma JG, E Jacobsen, T Bisseling and WJ Feenstra: A mutant of pea (*Pisum sativum*) possibly disturbed in the production of a compound required for the induction of nitrogenase activity in bacteroids. These proceedings, 1986.
18. Young JPW and P Matthews: A distinct class of peas (*Pisum sativum* L.) from Afghanistan that show strain specificity for symbiotic *Rhizobium*. Heredity 48:203-210, 1982.

PLANT HOST GENETICS OF NODULATION INITIATION IN SOYBEAN

PETER M. GRESSHOFF, JANE E. OLSSON, DAVID A. DAY, KATHRYN A. SCHULLER, ANNE MATHEWS, ANGELA C. DELVES, ARNO KROTZKY, G. DEAN PRICE AND BERNARD J. CARROLL.
Botany Department, Australian National University, Canberra, ACT, 2600, Australia.

INTRODUCTION

Symbiotic nitrogen fixation as exemplified by the legume-Rhizobium (or Bradyrhizobium) root nodule interaction is a well researched phenomenom illustrating plant-microbe interaction. The functional nitrogen fixing symbiosis requires cooperation between the bacterium and the plant. The last decade has witnessed a rapid expansion of the definition of Rhizobium genes that are involved in the symbiosis. The plant's contribution, although always recognised as being important, recently received more attention through two major developments. The first was the application of DNA technology to the analysis of gene expression of legume symbiotic genes (Verma et al 1985) and the second was the realisation that existant plant variability may be insufficient in many legumes to permit the isolation of symbiotically defective germplasm. For this reason, research with Pisum sativum (Feenstra and Jacobson, 1985) (LaRue et al, 1985), Cicer arietinum (Davies et al, 1985) and Glycine max (our laboratory) has concentrated on the isolation of symbiotic mutants after induced mutagenesis (Carroll et al, 1985a,b). The approach has been highly successful, and our knowledge on the soybean mutants is summarised in the following pages:-

SUMMARY OF RESULTS

(a) Non-nodulation mutants

Three non-nodulation mutants of the soybean cultivar Bragg were isolated after ethyl methanesulphonate (EMS) mutagenesis (Mathews et al, 1987). Genetic complementation analysis shows that the mutants nod49 and nod772 are allelic with the naturally occurring soybean variant rj_1 (Mathews et al, in preparation). Mutant nod49 plants like rj_1 plants lack root hair curling and subepidermal cortical cell divisions. Mutant nod772, despite its allelism to nod49, is more "leaky" and shows some root hair curling and an increased level of "occasional nodulation". Mutant nod139 also lacks root hair curling and has very few "occasional nodules" (see Mathews et al, this volume).

"Occasional nodules" occur on all three mutant isolates, as they occur also on rj_1 plants. These nodules are of normal structure and fix nitrogen efficiently. Nodule occupancy tests reveal that the resident bacterium is identical in its symbiotic properties to the original inoculant. Thus the plant did not select a variant bacterium, which is capable of suppressing the non-nodulation response.

A large range of unrelated Bradyrhizobium japonicum isolates were tested for their ability to elicit the mutant phenotype, which they all did with varying degrees of "occasional nodulation". Non-nodulation was always observed under field conditions. "Occasional nodulation" was influenced by the inoculum level. Increased bacterial densities resulted in increased mean nodule numbers and mass as well as nodulation frequencies, when bacteria were inoculated into Leonard jars containing vermiculite. However, despite the presence of some large nodules, their activity was insufficient to support good nitrogen fixation dependent plant growth.

By grafting shoots onto roots it was revealed that the non-nodulation phenotype (for all of the three mutants resembling rj_1 plants) was controlled by the genotype of the root (Delves et al, 1986). By grafting the shoot of a supernodulation mutant (like nts382; see later discussion), which usually confers the supernodulating phenotype on both other nts mutant roots as well as wildtype roots, onto a root of a nod49 plant, it was demonstrated that the shoot control of supernodulation was not permitted expression when the root was that of the nod49 mutation.

This interaction at the tissue level was further investigated by two other means.

Plants from nod49, nts382 and Bragg were cocultured at high plant densities (resulting in intimately intergrown root systems) in pots and Leonard jars. Under all conditions nodulation was as expected for the "control" planting. In other words, nod49 plants (like rj_1 plants) failed to produce a diffusable rhizosphere factor that inhibited nodulation of control plants, similarly, control plants failed to produce an exudate substance which could suppress (cross-feed) the mutational block of mutant nod49. Thus we believe that nod49 (like rj_1 plants) exhibits a "resistance to Bradyrhizobium") manifested through an inability to interpret, or receive, a bacterial signal necessary for the initiation of the nodulation sequence. "Occasional nodules" are viewed as an escape from that resistance facilitated by either an alternative infection mechanism, random chance, or possibly somatic mosaicism.

Reciprocal crosses between nts382 and nod49/nod139/nod772 resulted in wildtype F1 progeny, indicating that both nts382 and the non-nodulation mutations are caused by recessive genes. Instead of the 9:3:3:1 ratio expected in F2 plants from the above selfed F1 plants for two unlinked non-epistatic recessive genes, a 9:3:4 (wildtype:nts:nod^-) ratio was observed. This is explained by the postulated epistasis, which prevents the expression of the homozygous nts condition in the homozygous recessive nod^- mutant. Thus the homozygous recessive double mutant has a non-nodulation phenotype. In order to verify this hypothesis F3 plant material was raised. Likewise shoots from F2 nod^-plants were grafted onto wildtype roots. Such roots serve as an "indicator" for the presence of a double recessive nts conditions in the grafted scion and should express the supernodulating phenotype.

(b) Supernodulation mutants

As described by Carroll et al (1985a,b) it was possible to isolate soybean mutants which form an increased number of nodules. This supernodulation is also expressed in the presence of nitrate. Table 1 gives an indication of the symbiotic parameters of some soybean mutants and the wildtype cultivar Bragg.

TABLE 1. Symbiotic parameters of some supernodulation mutants.

Genotype	Nodule number $-NO_3$	Nodule number $+NO_3$	nodule dry weight $-NO_3$	nodule dry weight $+NO_3$	nitrogenase activity $-NO_3$	nitrogenase activity $+NO_3$
	/plant		mg/plant		nmol $C_2H_4.min^{-1}$/plant	
Bragg	26 ± 6	19 ± 7	31 ± 10	5 ± 3	71 ± 13	1 ± 1
nts1007	334 ± 292	991 ± 231	92 ± 49	179 ± 35	98 ± 29	88 ± 20
nts1116	101 ± 26	74 ± 45	66 ± 12	30 ± 12	85 ± 17	23 ± 10
nts2062	123 ± 46	299 ± 69	120 ± 17	144 ± 13	169 ± 24	108 ± 53
nts382	576 ± 77	1007 ± 154	166 ± 9	193 ± 20	119 ± 35	69 ± 11

- harvested 26 days after planting. Inoculated with B. japonicum strain USDA110. (means of 3-6 plants ± standard deviation). Data as collected by Carroll et al, 1985a.

In all cases (12 mutants were studied in detail) supernodulation correlated with nitrate tolerance.

Supernodulating nts382 plants were analysed for their growth potential in controlled glasshouse conditions (Day et al, 1987). Plants were grown either inoculated in the absence of external nitrogen source (NO_3^-) or uninoculated in the presence of 5mM KNO_3. For both treatments, nts382 growth up to 13 days after planting was faster than that of the cultivar Bragg. Thereafter, supernodulation of inoculated nts382 occurred and growth of the cultivar Bragg was faster. Shoot and root dry weight increments were greater as was leaf area in Bragg, but the N content of nts382 was higher. Shoot growth of uninoculated plants was similar but roots were slightly smaller in nts382 than Bragg. Early lateral root formation (prior to nodule emergence) was greater in nts382 plants regardless of the culture regime. Thus it was concluded that although supernodulation is a "retardant" to shoot and root growth in mutant nts38, the mutant itself still shows an altered phenotype (albeit slightly) in the absence of any nodulation (Day et al, 1987). Mutant nts1007 which appears to be allelic with nts382, had a more prolific root system even with nodulation than mutant nts382. Such differences are expected in view of the independent genetic nature of the individual isolates.

Nodules of mutant nts382 are characterised by an altered morphology. Plant cells are smaller (less hypertrophy) and there is less leghemoglobin and a smaller infected area (G.D. Price, unpublished data).

Thus the bacteroid content per nodule and per cell is decreased relative to the wildtype. Nitrogenase activity per milligram bacteroid protein is identical in both Bragg and nts382. Kathryn Schuller and David Day found that if the nodule number of nts382 is reduced by low titre Bradyrhizobium inoculation, average nodule size increases and specific activity per nodule also approaches that of wildtype. Thus the nts condition results in a large series of pleiotropic changes ranging from altered nodule and plant structure to changes in flowering times (see Gresshoff and Delves, 1986).

Genetic analysis supports the conclusion that the nts382 mutant is caused by a single gene. Nts382 was backcrossed to wildtype cultivars Willams and Clark, F1 (wildtype phenotype) plants were raised and allowed to set F2 seeds. These segregated with a 3:1 ratio as predicted for simple single gene Mendelian inheritance. The F2 nts plants were raised to obtain the F3 (self fertilised). All F3 plants derived from nts F2 plants showed the nts phenotype (Delves, unpublished data).

The supernodulation phenotype gives a moderate tolerance to delayed nitrate application (Schuller, 1986). When symbiotic plants are exposed to 10mM nitrate, acetylene reduction and ^{15}N incorporation fall dramatically in Bragg plants, but not in nts382 plants. Prolonged exposure, however, removes this apparent tolerance as nodule senescence sets in. This apparent short term tolerance may be a function of the altered nodule anatomy which may result in altered gas diffusion into the nodule.

Grafting experiments (Gresshoff et al, 1985; Delves et al, 1986; Gresshoff and Delves, 1986) coupled with the application of shoot extracts and/or phytogrowth regulators suggest that the nts phenotype requires the presence of an nts shoot. This holds true for several nts mutants (Delves, unpublished data). It is clear now that the supernodulation response occurs because of a release of nodulation inhibition, which occurs in wildtype plants. We are presently trying to specify the chemical nature of the shoot factor which is directly involved in the autoregulation response, and which clearly is absent (or reduced) in the nts mutants.

(c) Commercial application

As so much of our funding is available because of the potential application of our research to agriculture, field tests were done using the mutant lines. The non-nodulation mutants are of extreme value in determining nitrogen fixation levels in field conditions. In addition they may allow the development of a specific symbiosis, which could eliminate Rhizobium competition in the soil if a specially constructed Bradyrhizobium strain is available to overcome the mutational blockage in the plant.

Yield data on the nts mutants are relatively variable. As the plants have an altered growth habit, new agricultural systems need to be developed to allow their full potential to be realised.

In a replicated field trial carried out in collaboration with Dr D. Herridge (Tamworth) it was found that nts1007 plants yielded 80-90% of Bragg. In contrast nts1116 plants gave seed yields between 80 and 115% of Bragg depending on the N content and Bradyrhizobium history of the field site. Nts mutants clearly had increased symbiotic dependency, suggesting that additional N was spared in the soil.

Thus the moderate nodulation mutant nts1116 appears to have commercial applicability. More complete agronomic trials require further back-crossing and progeny analysis. The possibility exists that supernodulation mutants such as nts382 have real growth potential in environmental conditions which otherwise restrict nodule development. Some preliminary experiments by B. Carroll in collaboration with researchers in Brisbane (Australia) indicated that nts382 plants grow better at acid pH than Bragg plants do at neutral pH. Such findings require further confirmation.

Whatever the agronomic applications of the soybean mutant material(here described), it will provide the source for expanded analysis into the plant genes which control infection, nodule initiation, and nodule growth.

Conclusion

Few advances have occurred in the host genetic analysis of plant-microbe interactions since the proposal of the gene-for-gene hypothesis by Flor in 1955. It is of interest to observe that both pea and soybean mutants fall into some "mutational" hot spots. Likewise the presence of "occasional nodules" is not yet explained.

As mutations in the "pathogen" change avirulence into virulence through a "loss" of function (i.e. a dominant allele becomes recessive), the plant "requires" to attain a new function through mutation (recessive to dominant). Plant-microbe interactions presumably have co-evolved, so how can one attain such frequency of "beneficial" mutations that generate a new function. The generation of diversity in immune response genes in mammals is well studied and understood. Is it possible that in plants the genes that are responsible for the maintenance of host-pathogen balance are also subject to a genic rearrangement mechanism such as seen with the immunoglobulin genes or the trypanosome surface antigen genes? Perhaps the isolated "resistances to Rhizobium" represent such alterations, just as "occasional nodules" may represent somatic rearrangements to sensitivity in a chimeral tissue. The analysis of the precise nature of nodulation control genes and their molecular surroundings will shed more light on such hypotheses.

REFERENCES

1) Carroll BJ, McNeil DL and PM Gresshoff: Islation and properties of soybean mutants that nodulate in the presence of high nitrate concentrations. Proc. Nat. Acad. Sci. (USA) 82, 4162-4166. 1985a.
2) Carroll BJ, McNeil DL and PM Gresshoff: A supernodulation and nitrate-tolerant-symbiotic (nts) soybean mutant. Plant Physiol. 78, 34-40. 1985b.

3) Davies TM, Foster KW and DA Phillips: Non-nodulation mutants of chick pea. Crop Sci. 25, 345-348. 1985.
4) Day DA, Lambers H, Bateman J, Carroll BJ and PM Gresshoff: Growth comparisons of a supernodulating soybean (Glycine max) mutant and its wildtype parent. Physiol. Plant (in press). 1987.
5) Delves AC, Mathews A, Day DA, Carter AS, Carroll BJ and PM Gresshoff: Regulation of the soybean-Rhizobium nodule symbiosis by shoot and root factors. Plant Physiol. (in press). 1986.
6) Feenstra WJ and E Jacobson: Pea mutants with an altered response to Rhizobium leguminosarum in: Analysis of the plant genes involved in the legume-Rhizobium symbiosis. 50-51, Paris OECD. Publ. 1985.
7) Flor HH: Host parasite interaction in flax rust, its genetic and other implications. Phytopathol. 45, 680-685. 1955.
8) Gresshoff PM and AC Delves: Plant genetic approaches to symbiotic nodulation and nitrogen fixation in legumes in: King PJ and Hohn TH (eds). Plant Gene Research, vol. 3, Springer Verlag (in press) 1986.
9) Gresshoff PM, Day DA, Delves AC, Mathews A, Olsson JE, Price GD, Schuller KA and BJ Carroll: Plant host genetics of symbiotic nitrogen fixation and nodulation in pea and soybean in: Nitrogen Fixation Research Progress, pp. 19-25 (eds) HJ Evans, PJ Bottomley an WE Newton. M. Nighoff Publ. Dardrecht, NL. 1985.
10) LaRue TA, Keen BA and E Gartside: Plant mutation defective in symbiotic nitrogen fixation in: Analysis of plant genes involved in the legume-Rhizobium symbiosis. 39-48, Paris OECD Publ. 1985.
11) Mathews A, Carroll BJ and PM Gresshoff: Characterisation of the non-nodulation mutants of soybean (Glycine max (L.) Merr.) J. Plant Physiol. (submitted). 1987.
12) Schuller KA: PhD dissertation, Botany Department, Australian National University, Canberra. 1986.
13) Verma DPS, Lee JS, Katinakis P and B Sutton: Nodule specific genes of soybean in: Analysis of plant genes involved in the legume Rhizobium symbiosis. 74-84, Paris OECD. 1985.

A MUTANT OF PEA (PISUM SATIVUM) POSSIBLY DISTURBED IN THE PRODUCTION OF A COMPOUND REQUIRED FOR THE INDUCTION OF NITROGENASE ACTIVITY IN BACTEROIDS

JENNE G. POSTMA[1], EVERT JACOBSEN[1], TON BISSELING[2] and WILLEM J. FEENSTRA[1]

[1]Dept. of Genetics, University of Groningen, 9751 NN Haren, The Netherlands

[2]Dept. of Moleculair Biology, Agricultural University, 6703 BC Wageningen, The Netherlands.

INTRODUCTION

Compared to our knowledge of the genes of Rhizobium species concerned in the symbiontic nitrogen fixation, data about the genetic information of the host plant pertaining to the interaction with the bacteria are relatively scarce. Surveys of the available data have recently been given (1, 2). For a genetic analysis mutant alleles of host plant genes are required. The effect of a mutation may either be a block in a step leading to the formation of nodules or interfere with the functioning of the bacteroids.

Pea (Pisum sativum) is genetically well-known and is a suitable species for the induction and selection of mutants. Moreover, nodulins, i.e. plant proteins which are specific for nodule formation and function, have been isolated from this species (3). Therefore it was chosen for the isolation of mutants which are disturbed in the symbiosis with Rhizobium leguminosarum.

Nodulation-defective mutants (sym^-) can be screened more easily than mutants having ineffective nodules (fix^-), since for the identification of the latter a biochemical test (acetylene reduction) has to be employed. Pre-selection for nitrogen-deficiency symptoms during growth reduces the amount of work to be done.

Usually, also non-mutant plants when completely dependent on symbiontic nitrogen fixation for their nitrogen supply show poor growth, especially early in their development. In order to ensure good growth of the non-mutant plants, we used an early and profusely nodulating mutant of pea, nod_3, which does not show deficiency symptoms when wholly depending on N-fixation from the beginning of its growth. The early nodulation also allows a better identification of non- or late nodulating mutants.

MATERIALS AND METHODS

Mutagenic treatment and growth of M1

Seeds of the mutant nod_3 (4), isolated from the variety 'Rondo' of Pisum sativum (L) were used for the induction of mutations. Mutagenic treatment and growth of M1 were essentially as given in (5), except that seeds were pre-imbibed for 6 h at room temperature.

Selection of mutants

Six to nine seeds per M2-family were germinated in vermiculite moistened with tapwater. After one week seedlings were inocculated with Rhizobium leguminosarum strain PF2 and placed on aerated liquid medium (standard mineral solution (5) supplemented with 1070 mg/l KCl). After two more weeks plants were visually screened for the presence of nodules on the roots and after another week for N-deficiency symptoms in the leaves. The acetylene-reducing capacity of abnormal plants which did have nodules was determined. Prospective mutants of both types (sym^- and fix^-) were transferred to liquid medium of the same composition, but with KNO_3 (1500 mg/l) substituted for KCl, and grown to maturity.

Due to the period of poor growth preceding the moment of transfer to complete nutrient solution, yield and quality of seeds from fix^--plants was low (1-5 seeds per plant).

Biochemical analysis

Preparation of mRNA from nodules, in vitro translation and 2-D gel electrophoresis were carried out according to (6). Analysis of nodule proteins by Western blotting and reaction with antisera against components I and II was performed as described in (3).

RESULTS

Isolation of mutants

305 M2-families were screened for plants without nodules. In 6 families one or more sym^--plants were found which, after selfing, yielded an M3-family which likewise showed the mutant character. Thus, at least 2% of the M1-plants carried a mutant allele conferring resistance to nodulation.

84 M2-families could be tested for the presence of fix^--plants. Due to contamination of the nutrient solution with unwanted bacteria not all batches of M2-plants could be grown for three weeks which was required for the detection of non-fixing plants. In 1 family 1 normally nodulating but non-acetylene-reducing plant (fix^-) was found. Upon selfing of this plant, 2 plants of the same phenotype were obtained which, in their turn, yielded small progenies in which fix^-- and sym^--plants were present. All sym^+-plants were fix^-.

A fix^--plant was crossed with the parent type; the F1 showed normal nodulation and fixation, indicating recessiveness of the mutant character.

Tentative description of the fix^--mutant

Supplementation of the nutrient solution with bound nitrogen (NO_3) restored growth to normal, showing that the blcok in the metabolism is in the assimilation and not in the use of nitrogen. Nodules have the normal pink colour but do not produce ethylene, even when exposed to acetylene for periods of up to 18 h.

2-D gel electrophoresis of the products of in vitro translation of mRNA from nodules showed the presence of all nodulins which can be detected in the wildtype, a.o. leghemoglobin. Bacteroids, which were isolated from the nodules, looked normal. However, when proteins were extracted from the nodules and analysed as described in (3) no bands were found reacting with antisera against components I and II of nitrogenase.

DISCUSSION

Sym^--mutants were found in a rather high proportion of the M2-families: 2%. In an earlier study (5) chlorophyll mutants were found in 3 to 4% of the M2-families. Probably, the fix^--plant which was originally detected was heterozygous for a sym^--allele, as was indicated by the segregation of nodule-free plants in later generations obtained by selfing. The simultaneous induction of a sym^-- and a fix^--mutation may have been purely coincidental; however, a mutant plant in which both a fix- and a sym-gene had mutated was also found by Kneen et al. (these proceedings). Up to now, progenies of fix^--plants have been too small to allow a more extensive genetic analysis of the fix^--character and its possible relation to the inhibited nodulation.

The preliminary results of the biochemical analysis showed the absence of a compound of bacterial origin in the ineffective nodules. Since the primary effect of the mutation must be in the plant, this observation might suggest that a signal compound required for the activation of one or more bacterial genes is lacking in the mutant. This activation might apply to the nif A-gene which, in its turn, activates the nif HDK-genes, the structural genes for nitrogenase (7). The signal compound itself may be present in low concentration in the wildtype; enzyme(s) involved in its synthesis then will also be present in low quantities. It even needs not to be a nodulin, since the production of the signal compound does not necessarily need to be confined to the nodules but may take place throughout the plant. Therefore, the mutation, if it were to affect noticeably the presence of an enzyme, does not need to show up in the nodulin pattern.

Further study will reveal whether our conjecture of an effect on the activation of the bacterial nif A-gene will come true.

REFERENCES

1. Gresshoff PM, DA Day and AC Delves. In: Nitrogen fixation research progress. HJ Evans, PJ Bottomley and WE Newton (eds); Martinus Nijhoff Publishers, Dordrecht, 1985 pp 19-25.
2. LaRue TA, BE Kneen and E Gartside. In: Analysis of the plant genes involved in the legume-Rhizobium symbiosis. Report OECD, Paris, 1985 pp 39-48.
3. Bisseling T, C Been, J Klugkist, A van Kammen and K Nadler. EMBO J. 2, 1983: 961-966.
4. Jacobsen E and WJ Feenstra. Plant Sci Letters 33, 1984: 337-344.
5. Feenstra WJ and E Jacobsen. Theor.Appl.Genet. 58, 1980: 39-42.
6. Govers F, T Gloudemans, M Moerman, A van Kammen and T Bisseling. EMBO J. 4, 1985: 861-867.
7. Schetgens TMP. Thesis. Wageningen, 1986.

NON-NODULATION MUTANTS OF SOYBEAN

ANNE MATHEWS, BERNARD J. CARROLL AND PETER M. GRESSHOFF
Botany Department, Australian National University, Canberra, ACT, 2600

INTRODUCTION

Three non-nodulating mutants (nod49, nod139 and nod772) were isolated from mutagenized soybean populations (Carroll et al, 1986). The genetic, anatomical and physiological analysis of these mutants as well as the naturally occurring non-nodulation (rj_1) mutant (Williams and Lynch, 1954) was carried out.

RESULTS AND DISCUSSION

Complementation tests indicate that mutant nod139 is not allelic to the remaining non-nodulation mutants (nod49, nod772 and rj_1). We propose the new gene 'rj_6' for the complementation group identified by nod139.

TABLE 1. Complementation analysis of soybean mutant lines

	Bragg	nod49	nod139	nod772	rj_1	nts382
Bragg	+	+	+	+	+	+
nod49	+	-	+	-	-	+
nod139	+	+	-	+	+	+
nod772	+	-	+	-	-	+
rj_1	n.d.	-	+	-	-	n.d.
nts382	+	+	+	+	n.d.	+

+ = complementation giving normal nodulation;
- = lack of complementation.

Nod49 and nod772 are allelic to rj_1. However, nod772 has a leaky non-nodulation phenotype, showing a few curled root hairs. It also has slightly increased "occasional" nodulation (2.0 nodules per plant as compared to 0.5 nodules per plant for nod49) when inoculated with *Bradyrhizobium japonicum* USDA110. Mutant nod49 has no curled root hairs and preliminary tests have failed to shown subepidermal cell divisions (W.D. Bauer, pers. comm.)

The non-nodulation mutants are resistant to nodulation by most strains of *B. japonicum* at medium (10^7 - 10^8 cells/ml) cell numbers but do form a few nodules at high (10^8 - 10^9 cells/ml) cell numbers. Occasional nodules on nod49 are normal in development and similar to Bragg nodules when observed by light microscopy. They fix nitrogen.

TABLE 2. Total root hairs and percent curled root hairs of soybean mutants

	Age of plants	Plant genotypes Bragg	nod49	nod139	nod772	rj_1	nts382
Total root hairs	1 week	2220	1450	1732	735	704	1543
	2 weeks	1197	1072	288	412	265	537
% curled root hairs	1 week	10.1	0	0	0.6	0	10.9
	2 weeks	11.7	0	0	2.3	0	15.8

Root hairs on the left hand side of the root were counted. The figures indicate markedly curled root hairs. The terminal 8 cm of the root was observed.

Crosses between the supernodulation mutant nts382 (Carroll et al, 1985) and the non-nodulation mutant nod49 gave F_1 plants with wild-type nodulation pattern. The F_2 plants segregated into three phenotypic classes in the ratio of 9 wild-type:3 supernodulators:4 non-nodulators. Thus non-nodulation is epistatic on supernodulation. True double recessive mutants ($nod^-/^-nts^-/^-$) can be detected by shoot grafts onto wild-type roots. Nod49 and nts382 are unlinked and segregate independently. Soybean behaves as a true diploid organism.

Non-nodulation mutants, supernodulation mutant nts382 and Bragg were cocultured in pots or Leonard jars so that root systems were in close proximity. No inhibition of nodulation of Bragg or nts382 was observed, indicating that no significant amounts of inhibitory substance(s) was present in the exudate of these mutants. Likewise, the supernodulation mutant or Bragg had no effect on nodulation of the non-nodulation mutants, indicating that there is no apparent rhizosphere deficiency in the non-nodulating mutants which could be supplemented by wild-type exudates. Furthermore, no significant reduction in nodulation frequency on Bragg was observed when B. japonicum strain USDA110 was incubated in the exudate of the non-nodulation mutants.

Grafting experiments indicate that the inability of the non-nodulation mutants nod49 and nod139 to nodulate is strictly determined by the genotype of the root tissue (Delves et al, 1986). The supernodulation nts382 and Bragg shoots were unable to overcome the non-nodulation of the nod49 and nod139 root stocks.

REFERENCES

1. Carroll BJ, McNeil DL and Gresshoff PM : A supernodulation and nitrate-tolerant symbiotic (nts) soybean mutant. Plant Physiol. 78, 34-40, 1985.
2. Delves AC, Mathews A, Day DA, Carter AS, Carroll BJ and Gresshoff PM: Regulation of the soybean-Rhizobium nodule symbiosis by shoot and root factors. Plant Physiol (in press) 1986.
3. Carroll BJ, McNeil DL and Gresshoff PM: Plant Science (submitted) 1986.
4. Williams LF and Lynch DL: Inheritance of a non-nodulating character in the soybean. Agron. J. 46, 28-29, 1954.
5. Gresshoff PM and Delves AC: Plant genetic approaches to symbiotic nodulation and nitrogen fixation in legumes. In: King PJ, Hohn TH (eds) Plant Gene Research vol. 3, Springer Verlag, Vienna (in press), 1986.

EARLY NODULINS IN ROOT NODULE DEVELOPMENT

Jan-Peter Nap, Marja Moerman, Albert van Kammen, Francine Govers, Ton Gloudemans, Henk Franssen and Ton Bisseling

Department of Molecular Biology, Wageningen, The Netherlands

In recent years the occurrence of nodulins in nodules of leguminous plants has been firmly established and it has become evident that a differential expression of nodulin genes accompanies the development of a root nodule (e.g. Govers _et al._, 1985). As can be anticipated, the developmental decisions in this plant differentiation process will take place relatively early in development. Therefore our present interest of research focusses on the early events that are related to nodule development. In this paper we will describe the isolation of cDNA clones representing nodulin genes expressed early in development and we will outline the strategies for the elucidation of both the functions that these nodulins may serve in root nodule development and the mechanism by which the expression of these genes is regulated.

Nodulins involved in root nodule development

Nodule development was studied by the analyses of RNA populations by _in vitro_ translation of RNA followed by two dimensional (2D) electrophoresis and by Northern blot analyses. By these methods the majority of the nodulin mRNAs in pea nodules, including the leghemoglobin (Lb) mRNAs, are first detectable at 13 days after sowing and inoculation (Fig. 1B). Histological examination of longitudinal sections of infected roots of 8, 10 and 13 day old pea plants (Fig. 1A) showed that an entire nodule structure is already present at day 10. From the apical meristem infected as well as uninfected cells, the two cell types that are present in the symbiotic zone of the mature nodule, have been formed (Fig. 1A). Vascular bundles are present at the periphery of the nodule, an uninfected cortex surrounds the central part of the nodule and in the zone with the differentiating cells a large number of infection threads is present. At day 8 this differentiation process has started and in some cells a few bacteria are present (Fig. 1A); at 13 days the infected cells are fully packed with rhizobia (Fig. 1A). Since most nodulin mRNAs are expressed a good 3 days after the full nodule structure has been formed, we hypothesize that only the nodulin genes that are expressed before the Lb genes, which will be called early nodulin genes, can be involved in nodule morphogenesis. A conclusion that also follows from our studies on soybean nodule development (Gloudemans _et al._, 1986).

Early nodulins are involved in nodule morphogenesis

To study the part of the developmental program that leads to the formation of a root nodule structure we have isolated by differential hybridization cDNA clones from a soybean cDNA library that represent early nodulin genes (Franssen et al., 1986). One of these clones, pENOD2, showed strong homology to pea, alfalfa and common vetch (Vicia sativa) nodulin mRNA on Northern blots, suggesting that the ENOD2 gene is conserved and that its

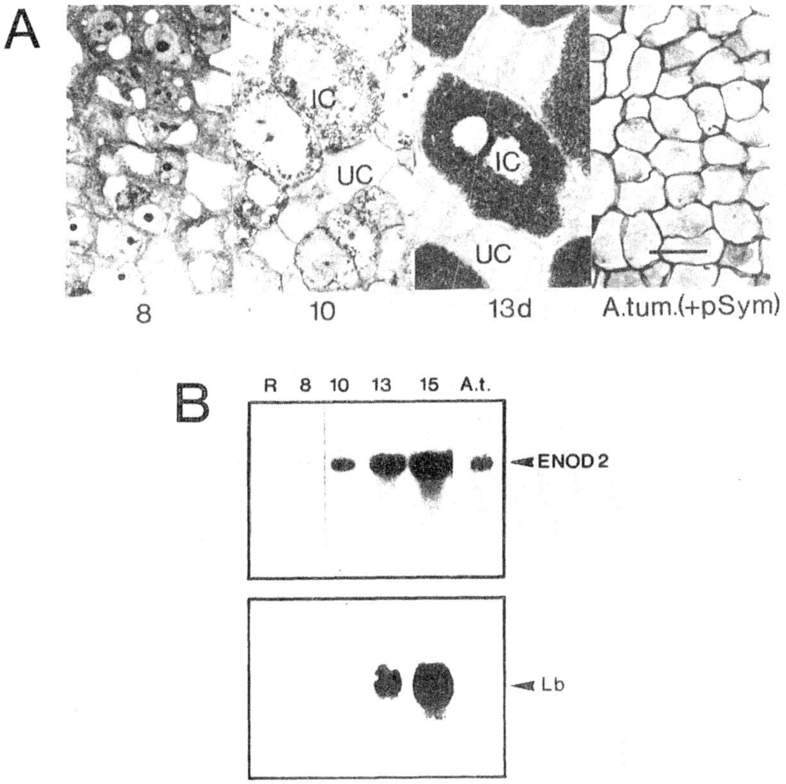

Figure 1. Cytology of pea nodules and the expression of the ENOD2 and Lb genes during root nodule development.
(A) Light micrographs of parts of the nodule tissue, containing differentiated cells from 8, 10 and 13 day old pea plants and pea nodules formed by an Agrobacterium transconjugant (LBA2712) carrying a Sym plasmid from R.leguminosarum (A.t.). (IC) infected cells, (UC) uninfected cells. Bar = 600 μm.
(B) Autoradiographs of Northern blots containing RNA from 8 day old uninfected roots (R) and nodules from 8, 10, 13 and 15 day old pea plants and 21 day old nodules formed by the Agrobacterium transconjugant. The blots were hybridized with pENOD2, the soybean early nodulin cDNA clone and pPsLb102, a pea Lb cDNA clone, respectively.

protein product presumably has the same function in different leguminous plants and different types of nodules.

From the appearance of early nodulins during development it can be assumed that they will be involved in nodule morphogenesis. However, since normally nodule morphogenesis and the infection process go hand in hand, the early nodulins can be involved in the infection process as well. To discriminate between these possibilities we have studied the expression of the ENOD2 gene in soybean nodules formed by R.fredii (USDA 257) (Franssen et al., 1986) and alfalfa nodules formed by R.meliloti mutants that have lost the ability to produce exopolysaccharides (Finan et al., 1985). In both types of nodules the absence of the infection process is shown by the lack of intracellular bacteria as well as the lack of infection threads, but in both types of nodules the ENOD2 RNA was found by Northern blot hybridization. On the other hand, ENOD2 RNA is not detected in root hairs from infected plants (Franssen et al., 1986). Therefore we conclude that ENOD2 will not be involved in the infection process, but more likely in nodule morphogenesis.

pENOD2 hybrid selects a soybean RNA that encodes for a nodulin with a molecular weight of 75.000 (indicated in Fig. 2A by an arrowhead) that was previously identified as N-75 (Franssen et al., 1986). The appearance of ENOD2 RNA during soybean root nodule development was studied by in vitro translation followed by 2D gelelectrophoresis rather than by Northern blot analyses, because only minute quantities of RNA could be obtained from the younger stadia of development. The ENOD2 RNA was already detectable in RNA from the youngest nodules we could excise, 7 days after sowing and inoculation (Fig. 2C). In this stage of development only a globular meristem is present and no differentiation into infected and uninfected cells has taken place yet (Fig. 2B). The 2D patterns of in vitro translated nodule RNAs showed that in addition to ENOD2 at least 3 other early nodulin genes are expressed in the nodule meristems of 7 day old plants. In root segments of younger infected soybean plants no nodulin gene expression was detected (Gloudemans et al., 1986).

The failure to detect nodulin RNAs in root segments of soybean plants younger than 7 days might be due to the fact that only a relatively low number of plant cells are involved in the plant-bacterium interaction. Since the concentration of the ENOD2 RNA, as well as of the other early nodulin RNAs (not shown), markedly increases from 7 till 10 days (Fig. 2C), we reason that the early nodulin genes are not expressed during the first cortical cell divisions but more likely shortly before day 7. This would implicate that the meristems formed upon interaction with Rhizobium may be similar to other plant meristems up to a certain stage. Detailed cytological observations have shown that in soybean roots many more infection loci are formed than the final number of nodules (Calvert et al., 1984) and up to the stage where the nodule emerges through the epidermis it can be aborted. When it reaches the "emergence" stage at day 7 the nodule will continue to develop. In soybean root nodule development the first expression of the ENOD2 gene coincides with the moment the nodule reaches this "emergence" stage. It is conceivable that due to the expression of early nodulin genes the general meristem of an infection locus becomes a "nodule meristem" committed to develop into a nodule structure.

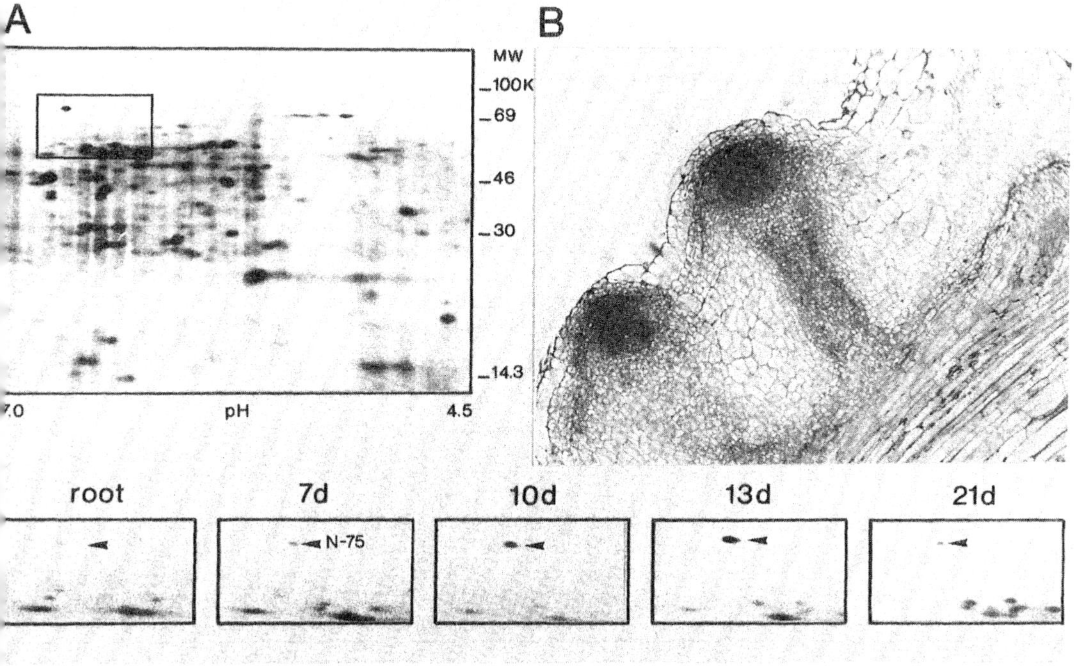

Figure 2. The appearance of ENOD2 during soybean root nodule development.

A. Fluorograph of a 2D-gel of in vitro translation products from total RNA from soybean nodules (20d).

B. Light micrograph of a longitudinal section through an infected soybean root 7 days after sowing and inoculation, stained with toluidine blue.

C. Fluorographs of 2D-gels of in vitro translation products from RNA of inoculated roots and nodules from 7, 10, 13 and 21 day old soybean plants inoculated at day 0 with Bradyrhizobium japonicum USDA110. Only the part of the gels indicated in (A) is shown.

Rhizobium genes involved in early nodulin gene induction

As a first approach to unravel the mechanism by which early nodulin genes are regulated we identified the Rhizobium genes involved in the induction of these plant genes. It was shown previously that Rhizobium strains cured of their own Sym plasmid regain the ability to form root nodules after the introduction of cloned DNA containing the nod-region (Downie et al., 1983). Downie et al. (1983) showed by deduction that 10 kb of the Sym plasmid, in a R.phaseoli chromosomal background, is sufficient for the induction of nodules on pea roots. Using the same Rhizobium strains we have shown that in these nodules all nodulin genes are expressed (Govers et al., 1986). Hence it was concluded that besides the nod-genes all other genes encoded by the Sym plasmid cannot be involved in the induction of the expression of nodulin genes. This result indicates that the nod-genes are involved in the induction of nodulin genes, although it does not exclude a role of the Rhizobium chromosome. However, in an Agrobacterium chromosomal background the total Sym plasmid (LBA 2712) gives rise to nodules on pea in which we have demonstrated that the ENOD2 gene (Govers et al., 1986) is expressed (Fig. 1B) but the second pea early nodulin gene N-40' (Govers et al., 1985, 1986) is not. The nodules formed by the Agrobacterium transconjugant (LBA 2712) lack intracellular bacteria (Fig. 1A), although infection thread like structures are present (Govers et al., 1986). These nodules have the same organization as normal nodules: a central tissue surrounded by uninfected cortex cells and peripherically located vascular bundles. However, the central tissue contains only one cell type. This are relatively small non-dividing cells that seem to represent neither the infected nor the uninfected cells that are found in normal root nodules (Fig. 1A). Since the N-40' gene is not expressed in these "empty" nodules in contrast to the ENOD2 gene, the two early nodulin genes we have identified represent different steps in the developmental program.

A.tumefaciens itself is unable to induce the expression of the ENOD2 gene in tumors formed on the stem of pea plants, so the Agrobacterium genome does not harbour genes involved in the induction of the expression of the ENOD2 gene. Hence these genes are most likely located on the Sym plasmid and since only 10 kb nod region of this Sym plasmid is essential, it can be deduced that the Rhizobium genes involved in the induction of expression of the ENOD2 gene have to be located in that nod region of the Sym plasmid (Govers et al., 1986).

Because the early nodulin N-40' is not expressed in the "empty" nodules on pea, one might argue that the nod region is not sufficient for the induction of this nodulin gene. However, a similar early nodulin gene N-40 in nodules formed on V.sativa by the same Agrobacterium transconjugant is expressed (Moerman et al., in preparation). It is not found in tumors on V.sativa, so the expression of N-40 (V.sativa) is also regulated by the nod genes. Assuming that N-40 (V.sativa) and N-40' are similar, we speculate that the expression of the N-40' gene is regulated by the nod genes as well. Since the outside of an Agrobacterium differs from Rhizobium it can be anticipated that the Agrobacterium is recognized by the plant in a different way than Rhizobium and the Agrobacterium might be subject to plant defence mechanisms. The striking differences between the V.sativa and pea nodules formed by the Agrobacterium transconjugant indicate that the presence of genetic information for the induction of a part of the developmental program is apparently no guarantee for the actual realization of that induction. The failure to induce a part of the developmental program can be due not only to the absence of the genetic information, required in a certain host, but also to the blockade of that program by the defence response of the host plant.

References

1. Calvert HE, Pence M, Pierce M, Malik NSA, Bauer WD (1984) Anatomical analysis of the development and distribution of Rhizobium infections in soybean roots. Can J Bot 62:2375.
2. Downie JA, Ma QS, Knight CD, Hombrecher G, Johnson AWB (1983) Cloning of the symbiotic region of Rhizobium leguminosarum: the nodulation genes are between the nitrogenase genes and a nifA-like gene. The EMBO J 2:947-952.
3. Finan TM, Hirsch AM, Leigh JA, Johansen E, Kuldan GA, Deegan S, Walker GC, Signer ER (1985) Symbiotic mutants of Rhizobium meliloti that uncouple plant from bacterial differentiation. Cell 40:869-877.
4. Franssen HJ, Gloudemans T, Stiekema W, Van Dam H, Govers F, Louwerse J, Bisseling T (1986) Molecular cloning of cDNA for Nodulin-75: a gene product involved in early stages of nodule development. Submitted.
5. Gloudemans T, De Vries SC, Bussink H-J, Franssen HJ, Louwerse J, Bisseling T (1986) Nodulin gene expression during soybean (Glycine max) nodule development. Submitted.
6. Govers F, Gloudemans T, Moerman M, Van Kammen A, Bisseling T (1985) Expression of plant genes during the development of pea root nodules. The EMBO J 4:861-867.
7. Govers F, Moerman M, Downie JA, Hooykaas P, Franssen HJ, Louwerse J, Van Kammen A, Bisseling T (1986) Rhizobium nodulation genes are involved in expression of an early nodulin gene. Submitted.

PERIBACTEROID MEMBRANE NODULINS OF SOYBEAN

MARC G. FORTIN and DESH PAL S. VERMA
Centre for Plant Molecular Biology, Department of Biology, McGill University, 1205 Docteur Penfield Avenue, Montreal, Canada H3A 1B1

1. INTRODUCTION

The symbiotic relationship between a legume plant and *Rhizobium* requires the expression of new sets of genes from both the host and microsymbiont. The products of these genes are neccessary for the formation of the nodule structure and nitrogen reduction and assimilation. One of the essential events in this interaction is the segregation of *Rhizobium* inside the host cell as an organelle-like structure delimited by the peribacteroid membrane (pbm) (1). The pbm plays a crucial role in the symbiosis, since it mediates all the molecular exchanges between the two organisms.

Pbm originates from the plant cell plasma membrane that surrounds the infection thread at the time of release of bacteria from the thread into the nodule cell (2,3). The pbm must, however, acquire new proteins that will be required to fullfill the new roles imposed on this membrane by the symbiosis. Nodule-specific proteins (nodulins) have been identified in the pbm of soybean nodules (4). These proteins are found exclusively in pbm and are not incorporated into the plasma membrane of infected cells (4) despite the high level of homology between these two membranes.

One way this specific targetting can be achieved is through the use of differentiating features on the protein that can be recognized as pbm-specific. A sub-library of pbm and peribacteroid fluid (pbf) specific cDNA clones was created from a nodule-specific library (5). We determined the primary structure of a pbm nodulin (nodulin-26) in order to compare it to other previously characterized pbm nodule-specific proteins.

2. MATERIALS AND METHODS

Plants (*Glycine max* L. cv. Prize) were grown and inoculated with *Rhizobium japonicum* (strain 61A76) as described (6).

To obtain a probe for screening pbm nodulins, polysomes from three-week-old nodules were purified (6) and reacted with anti-pbm and anti-pbf sera as in (7). The antigen-antibody complexes were purified on a Protein A-Sepharose column (7). The polyA$^+$ RNA was recovered from these immuno-purified polysomes by EDTA treatment and oligo-dT chromatography (8). cDNA was synthesized using oligo-dT priming and used to screen the nodule-specific cDNA library (5).

Hybrid selection was performed as in (8) using gel-purified insert DNA, and the released RNA was translated *in vitro* in rabbit reticulocyte lysate (5). The labelled translation products were immunoprecipitated with different sera as in (5), and separated by SDS-PAGE (9). The secondary structural analysis was carried out as described (10).

3. RESULTS

The nodule-specific cDNA library depleted of nod A, B, C and D sequences (5) was found to contain approximatively 200 clones hybridizing with the probe prepared against mRNA from immunoprecipitated polysomes with pbm/pbf antibody.

Clone plB1 (nodulin-26) contains a 940 bp insert and codes for a 26.5 kDa protein which is immunoprecipitated specifically by the pbm antiserum. No reaction is obtained when antisera raised against pbf or cytoplasmic fractions are used (Figure 1). Clone plB1 was verified to be nodule-specific by probing a Northern blot containing nodule and root polysomal RNA (Figure 2) with the plB1 insert. The plB1 probe hybridized with a 1180 nucleotide message in nodule RNA only.

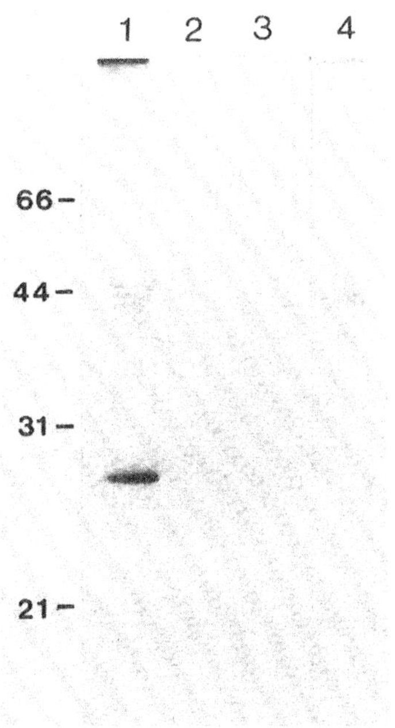

FIGURE 1. Immunoprecipitation of [^{35}S]-methionine-labelled in vitro translation products of mRNA hybrid-released from purified plB1 insert DNA. Proteins were incubated with sera raised against purified pbm (1), pbf (2) or nodule cytoplasmic fraction (3). Lane 4: pBR322 control. Protein markers are shown as x 10^{-3} kDa.

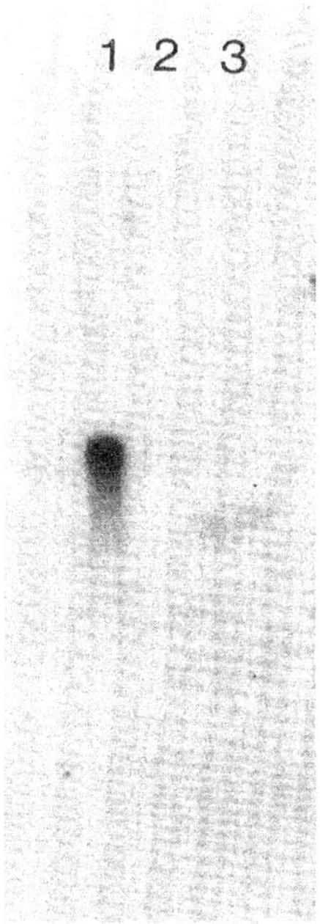

FIGURE 2. Northern blot of nodule (1), root (2), leaf (3), polysomal RNA probed with nick-translated plB1 insert DNA. 20 ug of polysomal RNA was applied to each lane and electrophoresed in formadehyde containing gels.

The nucleotide sequence of plB1 was determined and a single open-reading frame was found (Fortin et al, in preparation). Consistent with the fact that nodulin-26 is immunoreactive with pbm antiserum, it possesses a signal sequence at its amino terminus having all the characteristics of an eucaryotic signal sequence as established by von Heijne (10). The hydrophobic profile of the protein shows that it has four regions of sufficient hydrophobicity and length to span the membrane (domains a, b, c and d, Figure 3A). Domain b can be folded in a stable α-helix (using the algorithm described in (11)) which is designated as a transmembrane α-helix using the hydrophobic moment plot of Eisenberg (13). Secondary structure and hydrophobicity considerations allow us to propose the model shown in Figure 3B for the topology of nodulin-26 in pbm.

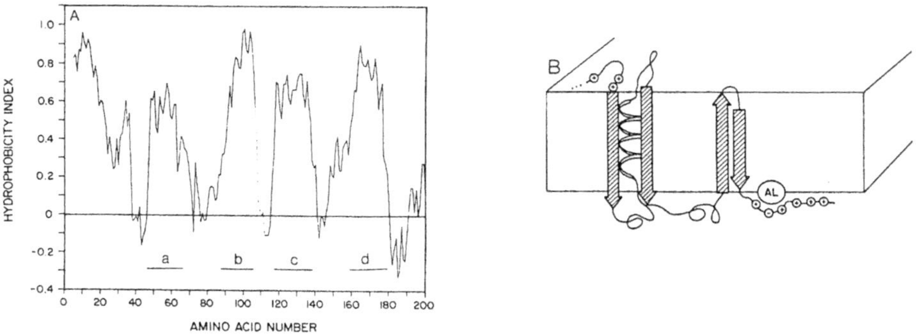

FIGURE 3. (A) Hydrophobicity analysis of nodulin-26. Hydrophobicity values for individual amino acids are those of the normalized consensus scale of Eisenberg (13). Domains a, b, c and d represent hydrophobic regions of sufficient hydrophobicity (>0.42) and length to traverse a membrane. (B) Proposed secondary structure and topology of nodulin-26 in pbm. AL: acidic lipids.

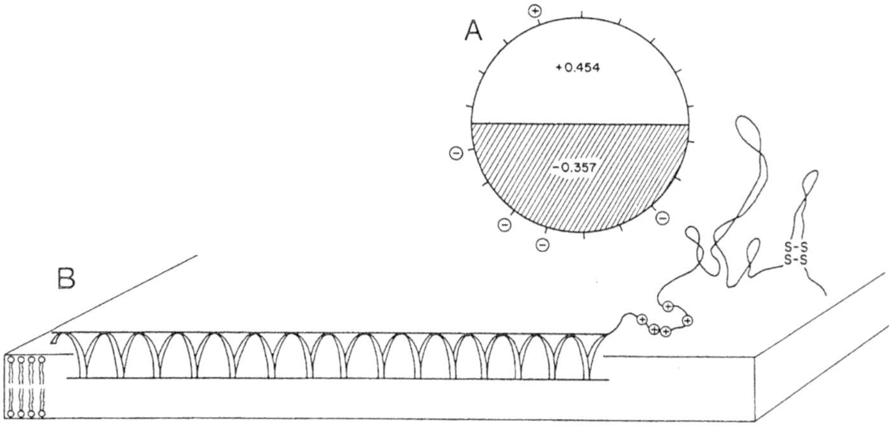

FIGURE 4. (A) Helical wheel diagram of the α-helix from residue 31 to 86 of nodulin-24. Values in each half of the helical wheel represent the average hydrophobicity of their respective half-helix. (B) Proposed topology of nodulin-24 at the surface of pbm. The upper surface faces the bacteroid.

How do these features of nodulin-26 compare with the ones of nodulin-24? Nodulin-24 does not have any transmembrane segment. It forms a 55 amino acid long (residues 31 to 86) α-helix of high amphiphilicity (Figure 4A). Since this domain contains three 18 amino acids repeats, the helix exposes exactly the same residues (except for two conservative substitutions at positions 44 and 48) at exactly the same angle (relative to the helix axis) over the entire length of the helix. The high amphiphilicity of this domain allow nodulin-24 to associate with the surface of pbm (Figure 4B).

No primary or secondary structure homology is found between nodulin-24 and nodulin-26, which suggest that pbm-specific features reside in other properties of these proteins. The models proposed for the topology of these nodulins in the pbm place clusters of four to five basic amino acids close to the membrane surface. It is interesting to note that Mellor et al. (14) have described strongly acidic lipids specifically associated with pbm in nodule cells. Interaction of these basic amino acids with acidic membrane components may be involved in segregating membrane nodulins away from plasma membrane proteins.

4. DISCUSSION

The symbiotic relationship requires that not only gene expression be modified, but also that the pathways of membrane/organelle biogenesis be altered. Vacuoles disintegrate following the invasion of nodules cells by bacteria (15), and a new subcellular compartment is created where nitrogen fixation occurs. A number of nodule-specific transcripts have been characterized in our laboratory. The products of these genes have different subcellular locations (see Table 1).

TABLE 1. Characterization of soybean nodulins.

Nodulin	Apparent MW kDa[1]	Actual MW kDa[2]	Signal Sequence[3]	Subcellular Location	Function
23	23.5	24.3	yes	pbm	?
24	24	15.1	yes	pbm	?
26	26.5	22.5	yes	pbm	?
26b	25.5	23.5	no	cytoplasmic	?
27	27	22.4	no	cytoplasmic	?
35	33	35.1	no	peroxisome	uricase II
(44) E-27*	42	39	yes	?	?
1ØØ	9Ø	?	?	cytoplasmic	sucrose synthetase

[1] by SDS-PAGE (see Jacobs et al., these Proceedings)
[2] derived from the amino acid sequence
[3] according to the criteria of von Heijne (11)
* Data from Ref. 16

Despite a high level of homology between several nodulins, some are part of the pbm whereas others are found in the cytoplasmic fraction (see Jacobs et al., these Proceedings). This illustrates the fact that targetting nodulins to pbm is a complex process we are just beginning to understand. The first requirement for a protein to be transported to pbm is the presence of a functional signal sequence. This will allow the protein to be translated on rough endoplasmic reticulum and to be subsequently transported to the Golgi apparatus. The processing events occurring in the Golgi are not yet fully understood, but N-linked glycosylation is certainly not involved in the pbm-specific targetting process since no Asn-X-Ser/Thr signal is present in nodulin-24.

The subsequent transport from the Golgi apparatus to pbm can occur via either a de novo pbm-specific biosynthetic pathway, in which case specific sorting signals are required on the proteins, or via a constitutive plasma membrane biosynthetic pathway. Since pbm nodulins are not found in the plasma membrane, the latter possibility implies that vesicles transporting pbm nodulins fuse preferentially with membrane surrounding bacteria, possibly due to a factor coming from the bacteria, or that the turnover rate for pbm nodulins in the plasma membrane is much higher than in pbm. Recent studies (10) suggest that specific signals may be involved in the induction of particular pbm nodulins during the formation of the endosymbiotic compartment.

REFERENCES

1. Verma DPS and Long S: Molecular biology of legume-Rhizobium symbiosis. in: International Review of Cytology (Jeon K, ed), Academic Press, Supp 14, pp 211-245, 1983.
2. Verma DPS, Kazazian V, Zogbi V and Bal AK: Isolation and characterization of the membrane envelope enclosing the bacteroids in root nodules. J Cell Biol 78, 919-936, 1978.
3. Robertson JG, Warburton MP, Lyttleton P, Fordyce AM, Bullivant S: Membranes in lupin root nodules. II. Preparation and properties of peribacteroid membranes and bacteroid envelope inner membrane from developing lupin nodules. J Cell Sci 30, 151-174, 1978.
4. Fortin MG, Zelechowska M and Verma DPS: Specific targetting of the membrane nodulins to the bacteriod enclosing compartment in soybean nodules. EMBO J 4, 3041-3046, 1985.
5. Fuller F, Kunstner PW, Nguyen T and Verma DPS: Soybean nodulin genes: Analysis of cDNA clones reveals several tissue-specific sequences in nitrogen-fixing root nodules. Proc Natl Acad Sci USA 80, 2594-2598, 1983.
6. Verma DPS, Nash DT and Schulman HM: Isolation and in vitro translation of soybean leghemoglobin mRNA. Nature (London) 251, 74-77, 1974.
7. Korman AJ, Knudsen PJ, Kaufman JF and Strominger JL: cDNA clones for the heavy chain of HLA-DR antigens obtained after immunopurification of polysomes by monoclonal antibody. Proc Natl Acad Sci USA 79, 1844-1848, 1982.
8. Maniatis T, Fritsch EF and Sambrook J: Molecular Cloning: A Laboratory Manual, Cold Spring Harbor, 1982.
9. Laemmli UK: Cleavage of structural proteins during the assembly of the head of bacteriophage T4. Nature New Biol 227, 680-685, 1970.
10. Fortin, MG, Morrison, NA and Verma DPS: (in preparation)

11. von Heijne G: A new method for predicting signal sequence cleavage sites. Nucl Acids Res 14, 4683-4690, 1986.
12. Argos P, Nayarana SVL and Neilsen NC: Structural similarity between legumin and vicilin storage proteins from legumes. EMBO J 4, 1111-1117, 1985.
13. Eisenberg D: Three-dimensional structure of membrane and surface proteins. Ann Rev Biochem 53, 595-623, 1984.
14. Mellor RB, Christensen JME, Bassarab S and Werner D: Phospholipid transfer from ER to the peribacteroid membrane in soybean nodules. Z Naturforsch 40c, 73-79, 1985.
15. Kijne JW: The fine structure of pea root nodules. I. Vacuolar changes after endocytic host cell infection by <u>Rhizobium leguminosarum</u>. Physiol Plant Pathol 5, 75-79, 1979.
16. Sengupta-Gopalan S, Pitas JW, Thompson DV, Hoffman LM: Expression of host genes during root nodule development in soybeans. Mol Gen Gent 203, 410-420, 1986.

ISOLATION OF NODULE SPECIFIC c-DNA CLONES FROM MEDICAGO SATIVA

György B. KISS, Éva VINCZE and Zoltán VEGH

INSTITUTE OF GENETICS, BIOLOGICAL RESEARCH CENTER
HUNGARIAN ACADEMY OF SCIENCES, SZEGED, P.O.B. 521, HUNGARY

1. SUMMARY

Two nodule specific genes have been isolated from nodule c-DNA clone bank of Medicago sativa using differential hybridization. One of the two genes was leghemoglobin which showed high level of sequence homology to leghemoglobin from other legumes. The second c-DNA gene, nodulin-1 was also sequenced and revealed two direct repeats RA and RB. Repeat RB consists of three tandem copies of 18 amino acids. From the genomic library of Medicago sativa constructed in our lab, nodulin-1 was isolated and partially characterised. The nucleotide sequence showed that this gene contains several introns. The function of nodulin-1 in symbiotic nitrogen fixation is not known at present.

2. INTRODUCTION

Dinitrogen is converted most effectively to ammonia by symbiotic nitrogen fixation. This biological process takes place in nodules of different legumes. Rhizobia are able to induce nodules on their host, to enter plant cells and to reduce dinitrogen to ammonia, wich is in turn used by the plant as nitrogen source. The micro-environment, energy supply, as well as other conditions necessary for Rhizobial activities are given by the plant. Nodule development and persistance necessitate coordinate interaction of functions coded by genes of both partners. Genes were identified to some extent from both Rhizobia and legumes but at presence the whole biological "pathway" is far from understanding.

In this study, we present the isolation and partially characterization of nodule specific genes from Medicago Sativa.

3. RESULTS AND DISCUSSION

3.1. Construction of c-DNA library

Poly-A+ mRNA was isolated from nodule tissue of Medicago sativa. After reverse transcription and second strand synthesis, double stranded c-DNA was digested by S1 nuclease followed by EcoRI methylation, ligation to EcoRI linkers and digestion with EcoRI . Liberated linker residues were separated from c-DNAs by chromatography on Sepharose 4B column. Cloning was performed in EcoRI cut lambda 1149 vector. After in vitro packaging, plate lysates were prepared on Escherichia coli Pop 13 strain. Phages were collected and stored.

3.2. Isolation of nodule specific clones

Independent phage plaques were picked from the c-DNA library and their DNA was transferred to duplicate nitrocellulose filters. Poly-A+ mRNA was isolated from nodule and root tissue, labelled using reverse transcriptase and used as hybridization probes. Two recombinant clones could be identified after the differential hybridization. The insert of both clones were used as probes to re-screen the c-DNA library and to isolate more similar c-DNA clones.

3.3. Leghemoglobin genes from Medicago sativa

Two independent c-DNA clones (clone 460 and 560) were isolated and sequenced (7). The DNA sequence of clone 560 revealed 384 bases long coding and 138 bases long untranslated region. It contained two sequences, AAATTAT and AAATATT 38 and 25 bases, respectively upstream from the 3 prime end. These sequences are similar to the poly-A+ signals of the eucaryotic mRNAs.

The sequence of clone 460 was identical to that of clone 560 except it was shorter by 26 bases and had only one potencial poly-A+ additional signal (AAATTAT) 22 bases away from the 3 prime end.

The amino acid sequence deduced from the nucleotide sequence of these two c-DNA clones were identical and revealed high percent of homology to leghemoglobins of different legumes. The percent of homology is 80 % to Lb1 of Pisum sativum (5), 77 % to Lb1 of Vicia faba (5), 65-67 % to Lba and Lbc of Glycine max (1,2,5), 65 % to Lba of Phaseolus vulgaris and 56-60 % to Lb1 and Lb2 of Lupinus luteus (5). According to the amino acid sequences and their analyses an evolutionary tree could be suggested for the evolution of Lb genes (see Figure 1).

Evolutionary tree for Lb genes

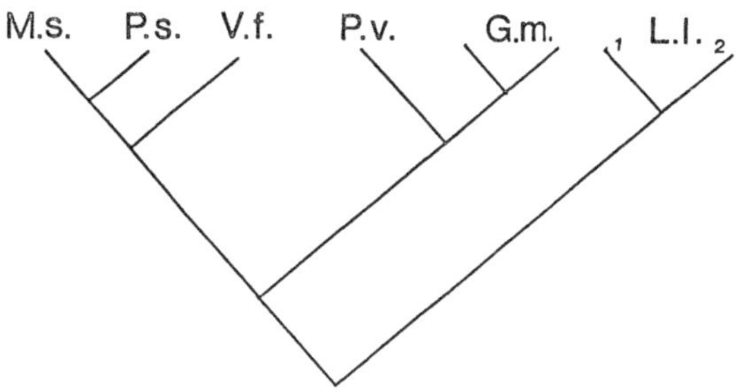

Figure 1. Possible evolution of leghemoglobin genes

M.s. = Medicago sativa, P.s. = Pisum sativum, V.f. = Vicia faba, P.v. = Phaseolus vulgaris, G.m. = Glycine max, L.l. = Lupinus luteus

3.4. Nodulin-1 gene from alfalfa

Three independent c-DNA clones were isolated after differential hybridization and re-screening (clone 385, 780 and 910). All three clones were identical according to sequence analyses, therefore they probably represent transcripts of one genomic gene. This gene is denoted as nodulin-1 gene of Medicago sativa (nod-1). "Northern" hybridization experiments (6) showed that this gene is expressed only in nodules. "Southern" hybridization showed at least one 20 kb EcoRI fragment and at least six BclI fragment of 2.4, 4.0, 5.0, 5.4 6.6 and 8.0 kb fragments. This result might suggest that there is no more than one or two copies of nod-1 gene in the genom.

3.4.1. Nucleotide and amino acid sequence analyses of nod-1

The insert of the three c-DNA clones were sequenced by Maxam-Gilbert and dideoxy chain termination procedures. From these data, 770 base pairs of nod-1 c-DNA could be abtained. There is an open reading frame of 179 amino acids followed by stop codon, by 215 nucleotides untranslated region and 14 nucleotide long poly-A+ tail. The potencial poly-A+ additional signal (AAATTAT) is 17 nucleotides away from the 3 prime end of the transcripts.

The hydropathy plot (4) of the known protein region showed extremly hydrophilic nature suggesting cytoplasmic location.

3.4.2. Repeats in the nod-1 gene

According to homology matrix analyses of the known amino acid sequence of nod-1, two repeats could be identified, repeat RA and RB, respectively.

Repeat RA contains 12 amino acids and are present in two copies (RA1 and RA2). RA1 and RA2 are separeted by 22 amino acids and there is only one difference at the sixth position. The sequence of the two repeats is shown below using sigle letter amino acid code:

Repeat RA: CPPGPDTYLELS
RB: CPPGPHTYLELS

Repeat RB contains 18 amino acids and is present in three tandem copies (RB1, RB2 and RB3). The first repeat (RB1) is started downstream by 37 amino acids from the last amino acid of RA2. RB1 and RB3, respectively differs in three amino acids as compared to RB2 and there are six amino acids difference between them. From the comparison analyses of both nucleotide and amino acid sequence of repeates RB1, RB2 and RB3 one can speculate that repeat RB2 has not mutated since the point of divergence. The amino acid sequence of the repeats are shown below using single letter code:

Repeat RB1: DGPIEISTRTENENFIPS
RB2: DEPIEISTGTEKENFIPS
RB3: DEPSEISSGAEKENFIPS

18 base pairs repeats were found by Katinakis and Verma (3) in the nodulin-24 gene of soybean, but no amino acid and nucleotide sequence homology was found between nodulin-24 and nodulin-1 genes as concluded from homology matrix analyses and DNA-DNA hybridization .

3.4.3. Isolation of the genomic nod-1 sequence from alfalfa

Genomic library was constructed from total DNA of Medicago sativa in a lambda phage vector. This gene bank was used to isolate the genomic counterpart of the nod-1 c-DNA gene. One out of four hybridizing recombinants was recloned in pBR322 and partially characterised. DNA-DNA hybridization experiments and sequence analyses revealed that this gene containes several introns of which one was precisely localised (see Figure 2).

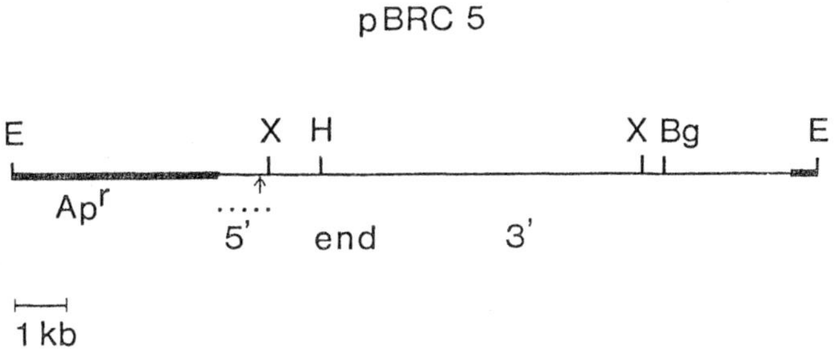

Figure 2. Genomic clone of nod-1 gene from alfalfa.

The insert of the original recombinant lambda clone hybridizing to nod-1 was recloned into pBR322 (thick bars) giving rise pBRC 5. The size of the insert (thin bar) is 11.5 kb. Dots show hybridization to the 5 prime end of the nod-1 c-DNA. Arrow indicates intron/exon junction. E = EcoRI, X = XhoI, Bg = BglII and H = HindIII.

4.REFERENCES

1. Brisson N. and D.P.S. Verma.1982. Soybean leghemoglobin gene family: normal, pseudo, and truncated genes. Proc.Natl.Acad.Sci. USA. 79:4055-4059
2. Hylding-Nielsen, J.J.,E.O. Jensen, K. Paludan, O. Wiborg, R. Garrett, P. Jorgensen and K. Marcker.1982. The primary structures of two leghemoglobin genes from soybean. Nucl.Acid Res. 10:689-701
3. Katinakis, P. and D.P.S. Verma.1985. Nodulin-24 gene of soybean codes for a peptide of the peribacteroid membrane and was generated by tandem duplication of a sequence resembling an insertion element. Proc.Natl.Acad.Sci. USA. 82:4157-4161
4. Kyte, J., and R.F. Doolittle.1982. A simple method for displaying the hydrophatic character of a protein. J.Mol.Biol. 157:105-132
5. Lehtovaara, P., A. Lappalainen and N. Ellfolk.1980. The amino acid sequence of pea (Pisum sativum) leghemoglobin. Biochem. Biophys. Acta 623:98-106 (1980)
6. Nobrega, F.G., C.L. Dieckman and A. Tzagoloff.1983. A rapid simple method for detecting specific RNA transcripts by hybridization to DNA probes in solution. Anal. Biochem. 131:141-145
7. Vegh et al. (in preparation)

ANALYSIS OF NODULE-SPECIFIC GENE EXPRESSION IN INEFFECTIVE ALFALFA ROOT NODULES AND CALLUS CULTURES DERIVED FROM INEFFECTIVE ROOT NODULES

Joanna F. Hanks[1], Lisa A. Macol[1], Jonathan Goldthwaite[2], and Ann M. Hirsch[1]
(1) Department of Biological Sciences, Wellesley College, Wellesley, Massachusetts 02181 and (2) Department of Biology, Boston College, 140 Commonwealth Avenue, Newton, Massachusetts 02167

INTRODUCTION

Nodule-specific proteins, commonly termed nodulins, have been detected in soybean (9), pea (1), and alfalfa root nodules (8,11) by means of nodule-specific antisera. Several of the genes encoding soybean nodulins have also been cloned (5). While the identity of some abundant nodulins such as leghemoglobin (Lb), nodule-specific glutamine synthetase (2), and, most recently, a polypeptide of the peribacteroid membrane (7) is known, the role of most of the nodulins in nodule development remains an open question.

The expression of plant genes involved in the pea-*Rhizobium* symbiosis has been analyzed by *in vitro* translation of mRNA followed by two-dimensional gel electrophoresis (6). These results indicate differential expression of nodulin genes during root nodule development. Although most of the nodulin genes are expressed concomitantly with the Lb genes, the gene encoding N-40' is expressed several days earlier. We have used a similar experimental approach to analyze the expression of nodule-specific genes of alfalfa. In parallel with the mRNA experiments, we have also employed antinodule sera to detect nodulins from the same tissues on two-dimensional gels.

EXPRESSION OF NODULE-SPECIFIC GENES

Total RNA isolated from both wild-type nitrogen-fixing alfalfa root nodules and non-infected root tissue by LiCl precipitation (3) was translated *in vitro*, followed by separation of the translation products by two-dimensional gel electrophoresis (10), essentially as described by Govers *et al.*(6). Comparison of the two-dimensional patterns of translation products from nodules with those from root mRNA indicated the majority of the polypeptides were present in both tissues. At least seventeen spots appear to be nodule-specific and therefore represent the expression of presumed nodulin genes (Figure 1). As several of these are minor spots, we cannot exclude the possibility that they are nodule-stimulated, with a very low level of expression in the root not detectable in our experimental system. For the purposes of this paper we have designated nodulins identified as translation products with a lower case 'n' and nodulins identified using antiserum with an upper case 'N', followed by the molecular weight in kd. The molecular weights of the nodulin translation products range from 14kd to 64kd, with 13 major spots designated n-46, n-38, n-24, n-18, n-17a, n-17b, n-17c, n-17d, n-15a, n-15b, n-15c, n-14a and n-14b. Nodulins n-15a, n-15b, n-15c, n-14a and n-14b are probably Lb expression products judging from their molecular weight, intensity, and relative position to spots from pea nodules (6).

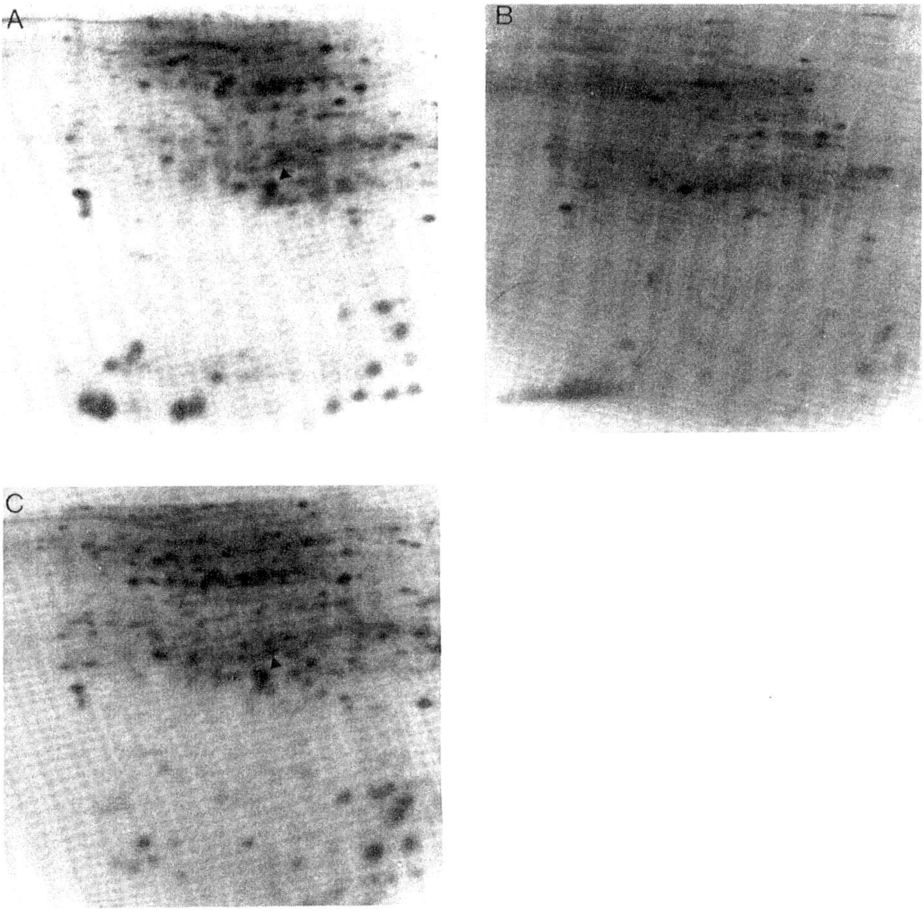

Figure 1. In vitro translation products from Fix^+ nodules (A), root (B), and ineffective nodules induced by fix21 mutant (C). n-38 is indicated.

Expression of nodule-specific genes was also investigated in ineffective root nodules elicited by infection with the spontaneous mutant of Rhizobium meliloti designated fix21 (see Clover, Finan, and Signer, this volume). These Fix^- nodules have aborted infection threads and contain very few bacteria. Several nodule-specific translation products were observed in experiments with fix21 nodule RNA, including n-51, n-40, n-38, n-32, n-24, n-18, n-17c and n-17d. n-38 is the major nodulin present in nodules induced by the fix21 mutant. Other nodulins were also present at low levels including n-14a, n-14b, n-15b and n-15c, the putative Lb spots (Figure 1).

Callus tissue cultures have been generated in our laboratory from ineffective root nodules induced by mutant R. meliloti, including fix21 and several of the mutants in the exo locus (4). RNA was isolated from the various callus cultures, including Iroquois root callus as a control, and the translation products were analyzed on two-dimensional gels as above for nodulin gene expression. A spot of the same molecular weight as n-38, the major nodulin seen in fix21 nodules, but slightly more basic in isoelectric point, appears on gels from callus derived from fix21 nodules and callus derived from exoB nodules. This second spot often appears as a doublet with n-38 on gels from fix21 nodules, but seldom on wild-type nodule gels, which have a single spot for n-38, and neither spot at 38kd is ever present on root gels. It is possible that the second 38kd spot represents an isomeric form of n-38. Callus cultures derived from fix21 and exoB nodules also appear to express n-32. ExoB derived callus also expresses n-51 as well.

NODULINS DETECTED WITH ANTINODULE SERUM

Many nodulins have also been detected in our laboratory by a peroxidase-linked immunoassay using antinodule serum to compare nodule and root proteins separated on two-dimensional gels. The relative abundance of proteins in the injection extract as well as the individual antigenicity of the proteins affects the intensity of spots seen by this method. We routinely load 70ug of root protein per gel to compare with 40ug nodule protein per gel. Using our antinodule serum we observe a total of 34 nodulins in Fix^{+} nodules. By far the most intense nodulins detected with our antiserum are N-43a and N-43b with other abundant nodulins designated N-76a, N-76b, N-42, N-41, N-39, N-38, N-35, N-28a, N-18 and N-17. These nodulins should not be confused with those detected as RNA translation products discussed above.

Several nodulins were observed on protein gels from nodules induced by the fix21 mutant, including N-14, N-17, N-30, N-34, N-28b and N-37. Since our antiserum has a relatively weak reaction to Lb, we have alternatively detected Lb on our gels with anti-Lb serum generously provided by T. Bisseling. Lb, however, was not detected in fix21 nodules using either our antiserum or the Bisseling anti-Lb serum. Thus it appears that the mRNA for Lb present in nodules induced by fix21 is not translated into protein *in vivo*.

REFERENCES

1. Bisseling T, Been C, Klugkist J, Van Kammen A and K Nadler. (1983) EMBO J. 2, 961-966.
2. Cullimore JV, Lara M, Lea PJ and BJ Miflin. (1983) Planta 157, 245-253.
3. DeVries SC, Springer J and JGH Wessels. (1982) Planta 156, 129-135.
4. Finan TM, Hirsch AM, Leigh JA, Johansen E, Kuldau GA, Deegan S, Walker GC and ER Signer. (1985) Cell 40, 869-877.
5. Fuller F, Kunstner PW, Nguyen T and DPS Verma. (1983) PNAS 80, 2594-2598.
6. Govers F, Gloudemans T, Moerman M, VanKammen A and T Bisseling. (1985) EMBO J. 4, 861-867.
7. Katinakis P and DPS Verma. (1985) PNAS 82, 4157-4161.
8. Lang-Unnasch N and FM Ausubel. (1985) Plant Physiol. 77, 833-839.
9. Legocki RP and DPS Verma. (1980) Cell 20, 153-163.
10. O'Farrell PH. (1975) J. of Biol. Chem. 250, 4007-4021.
11. Vance CP, Boylan KLM, StadeS and DA Somers. (1985) Symbiosis 1, 69-84.

NODULE SPECIFIC GENES IN *Phaseolus vulgaris*.

F. CAMPOS, M. VAZQUEZ, J. PADILLA, C. ENRIQUEZ AND F. SANCHEZ.

Depto. de Biología Molecular de Plantas,
Centro de Investigación sobre Fijación de Nitrógeno, UNAM.
Apdo. Postal 565-A. Cuernavaca, Morelos, MEXICO.

1. INTRODUCTION

During the symbiotic association between bacteria of the genus *Rhizobium* and the root of its host legume, the formation of a novel organ is induced: the nitrogen-fixing root-nodule.

A coordinated gene expression from both organisms is required for the accurate nodule development and function. The root-nodule is not only the interphase of two interacting genomes but also the forum of a continuos metabolic dialogue between a group of highly specialized plant- and bacterial cells, contributing to promote plant growth.

Nodulins (1) are a group of plant proteins that are induced specifically during the legume nodule development. Some of the abundant nodulins are associated to metabolic functions. Such is the case of the leghemoglobins (2), nodule uricase in soybean (3), and the gamma polypeptide of glutamine synthetase in Phaseolus vulgaris (4).

2. PROCEDURES

Material and methods. *Phaseolus vulgaris* L.cv. Negro Jamapa seedlings were inoculated with *R.phaseoli* CIAT899 (5), and other symbiotical-altered mutants, and grown in glasshouse as previously reported. Nodules were harvested at different developmental stages.

RNA isolation and Northern blot analysis. Total polysomal RNA was extracted following procedures described by Christofferson and Laties (6). Northern blot were prepared by electrophoresis of polysomal RNA according to Maniatis (7). and hybridized with the indicated nodule-specific cDNA clones.

in vitro translation. Total polysomal mRNA was *in vitro* translated in a rabbit reticulocyte lysate system with ^{35}S - methionine, and labeled products were separated in PAGE-SDS or 2-D gel systems (8).

3. RESULTS AND DISCUSSION

We have chosen to invetigate the expression of nodule-specific genes in *Phaseolus vulgaris* with the aim to comprehensively study the root-nodule symbiosis as a biochemical, genetical and developmental plant-bacteria interaction model. Our goal is to correlate those genes which are essential for the symbiotic association, with the proteins they encode for, and the function that these perform. This correlation should be framed in a temporal and spatial context. Our initial approach was to characterize the nodule-specific proteins (Fig. 1), nodule-specific transcripts, inferred from in vitro translation products (Fig. 2), and nodule-specific clones from a cDNA library by using molecular and immunological techniques.

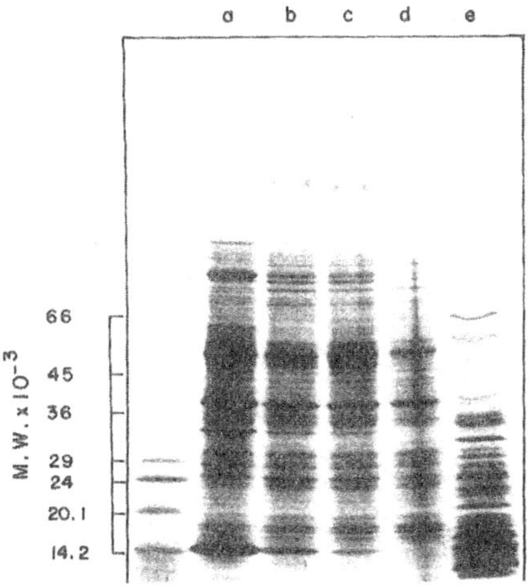

Figure 1. SDS-PAGE of nodule fractions. Lanes a-c soluble proteins from nodules formed by strains: a:CIAT899(Fix$^+$); b:CE108 (Fix±); c:cli80(Fix$^-$). Lane d: uninfected roots; e: CIAT899 bacteroid.

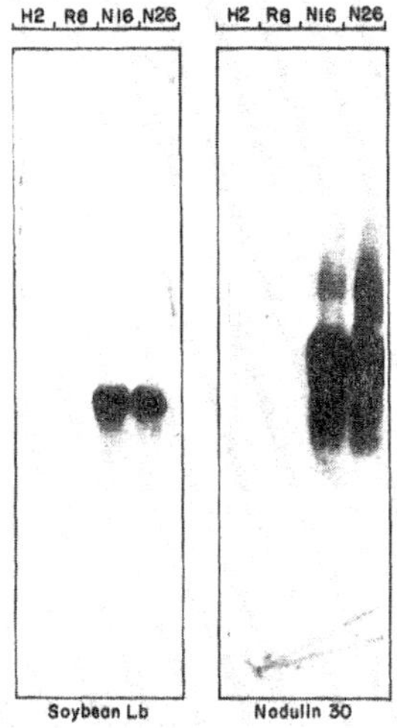

Figure 2. 2-D(O'farrell) from in vitro translation products from poly(A^+)mRNA. A:8 days old uninfected roots; B: 18 d nodules. (Arrows: specific polypeptides; β-γ: GS polypeptides.

At present, we know that Lb's, uricase II and GS-γ are induced at a transcriptional and translational level, and that these nodulins may serve as developmental markers, as they are expressed before the onset of nitrogen fixation, and their relative abundance varies depending on nodule age and the microsymbiont genotype. These variations suggested that sequential developmental stages and bacterial genomic traits are essential for the nodulins expression and accumulation.

In particular, we have described a group of in vitro translation products encoded by the most abundant transcripts in bean nodules. They have an unusual IEF pattern and 30 Kd MW range. They are absent in ineffective nodules and may account for a multigene family according to hybrid-selection and northern-blot analysis, (Fig. 3); however, the corresponding in vivo products have not yet been detected in soluble or membranal fractions.

With the aid of altered strains in symbiotic functions, we will be able to gain insight on the mechanism and factors involved in nodulin expression and accumulation in P.vulgaris root-nodules.

Figure 3. Northern-blot of polysomal RNA from: H2, 2 day-old hypocotiles; R8, 8d uninfected roots; N16-N26, 16 and 26d effective nodules hibridized to cDNA clones as indicated.

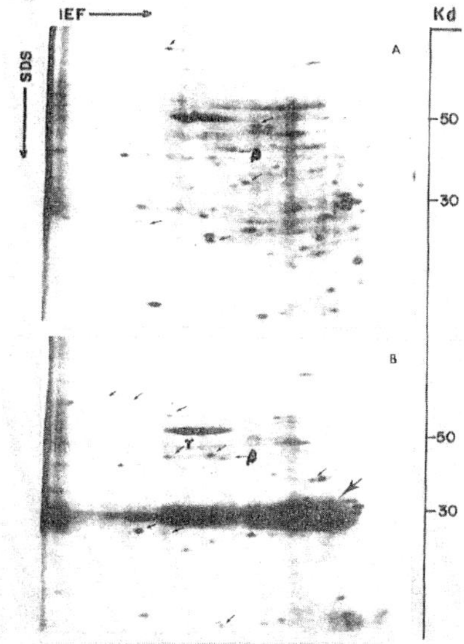

ACKNOWLEDGEMENTS:This work was supported by CONACyT-PCCBBNA-022632, Fondo de Est. e Inv. Ricardo J. Zevada. We thank D.P.S. Verma for providing leghaemoglobin and uricase cDNA clones from soybean.

REFERENCES

1. Legocki RP, and Verma DPS. Cell 20: 153-163, 1980.
2. Baulcombe D, and Verma DPS. NAR 5: 4141-4153, 1978.
3. Nguyen T, Zelechowska M, Foster V, Bergmann H, and Verma DPS. PNAS 82, 5040-5044, 1985.
4. Lara M, Porta H, Padilla J, Folch J, and Sánchez F. Plant Physiol 76, 1019-1023, 1984.
5. Martínez E, Pardo MA, Palacios R, and Cevallos MA. J. Gen Microbiol 131: 1779-1786, 1985.
6. Christofferson RE, and Laties GG, PNAS 79, 4060-4063, 1980.
7. Maniatis, T., Fritsch, E.F., and Sambrook, J. Molecular Cloning: A Laboratory Manual. CSH, New York. 1982.
8. Ortega JL, Campos F, Sánchez F, and Lara M. Plant physiol 80: 1051-1054, 1986.

INVESTIGATION OF PLANT GENES EXPRESSED DURING SYMBIOTIC NITROGEN FIXATION

S.G. Gottlob-McHugh and D.A. Johnson, Department of Biology, University of Ottawa, Ottawa, Canada

INTRODUCTION AND RESULTS

The development of the legume root nodule results from a complex interaction between *Rhizobium* and its host, requiring the co-ordinated expression of both bacterial and plant genes. To understand the molecular processes involved, an analysis of the genes of both partners is necessary. A great deal of progress has been made in identifying numerous *Rhizobium* genes (*nif*, *nod*, *fix*, etc.) involved in the symbiosis; relatively less progress has been made in identifying plant genes. In recent years, nodule specific plant proteins, nodulins, have been detected in soybean (1), pea (2), and alfalfa (3); but except for a few nodulins, little is known about their roles. Isolation and analysis of the nodulin genes will lead to a better understanding of the role of the plant genes and how they are regulated. To this end we have isolated and characterized five nodule specific cDNA clones from a soybean nodule cDNA gene bank (A). The clones have also been used to probe mRNA from nodules derived from the *Frankia*-alder symbiosis with the hope that some homology may exist between soybean and alder nodulins (B).

A) TABLE 1. Characterization of Nodulin Clones

Nodulin clone	Size of mRNA (bp)	Homology to Published Sequences
6-9-F	1475	None
7-3-H	780	N.D.
9-11-B	1500,1000	Yes
15-9-A	900	N.D.
36-1-A	1050	None

We could detect no homology of these five clones to root RNA, even after extended exposure of northern blots (3 weeks). Sequence analysis of the clones indicates that at least two clones show no sequence homology to published sequences (Nodulin 23, Nodulin 24 and Nodulin 35 from D.P.S. Verma's group), although 9-11-B is homologous to Nodulin 23. However, the other clones we have isolated may represent genes that have not been previously described.

FIGURE 1. Nodulin mRNA Accumulation Following Inoculation

Total soybean RNA was isolated from root and nodule tissue collected following inoculation of 2 day old soybean seedlings with <u>Bradyrhizobium japonicum</u> strain 61A76. RNA was spotted onto Biodyne filters and hybridized to the clones described in Table 1. (1) 9-11-B; (2) 7-3-H; (3) 6-9-F; (4) 15-9-A; (5) 36-1-A; (Lb) leghemoglobin. A) first detection of acetylene reduction; B) first appearance of flowers.

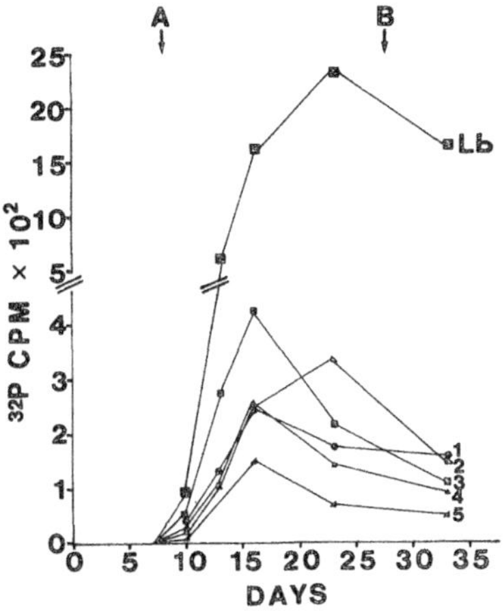

Our results indicate that nodulin mRNA begins to accumulate at the same time as leghemoglobin. We observed maximum levels of all but one of these nodulin messages at day 16 after which the levels decline. Leghemoglobin levels were at a maximum at day 23. This result suggests that perhaps four of the nodulins are necessary for the development and growth of the nodules but that high levels are not necessary for continued nodule function.

B) We have previously detectd the presence of leghemoglobin-like sequences in alder genomic DNA (4), therefore we wished to determine whether any other sequences homologous to nodulins could be detected. Southern blot analysis of the alder DNA indicates the presence of a sequence homologous to 6-9-F. Preliminary northern blot analysis suggests that a sequence homologous to 6-9-F may be expressed in alder nodules. Weak hybridization indicates that this sequence may be present in low levels in the alder nodule or have limited homology to 6-9-F. Perhaps 6-9-F represents a class of nodulins common to soybean and alder.

1. Legocki, R.P. and Verma, D.P.S. (1980) Cell 20, 153-163.
2. Bisseling, T. et al. (1983) EMBO J. 2, 961-966.
3. Lang-Unnasch, N. and Ausubel F. (1985) Plant Physiol. 77, 833-839.
4. Hattori, J. and Johnson,D.A. (1985) Plant Mol. Biol. 4, 285-292.

RHIZOBIUM INDUCED PLANT PROTEINS IN TARGET ROOT EPIDERMAL CELLS OF VIGNA UNGUICULATA

ARVIND BHAGWAT and JOSEPH THOMAS
Molecular Biology and Agriculture Division
Bhabha Atomic Research Centre
Bombay - 400085, India

1. INTRODUCTION

Recent studies have shown that invasion by Rhizobium is limited to a very narrow region above the root tip which does not possess root hairs at the time of inoculation (1,2). The cells in this region remain susceptible to infection for only a brief period of 3 to 4 h. At present, not much is known in biochemical, cytological or genetic terms regarding the events occurring at the root surface during such early stages of infection. An analysis of preinvasion events in cowpea (Vigna unguiculata) - Rhizobium symbiosis is reported here. Using rhizobia having different infection rates we have recently demonstrated the significant role of host in early intereactions. Cowpea roots under nitrogen deficiency, secrete substances which elicit faster nodulation response in rhizobia which otherwise possess slow infection status (3). Under the influence of host root exudate rhizobia show enhanced capsular polysaccharides (CPS) synthesis (4) and alteration in capsule topography and chemical composition (5). The newly synthesised host induced CPS (HI-CPS) increase the nodulation efficiency of Rhizobium. These observations prompted us to examine the effect of HI-CPS on the infectible zone of cowpea roots.

2. MATERIALS AND METHODS

2.1. Legume seedlings were grown in plastic growth pouches as described earlier (4). ^{125}I-labelling of fluorescein conjugated CPS (^{125}I.Fl.CPS) was done as described by Glabe et al (6). NRH zones were radiolabelled by addition of ^{14}C aminoacid mixture (0.37 MBq. ml^{-1}, with 50 ug chloramphenicol ml^{-1}; Isotope Division, B.A.R.C.) using droplet inoculation procedure (1).

3. RESULTS AND DISCUSSION

3.1. Nonspecific binding of HI-CPS Several experiments were performed to ascertain if there exists any fast acting lectin mediated recognition phenomenon in cowpea which would suggest a functional role for HI-CPS in the infection/nodulation process. Using ^{125}I.Fl.CPS, binding towards no root hair (NRH) zone and mature root hair (MRH) zone of cowpea which represent infectible and noninfectible areas respectively was examined. Neither HI-CPS nor native CPS (i.e., CPS isolated from Rhizobium grown in YEM medium,5) showed exclusive binding towards either of the root zones (Table 1) although HI-CPS showed enhanced binding to both root zones. The combined nitrogen status of the rooting medium also did not affect the binding of CPS (data not shown). Binding of both the CPS species to roots of Medicago sativa, Vigna radiata, Trigonella foenum-graecum was also examined. There was no clear cut specificity observed for binding of either of the CPS species to any of the legume roots. Moreover, binding of rhizobia to cowpea roots remained unaltered in the presence of both CPS species (7).

3.2. Root hair curling by HI-CPS HI-CPS and to much lesser extent native CPS, could induce root hair curling when tested on cowpea plants (Fig.1a).

TABLE 1. Nonspecific binding of ^{125}I-labelled fluorescein conjugated CPS of Rhizobium sp. strain 1001 to different regions of legume roots[@].

Host plant	Region of the root	μg CPS.cm root^{-1} HI-CPS	μg CPS.cm root^{-1} N-CPS
Vigna unguiculata	NRH	0.061	0.038
	MRH	0.073	0.041
Vigna radiata	NRH	0.055	0.042
Medicago sativa	NRH	0.034	0.021
Trigonella foenum-graecum	NRH	0.041	0.047

@ Legume seedlings were grown (3-7 d) in plastic growth pouches without addition of combined nitrogen. Radiolabelled, native CPS or HI-CPS (1.2 μg) were delivered to specific region of the legume root by the droplet inoculation procedure (1).

Significant curling was observed within 12 to 16 h of addition of HI-CPS. In general HI-CPS induced faster and more pronounced curling than whole cells of Rhizobium log phase cultures.

3.3. HI-CPS as an elicitor compound It appears attractive to postulate a role for rhizobial CPS in specificity/infection process beyond its immediate interaction with root lectins or mere binding to roots. The ability of HI-CPS to induce extensive root hair curling could be of significance in this context.

Interestingly , target root epidermal cells in the NRH (but not MRH) zone apparently recognised HI-CPS as an elicitor and in response initiated de novo synthesis of plant proteins (Fig.1b). The pattern of protein synthesis in the NRH zone of cowpea plants was analysed in the presence and absence of Rhizobium sp strain 1001, native CPS or HI-CPS for a period of 4.5 h to 8 h. Fig.1b shows the legume protein polypeptides synthesised de novo and visualised by SDS-polyacrylamide gradient (7.5% - 12.5%) gel autoradiography. Cowpea root epidermal cells in the NRH region synthesised 5-7 polypeptides of molecular weight 0.3×10^5 - 2.0×10^5 specifically in response to HI-CPS or Rhizobium sp strain 1001. Such polypeptides were not observed in uninoculated plants or with plants inoculated with Rhizobium but grown in Jensen medium supplemented with 15mM NH_4^+ or NO_3. The unique bands of MW > 95 K were also observed when the NRH zone was treated with other complimentary cowpea Rhizobium strains such as NC92, S24 and Arg2C. These polypeptides appear to be the earliest symbiosis specific plant proteins detected so far (8,9).

3.4. Induction of Rhizobium proteins by host root exudate Recently certain Rhizobium symbiotic genes have been shown to get induced by the root exudate of specific host legumes (10) and root exudate is required for the induction of nodC (11) belonging to the nodABC gene cluster. We have previously observed the appearance of new protein bands in Rhizobium sp. strain 1001 under the influence of cowpea root exudate (7). Several other slow growing Rhizobium strains which nodulate cowpea plants synthesised the specific proteins during incubation in root exudate and attained higher infection status, even at low inoculum density (12).

4. SUMMARY

The present data and our previous findings suggest the occurrence of a sequence of early interactions in the cowpea-Rhizobium symbiosis. These include (i) secretion of signal substances by the roots during nitrogen deficiency (ii) induction

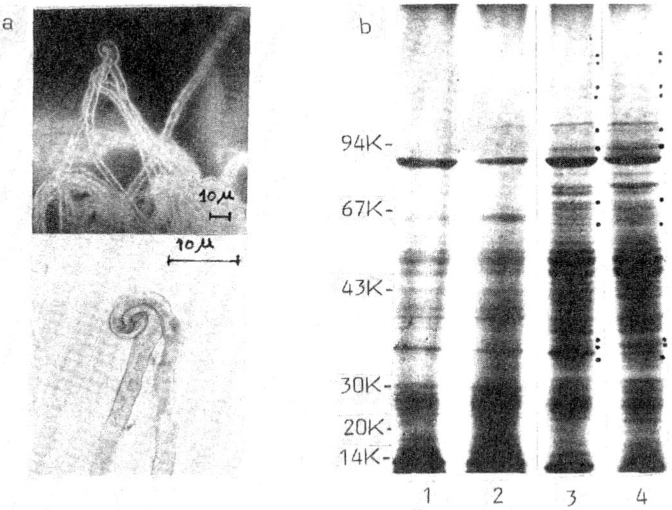

FIGURE 1. Induction of unique proteins and root hair curling by Rhizobium sp strain 1001 or by its HI-CPS. a) Light photomicrograph of the NRH region of cowpea plants 16 h after treatment with the Rhizobium (top), or its HI-CPS (1.2 μg plant^{-1};bottom). b) SDS-PAGE-autoradiograph showing symbiosis specific proteins (marked by dots) from NRH zone of cowpea plants treated with HI-CPS (lane 4) or Rhizobium sp strain 1001 cells (lane 3); same as in lane 3 except that plants were grown in the presence of 15mM NH_4^+(lane 1);uninoculated control (lane 2).

of specific Rhizobium proteins and CPS by the root secretions and (iii) elicitation by HI-CPS of symbiosis specific proteins in root target cells followed by root hair curling.

REFERENCES

1. Bhuvaneswari TV, Turgeon BG, Bauer WD (1980) Pl Physiol 66, 1027-1031
2. Bhuvaneswari TV,Bhagwat AA, Bauer WD (1981) Pl Physiol 68,1144-1149
3. Bhagwat AA, Thomas J (1982) Appl Env Microbiol 43, 800-805
4. Bhagwat AA, Thomas J (1983) Arch Microbiol 136, 102-105
5 Bhagwat AA, Thomas J (1984) Arch Microbiol 140, 260-264
6. Glabe CG, Harty PK, Rosen SD (1983) Anal Biochem 130, 287-294
7. Bhagwat AA (1985) Ph.D. Thesis, M.S. University of Baroda, Baroda, India
8. Fuller F, Verma DPS (1984) Pl Mol Biol 3, 21-28
9. Bisseling T, Franssen H, Govers F, Gloudemans T, Louwerse J, Moerman M. Nap J, Van Kammen A (1985) In Nitrogen Fixation Research Progress (eds: Evans HJ, Bottomley PJ, Newton WE) p 53-59. Martinus Nijhoff Pub. Dordrecht
10 Olson E, Sadowsky M, Verma DPS (1985) Bio/Technology 3, 143-149
11. Mulligan JT, Long SR (1985) Proc. Natn Acad Sci (USA) 82, 6609-6613
12. Philip G, Bhagwat AA, Thomas J (1986) M.S. in preparation

FOUR SOYBEAN NODULIN GENES EVOLVED FROM A COMMON ANCESTOR

F. JACOBS, M. ZHANG, M. FORTIN and D. P. S. VERMA
Centre for Plant Molecular Biology, Biology Department, McGill University, Montreal, Canada H3A 1B1

The nitrogen-fixing nodule is a differentiated tissue with unique structures and functions. A number of host-encoded proteins called 'nodulins' are specifically induced in this tissue, some of which have been identified and characterized ie. leghemoglobin (1), uricase II (nodulin-35) (2) and sucrose synthetase (nodulin-100) (3). The functions of most nodulins, however, remain to be determined. A subset of nodulin cDNA clones from a nodule-specific cDNA library (5) were observed to cross-hybridize under conditions of low stringency. Four of the most abundantly represented of these clones (encoding nodulin-23, -26b, -27 and -44) were further characterized. Nodulin-44 appears to be similar to the nodulin clone E27 recently described by Sengupta-Gopalan et al. (4).

Sequence comparisons of full length cDNA clones for nodulin-23, -26b and -27, revealed common structural features with nodulin-E27. The cDNAs are composed of three general 'domains', two of which have greater than 80% intermolecular homology. The first domain, which is approximately 250 bp long, begins at the initiation codon. The central domain is unique to each sequence and varies in length from 25 to 600 bp, with recursive elements occuring only in nodulin-E27. The third domain is less defined, both in its degree of homology and in its specific length. However, the homologous portions of this domain extend almost 200 bp into the 3' non-coding region of these clones.

The first and third domains also have a number of small insertions and deletions but in spite of this, the reading frames are conserved (Figure 1). However, approximately 30% of the conserved protein sequences consist of conservatively substituted amino acids. Searches through the National Biomedical Foundation Protein Data Base did not reveal any significant homologies with other known protein sequences.

An interesting feature of the alignment of these cDNAs is the coincident positions of the initiation and termination codons. In the latter case, the sequence of the immediate vicinity is unique to each cDNA, despite the surrounding homologies. The termination codons, therefore, appear to have been created in the same location by independent events, which suggests that these nodulins are subjected to similar functional constraints.

Closer examination of the amino terminal sequences of nodulin-23, -26b, -27 and -E27, which is the most highly conserved region in these nodulins, revealed properties of signal peptides (Figure 2). All four nodulins contain a hydrophobic sequence. Both nodulin-23 and -E27 have a potential cleavage site. The leader sequences of nodulin-26b and -27, however, appear not to meet all the criteria for a signal peptide. Nodulin-26b is missing a portion of the amino terminal hydrophobic and polar region while nodulin-27 has a lysine at position -5 of the only potential cleavage site, a residue never observed at this position (7). These findings lead us to predict that nodulin-23 and -E27 (or -44) are membrane associated proteins, while nodulin-26b and -27 are in the soluble fraction of the plant cell.

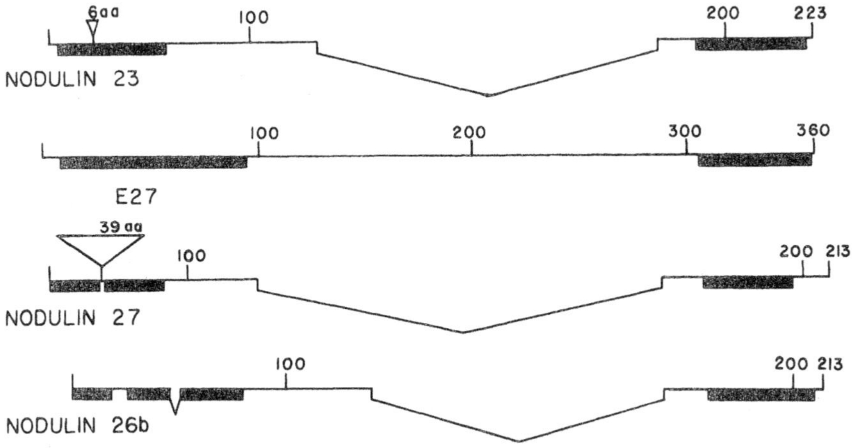

FIGURE 1. Protein homologies between nodulin-23, -26b, -27 and -E27. The protein sequences of nodulin-23, -26b, -27 and -E27, determined from their cDNA sequences, were analyzed. The nodulin protein sequences with >80% homology, including conservatively substituted residues, were aligned using the FastP computer program (6). The solid bars below the line represent either direct or conserved homologies of nodulin-23, 26b and 27 with nodulin-E27. For detailed sequences, see: nodulin 23 (8), nodulin-26b and -27 (9), nodulin-E27 (4).

In order to test their predicted subcellular locations, ^{35}S-labelled translation products of nodulin mRNAs were immuno-precipitated with either soluble nodule-specific (S-100) or peribacteroid-membrane specific antibodies (5,10). Immunoprecipitation of hybrid-released translation (HRT) products for nodulin-23 and -27 showed specific reactivity with the peribacteroid membrane and the S-100 antibodies respectively (Table 1). The strong reactivity of nodulin-23 and -27 suggest that despite similarities in their overall structures, these proteins are respectively found associated with the peribacteroid-membrane and the free cytoplasm of the soybean nodule cell, in agreement with our prediction.

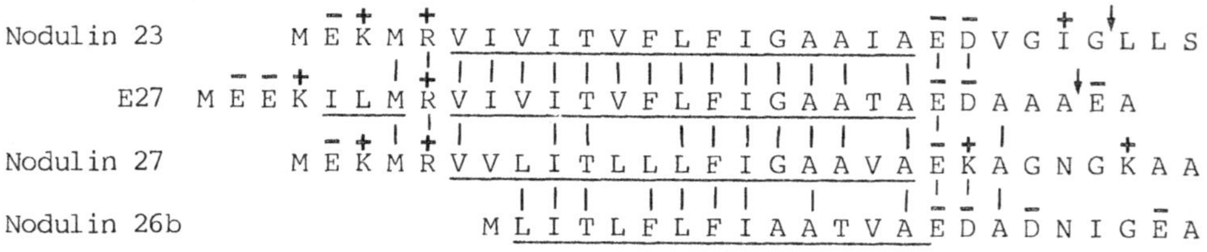

FIGURE 2. Amino terminal sequences of nodulin-23, -26b, -27 and -E27. Displayed are the amino terminal residues of nodulin-23, -26b, -27 and -E27. Vertical lines indicate direct amino acid homology with nodulin-E27. Basic and acidic amino acids are indicated by '+' and '-' respectively, while the hydrophobic regions are underlined. The arrows indicate putative sites for signal peptide cleavage (7).

TABLE 1. Reactivity of HRT nodulins with specific antibodies. ^{35}S-labelled hybrid-release translated (HRT) nodulins were immunoprecipitated with either soluble nodule specific (S-100) or peribacteroid membrane specific antibodies and subjected to SDS-PAGE. The strength of the reactivities were then scored.

Antibody	HRT nodulins			
	23	26b	27	44
soluble fraction	-	+	+++	+
Peribacteroid membrane	+++	+	-	+

Nodulin-26b and -44 reacted weakly with both antibodies, which did not permit deduction of their subcellular location. The weak reactions may be the result of imperfectly shared epitopes, since portions of these protein sequences are quite similar to nodulin-23 and -27.

CONCLUSIONS

We have found several structural similarities in four nodulins (nodulin-23, -26b, -27 and -44) of soybean. These nodulins appeared to have recently evolved from a common ancestor since they contain highly conserved coding and 3' non-coding nucleotide sequences. However, they showed diversity in both their structures and subcellular locations. Sequence analysis also revealed that the shared domains of homology exist at the amino acid level. Since the translational start and stop codons of these nodulins occur in the same relative positions in the homologous domains, these nodulins appear to be under a common functional constraint. The large number of conservative amino acid substitutions in the homologous domains and the variability in their overall structures, suggest that these nodulins probably do not have enzymatic functions. In addition, the amino terminal regions of these nodulins have properties of signal leader sequences, although two (nodulin-26b and -27) do not meet all the criteria for a cleavable signal peptide. Specific immunoprecipitation with nodulin hybrid-released translation products confirmed the predicted subcellular locations of nodulin-23 (peribacteroid membrane) and -27 (soluble). These nodulins have additional members in its family, several of which have been isolated.

REFERENCES

1. Sullivan D, Brisson N, Goodchild B, Thomas DY, Verma DPS: Molecular cloning and organization of two leghemoglobin genomic sequences of soybean. Nature 289:516-618, 1981.
2. Nguyen T, Zelechowska M, Foster V, Bergmann H, Verma DPS: Primary structure of the soybean nodulin-35 gene encoding uricase II localized in the peroxisomes of uninfected cells of nodules. Proc Natl Acad Sci 82:5040-5044, 1985.
3. Thummler F, Verma DPS: manuscript in preparation.

4. Sengupta-Gopalan C, Pitas JW, Thompson DV, Hoffman LM: Expression of host genes during root nodule development in soybeans. Mol Gen Gent, in press, 1986.
5. Fuller F, Kunstner PW, Nguyen T, Verma DPS: Soybean nodulin genes: Analysis of cDNA clones reveals several major tissue-specific sequences in nitrogen fixing root nodules. Proc Natl Acad Sci 80:2594-2598, 1983.
6. Lipman DJ, Pearson WR: Rapid and sensitive protein similarity searches. Science 227:1435-1441, 1985.
7. von Heijne G: A new method for predicting signal sequence cleavage sites. Nucl Acids Res 14:4683-4690, 1986.
8. Mauro V, Nguyen T, Katinakis P, Verma DPS: Primary structure of the soybean nodulin-23 gene and potential regulatory elements in the 5'-flanking regions of nodulin and leghemoglobin genes. Nucl Acids Res 13: 239-249, 1985.
9. Jacobs FA, Zhang M, Fortin M, Verma DPS: manuscript in preparation.
10. Fortin M, Zelechowska M, Verma DPS: Specific targeting of membrane nodulins to the bacteroid-enclosing compartment in soybean nodules. The EMBO J 4:3041-3046, 1985.

COORDINATED EXPRESSION OF NODULE-SPECIFIC AND ROOT GENES IN YELLOW LUPIN

M. SIKORSKI, U.SZYBIAK-STRÓŻYCKA, P. STRÓŻYCKI, B. GOLIŃSKA, C.J. MĄDRZAK, R. KAMP, B. WITTMANN-LIEBOLD, A.B. LEGOCKI

1. INTRODUCTION

The development of nitrogen fixing symbiotic system of leguminous plant and Rhizobium species is associated with modulation of gene expression of both partners. It is now well established that in the infected plant tissue a group of host-specific polypeptides - nodulins is synthesized. Although several nodulin encoding sequences have been cloned and their structure recognized, regulation of plant genes expression during symbiosis remains unknown.

In our previous studies using immunoelectrophoretic techniques we observed that the development of lupin nodules is accompanied by repression of some root proteins. We present here the data suggesting that coupled repression - derepression events occur during the nodule development in yellow lupin /Lupinus luteus/. Using HPLC separations and Western blotting we demonstrate that the synthesis of lupin leghemoglobins I and II is correlated with disappearance of 18 kD root protein.

2. EXPERIMENTAL

Yellow lupin leghemoglobins I and II were isolated from nodule tissue. Total cytoplasmic proteins were precipitated with ammonium sulphate. The fraction 50-80% of saturation was submitted to the Sephadex G-50 gel filtration. Crude leghemoglobin fraction was purified by high performance liquid chromatography on the Vydac TP reverse phase column, eluted with linear gradient of 0-50% isopropanol containing 0,1% trifluoroacetic acid and 0,01% 2-mercaptoethanol. Paralelly, identical procedures were applied to the total cytoplasmic proteins from 3 days old roots, 12 days old nodule zones, 23 days old and 35 days old nodules.

We have identified the 18 kD root protein that is absent from the nodule tissue. This protein was purified from 3 days old lupin seedlings by the same procedure as two leghemoglobins. The inspection of HPLC patterns indicated a great extent of similarity in the properties of these proteins. Root protein /R18/ is eluted from the HPLC column at the concentration of 42,5% of isopropanol, whereas leghemoglobins I and II at 43% and 43,5% respectively.

The analysis of protein patterns during the course of lupin nodule development indicates that protein R18 disappears at the phase when leghemoglobin reaches the level of the most abundant nodule protein i.e. at 23-rd day after

infection. On the other hand, in 23 days old uninfected roots protein R18 is present in high amount.
The electrophoresis of fractions obtained from HPLC separations and the calculation of relative concentrations of proteins during the course of nodule development is shown in Figure 1.

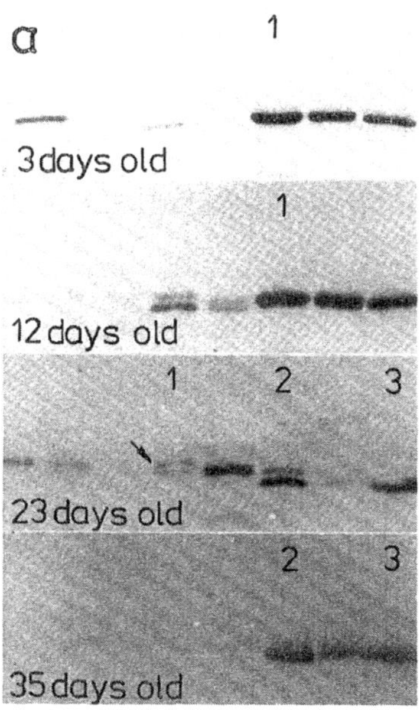

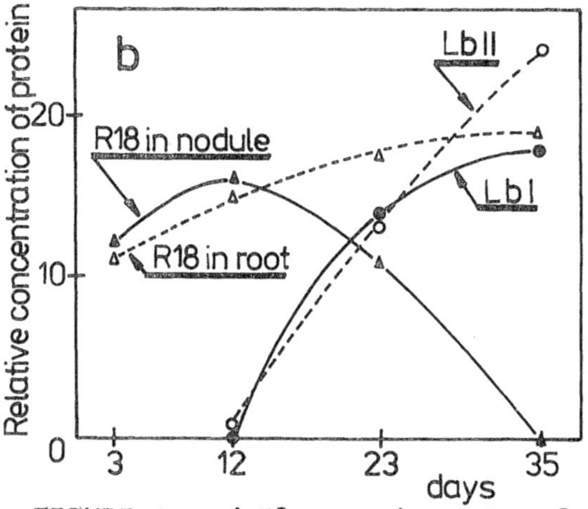

FIGURE 1. a/ Electrophoresis of fractions after HPLC separations. The fraction numbers are correlated with eluting concentration of isopropanol.
b/ The relative concentration of protein R18 and leghemoglobins in lupin roots and lupin nodules calculated from HPLC patterns and electrophoresis of purified fractions.

As it is shown in Fig. 2. Western blots do not reveal any immuno-crossreactivity between lupin leghemoglobins and protein R18.

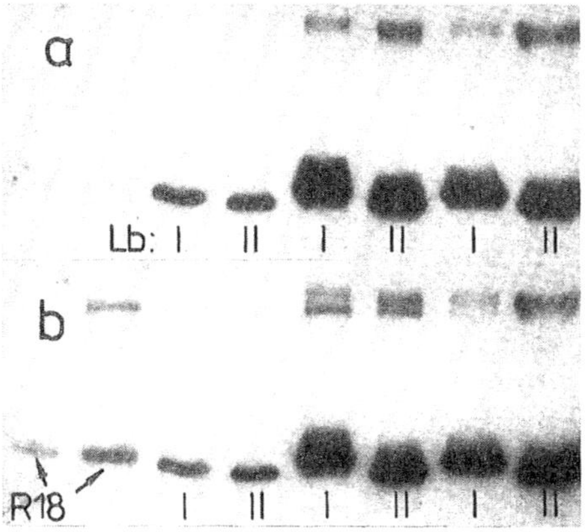

FIGURE 2. The Western blot of selected fractions contain-ing lupin leghemoglobins I and II and protein R18 reacted with anti-Lb antibody/a/. b/ the same blot was subsequently reacted with the anti body against proteins extracted from 23 days old uninfected roots.

Although the amino acid composition of leghemoglobins and R18 shows significant similarities /Table I/, the N-terminal sequences are different which indicates that these proteins are coded by different genes /Fig. 3/.

TABLE I

Amino acid residue	Lb I	Lb II	R18
Cys	0•	0•	0•
Asp	16.42	14.30	16.32
Glu	21.84	23.67	27.97
Ser	9.93	9.36	7.36
His	3.61	4.48	0
Gly	10.06	8.94	19.57
Thr	8.80	8.61	12.08
Arg	1.63	1.58	2.33
Ala	18.81	20.38	18.28
Tyr	3.73	3.60	6.04
Trp	0•	0•	0•
Met	0.53	0.50	0
Val	12.75	12.97	9.82
Phe	7.31	7.10	5.88
Ile	7.67	7.70	14.40
Leu	14.50	14.53	10.98
Lys	14.23	13.79	16.75
Pro	0•	0•	0•

•not detectable as OPA derivatives

Lb I Gly-Val-Leu-Thr-Asp-Val-Glu-Val-Ala-

Lb II Gly-Ala-Leu-Thr-Glu-Ser-Glu-Ala-Ala-

R18 Gly-Leu-Ala-Gly-Phe-Thr-
(Ile)

FIGURE 3. The comparison of N-trminal amino acid sequences of lupin leghemoglobins and root protein R18.

The fact that in lupin nodules leghemoglobin replaces root protein R18 that has several similar physico-chemical properties can indicate that the pool of proteins present in lupin nodule and in root maintain similar hydrophobicity. It may furthermore suggest that coupled repression - derepression system beeing a part of overall symbiotic regulatory mechanism, functions within the nodule.

PLANT GENE EXPRESSION DURING EFFECTIVE AND INEFFECTIVE NODULE DEVELOPMENT OF THE TROPICAL STEM-NODULATED LEGUME SESBANIA ROSTRATA

P. de Lajudie(1) and T. Huguet
Laboratoire de Biologie Moléculaire des Relations Plantes-Microorganismes, INRA-CNRS, BP 27, Auzeville, 31326 CASTANET-TOLOSAN CEDEX, FRANCE.
(1) Permanent Address : Laboratoire de Biologie des Sols, ORSTOM, BP1386 DAKAR, SENEGAL.

Sesbania rostrata is a tropical legume which develops nitrogen-fixing nodules on both stems and roots when infected by a specific Rhizobium (Dreyfus and Dommergues, 1981). This fixation is very efficient (Dommergues et al., 1985), and Sesbania rostrata has a real potential in agriculture as a green manure (Dreyfus et al., 1985 ; Rinaudo and Moudiongui, 1986). In agreement with what is described in other nitrogen-fixing symbioses i.e. root-nodulated plants like Glycine max (Legocki and Verma, 1980), Pisum sativum (Bisseling et al., 1983), Phaseolus vulgaris (Cullimore et al., 1983) and Medicago sativa (Lang-Unnasch and Ausubel, 1985), we have found plant genes specifically activated in stem and root nodules of Sesbania rostrata.

We compared two-dimensional (2D) polyacrylamide gel electrophoresis patterns of in vitro translation products of poly A^+-mRNA purified from uninfected roots and stems and from root and stem nodules induced by wild type Rhizobium sp. strain ORS571 (Dreyfus et al., 1983). 200-300 polypeptides could routinely be distinguished in each pattern, their molecular weights ranging from 10 000 to more than 100 000 daltons. The figure shows the 2D-patterns of the different tissues. Most of the polypeptides are present in all tissues. However at least 30-40 polypeptides exhibit different intensities in stems, roots, stem nodules and root nodules, reflecting tissue-related plant gene expression. Among these we found, according to nomenclature proposed by Van Kammen (1984), at least 7 "nodulins" (see figure : spots 28, 26, Lb1, Lb2, Lb3, Lb4, Lb5) and 9 "nodule-stimulated polypeptides" (see figure : spots 66, 54, 53, 52, 51, 38, 37, 25, 22) common to both root and stem nodules. In addition, whilst 44 is a genuine "root nodulin" (since it is not present in uninfected roots), it is a "stem nodule-stimulated polypeptide" (since there are trace amounts in uninfected stems). In the same way, 39 is a "stem nodulin" and a "root nodule-stimulated" polypeptide. In addition, by comparing uninfected stems to stem nodules, we found 3 "stem nodulins" (38, 33, 33') and 3 "stem nodule-stimulated" polypeptides (32, 32', 31), which are either not present or not stimulated in root nodules ; on the other hand, by comparison of root nodules with uninfected roots, there are 4 "root nodulins" which are not stimulated in stem nodules.

We used an antiserum raised against leghemoglobin purified from Sesbania rostrata stem nodules to immunoprecipitate the corresponding in vitro translation products. None of the uninfected root and stem poly A^+ RNA translation products reacted with the antiserum. On the other hand, five abundant polypeptides of molecular weights below 17kdal (belonging to the class of common root and stem nodulins mentioned above) were immunoprecipitated from both stem and root nodules. According to their decreasing isoelectric points, we named them Lb1 to Lb5 (see figure). In mature stem and root nodules, Lb2 and Lb5 are major polypeptides, Lb3 and Lb4 moderately abundant and Lb1 is a minor species.

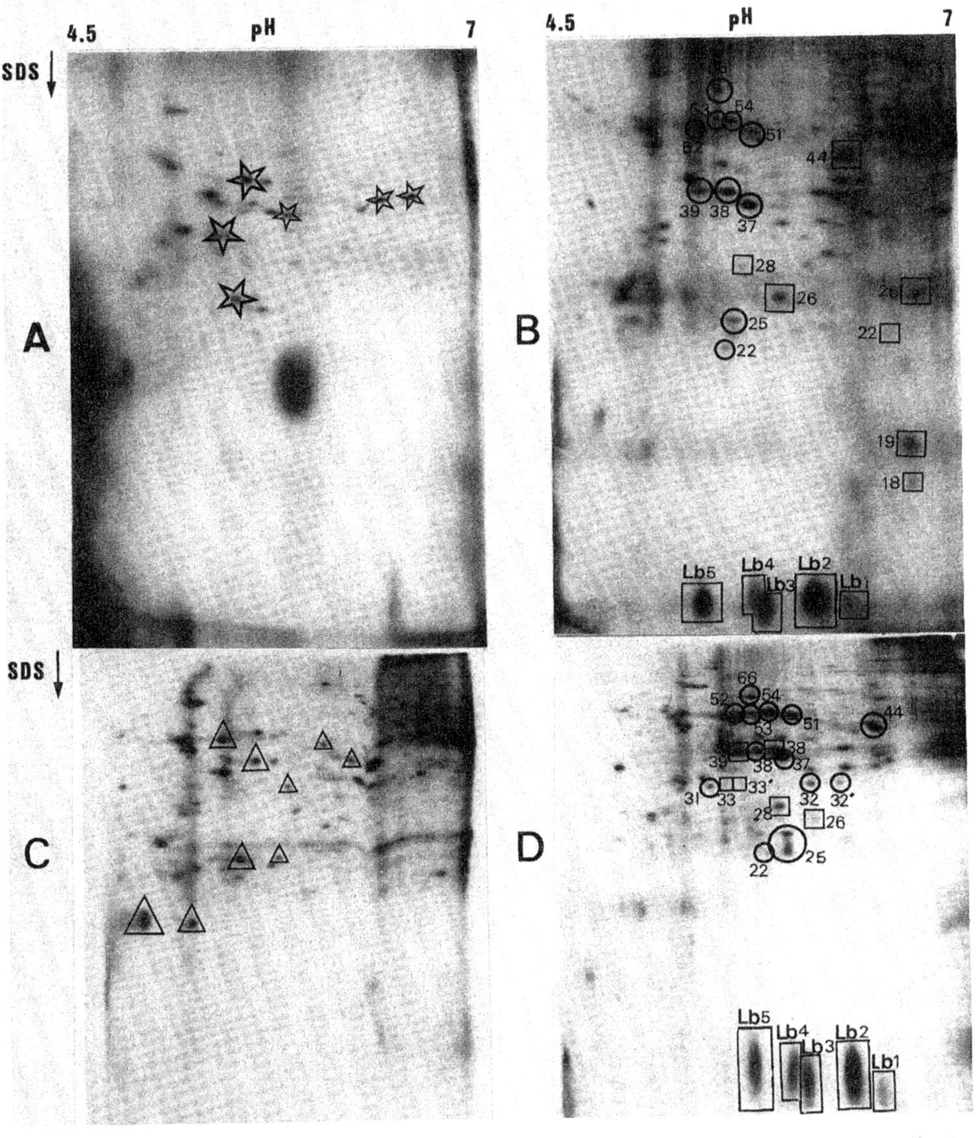

Differential plant gene expression in stem and root nodules compared to uninfected stems and roots. Fluorographs of two-dimensional polyacrylamide gel of *in vitro* translation products from poly A^{+}-mRNA isolated from :
(A) 15 days-old uninfected roots (B) effective root nodules induced by wild type *Rhizobium* sp. strain ORS571, 6 weeks after sowing (C) 15 days-old uninfected stems (D) 32 days-old effective stem nodules induced by wild type *Rhizobium* sp. strain ORS571.
Numbers indicate the polypeptides approximative molecular weights (Kdal).

□ nodule-specific polypeptide
○ nodule-stimulated polypeptide
☆ root-stimulated polypeptide
△ stem-stimulated polypeptide

We studied plant gene expression during stem nodule development. Leghemoglobin components can first clearly be detected at day 12 after inoculation, concomitantly with the majority of the other nodule-stimulated polypeptides and their intensities increase throughout our observation period (32 days after stem inoculation). Moreover their relative intensities vary during the same period : at day 12, Lb2, Lb3, Lb5 have more or less the same intensities while Lb4 is major and Lb1 is not clearly visible ; after day 16, Lb2 and Lb5 become majoritary, Lb3 and Lb4 are moderately abundant and Lb1 appears but remains minoritary. Apart from leghemoglobin, other nodule-stimulated genes are activated at different stages of development : some are stimulated early, like 44 and 38, detectable at day 6, or 28 and 25 detectable at day 7. Some polypeptides reach their maximum intensity and then remain more or less at the same level : this is the case for 38, 28 (plateau reached at day 12), for 31 (plateau reached at day 14), and for 54, 53, 51, 39, 32 and 32' (maximum intensity from day 18). Some others are transiently activated : 44 is very intense during the first stages and then decreases gradually ; 33, 33', 25 and 22 reach their maximum expression between days 12-16, then decrease. All these observations indicate that plant genes are sequentially expressed during stem nodule development in *Sesbania rostrata* like in pea (Bisseling *et al.*, 1983) and soybean (Fuller and Verma, 1984).

We also analysed ineffective stem nodules induced by a nif^- mutant *Rhizobium* sp. strain 5740 (Elmerich *et al.*, 1982). The resulting pattern is comparable with that of the effective stem nodules, except for the five leghemoglobin components which appear less abundant (data not shown). This indicates that in *S. rostrata*, like in other described systems (Fuller and Verma, 1984 ; Govers *et al.*, 1985 ; Lang-Unnasch and Ausubel, 1985), leghemoglobin genes are activated in ineffective nodules induced by nif^- *Rhizobium* sp. mutants, but are expressed at a reduced level compared to efficient nodules.

Acknowledgements :

We are thankful to D. Bogusz, from O.R.S.T.O.M. Dakar, Sénégal, for sending us antiserum raised against leghemoglobin purified from *Sesbania rostrata* stem nodules.

References :

Bisseling T., Been C., Klugkist J., Van Kammen A., Nadler K. 1983. EMBO J. 2, 961-966.

Cullimore J.V., Lara M., Lea P.J., Miflin B.J. 1983. Planta 157, 245-253.

Dommergues Y.R., Dreyfus B.L., Diem H.G., Duhoux E. 1985. La Recherche 16 (162), 22-31.

Dreyfus B.L., Dommergues Y.R., 1981. FEMS Microbiol. Lett. 10, 313-317.

Dreyfus B.L., Elmerich C., Dommergues Y.R., 1983. Appl. Envir. Microbiol. 45, 711-713.

Dreyfus B.L., Rinaudo G., Dommergues Y.R. 1985. MIRCEN Journal 1, 111-121.

Elmerich C., Dreyfus B.L., Reysset G., Aubert J.P. 1982. EMBO J. 1 (4), 499-503.

Fuller F., Verma D.P.S. 1984. Plant Mol. Biol. 3, 21-28.

Govers F., Goudemans T., Moerman M., Van Kammen A., Bisseling T. 1985. EMBO J. 4 (4) 861-867.

Lang-Unnasch N., Ausubel F.M. 1985. Plant Physiol. 77, 833-839.

Legocki R.P., Verma D.P.S. 1980. Cell 20 153-163.

Rinaudo G., Moudionqui A., 1986. Bull. Rech. Agr. Gembloux - In Press.

Van Kammen A. 1984. Plant Mol. Biol. Rep. 2 (2) 43-45.

EXPRESSION OF TWO ENZYMES INVOLVED IN UREIDE FORMATION IN SOYBEAN REGULATED BY OXYGEN

Knud Larsen and Bjarne Jochimsen
Department of Molecular Biology and Plant Physiology, University of Aarhus, DK-8000 Aarhus C, Denmark.

Introduction

Soybean (*Glycine max*) synthesize ureides (allantoin and allantoic acid) from currently fixed nitrogen in root nodules and use these compounds for transport of nitrogen. Ureides are formed as a result of purine *de novo* synthesis of inosine 5'-monophosphate (IMP) followed by oxidative catabolism (Triplett et al. 1980; Schubert 1981; Atkins et al. 1982; Boland and Schubert 1982).

In soybean root nodules the enzymes involved in the degradation of IMP: 5'-nucleotidase, purine nucleosidase, xanthine dehydrogenase, uricase and allantoinase have been identified and characterized (Tajima and Yamamoto 1975; Christensen and Jochimsen 1983). The properties of xanthine dehydrogenase (Triplett et al. 1982) and uricase (Bergmann et al. 1983; Lucas et al. 1983) have been determined with purified enzyme preparations. The subunit of one of these enzymes, uricase, has been shown to be identical with the nodule-specific nodulin-35 (Bergmann et al. 1983).

Appearance of purine catabolizing enzymes in soybean root nodules

In soybean root nodules 5'-nucleotidase, purine nucleosidase, xanthine dehydrogenase and uricase are expressed concomitantly with nodule development (Fig. 1 A,B,C,D). The last enzyme of the pathway, allantoinase, is unaffected by nodule development (Fig. 1 E). The initial expression of these enzymes does not depend on active nitrogen fixation, as demonstrated by comparison of effective (Fix^+) and ineffective (Fix^-) nodules. However with xanthine dehydrogenase, uricase (Fig. 1 C,D) and less pronounced with 5'-nucleotidase (Fig. 1 A) the maximal specific activities obtained are lower in Fix^- nodules. The specific activities of purine nucleosidase and allantoinase were not significantly different when Fix^+ and Fix^- nodules were compared (Fig. 1 A,E). From the patterns of appearance some similarities/differences in the mode of regulation of these enzymes would be expected.

Effect of oxygen on the expression of purine catabolizing enzymes in callus tissue

In order to identify possible bacterial signals that might control gene expression in the host plant, we assumed that invasion and proliferation of *Bradyrhizobium* within plant root cells might create a situation with an increased demand for oxygen for bacterial respiration. The lowered oxygen availability, within the plant cell, might then result in synthesis of enzymes characteristic of anaerobic or partial anaerobic conditions.

Our hypothesis was tested by incubating sterile callus tissue in serum bottles and adjusting the oxygen concentration in the gas phase . After 48 hours of incubation under such conditions, extracts were prepared and the specific activities of the different purine catabolizing enzymes determined. The positive results are shown in Figure 2.

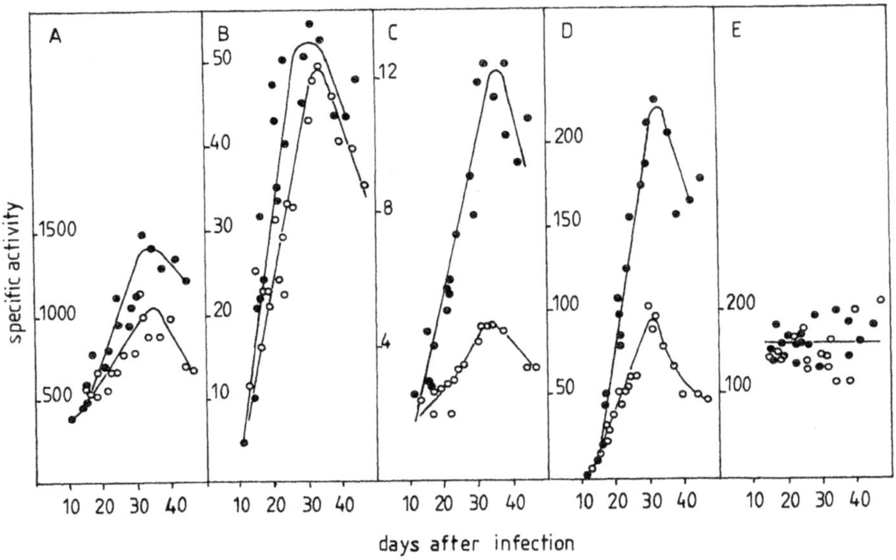

Figure 1. Specific activity (n mol/min·mg protein) of A, 5'-nucleotidase; B, purine nucleosidase; C, xanthine dehydrogenase; D, uricase and E, allantoinase in extracts from nodules collected at the indicated times after infection. Infections were with a Fix$^+$ and Fix$^-$ of B. japonicum : (●) B. japonicum 110 (spc4, nod$^+$, fix$^+$) and (o) B. japonicum A3 (spc4, nifD::Tn5, nod$^+$, fix$^-$)(Hahn and Hennecke 1984). Data from three consecutive experiments presented.

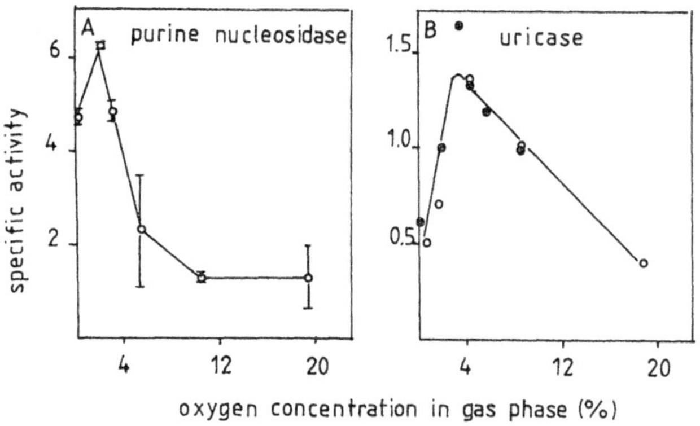

Figure 2. Specific activities (n mol/min·mg protein) of A, purine nucleosidase and B, uricase in extracts from callus tissue exposed to different oxygen concentrations as described (Larsen and Jochimsen 1986). Data with purine nucleosidase are the mean ± SE of three experiments. With uricase data are the average of duplicate experiments.

A four fold increase in specific activity of purine nucleosidase was observed, when the oxygen concentration was lowered to around 2% (Fig. 2 A) while a two fold increase in specific activity of uricase was obtained around 4-5% oxygen in the gas phase (Fig. 2 B). The increase in specific activity of uricase was demonstrated to be due to de novo protein syntheis using antibodies raised against highly purified nodule-specific uricase (Larsen and Jochimsen 1986). To determine whether the increase in purine nucleosidase activity is a result of elevated gene expression, or a modification (activation) of preexisting enzyme has to await the isolation of antibodies raised against purified purine nucleosidase.

The specific activities of 5'-nucleotidase, xanthine dehydrogenase and allantoinase did not respond to variations in oxygen concentration (Fig. 3 A,B,C).

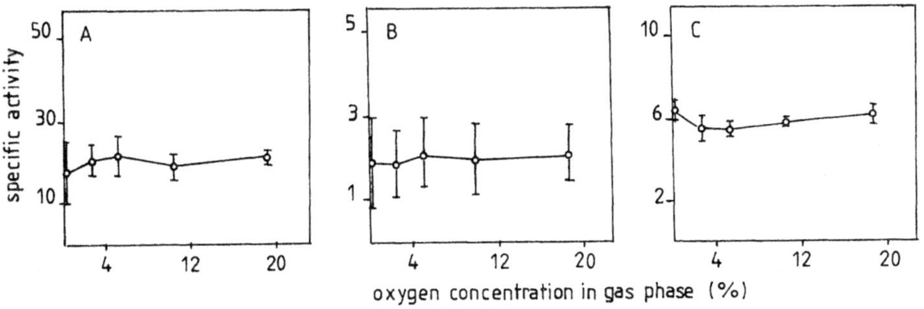

Figure 3. Specific activities (n mol/min·mg protein) of A, 5'-nucleotidase; B, xanthine dehydrogenase and C, allantoinase. Mean data ±SE of three experiments. Other details see legend to Fig. 2.

Since the pattern of appearance in Fix^+ and Fix^- nodules of glutamine synthetase resembles that of uricase (Werner et al. 1980), we tested whether the expression of two enzymes involved in amino acid metabolism was influenced by oxygen. Glutamine synthetase and aspartate aminotransferase did not respond to variations in oxygen concentration (data not shown).

Nodule-specific proteins

Uricase has been demonstrated to be nodule-specific (Bergmann et al. 1983) and the increase in specific activity is around 200 fold above root level (1 nmol/min·mg protein). The increase in specific activity observed with purine nucleosidase is only around 15 fold above root level (3-4 nmol/min·mg protein) and the enzyme can per se not be regarded as nodule-specific, unless a different form of the enzyme is expressed during nodule development, as has been demonstrated with glutamine synthetase from Phaseolus (Cullimore et al. 1983; Lara et al. 1983). In order to examine this possibility we have started to purify and characterize purine nucleosidase from Fix^+ nodules.

Two isoenzyme forms of purine nucleosidase can be separated by DEAE ion-exchange chromatography (Fig. 4). These two forms did not differ significantly with respect to kinetic parameters for the substrates: adenosine, inosine, xanthosine and uridine. Both forms showed a broad pH optimum around 8-9 (with uridine as substrate). Gelfiltration on Sephadex G150 columns revealed a molecular weight of 80.000 daltons for both forms.

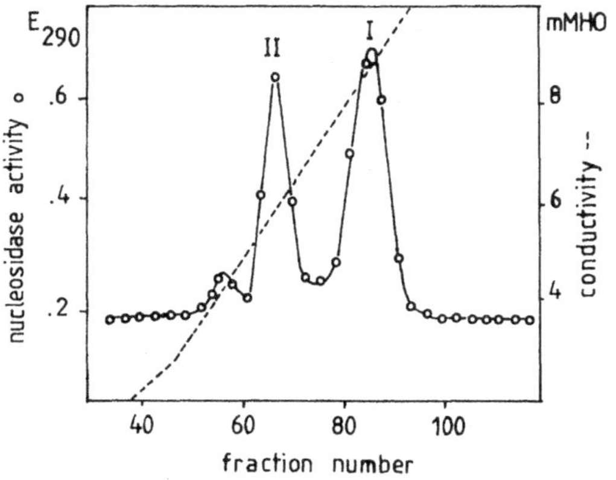

Figure 4. Elution profile of purine nucleosidase activity from a DEAE column (Sephadex A50) of a 35-55% ammoniumsulfate precipitate, prepared from five weeks old nodules (infected with B. japonicum 110). Fractions were assayed nucleosidase activity with uridine as substrate.

Tissue-specific appearance of purine nucleosidase

The separation of two isoforms of purine nucleosidase can easily be repeated by native polyacrylamide gelelectrophoresis. Nodule extracts contained equal amounts of activity of the two forms, both with Fix^+ and Fix^- nodules, while only minor activity of one form (I) is observed in leaves, (stem not shown) and roots (Fig. 5). This indicates a nodule-specific expression of form II, but also a tissue-specific increase in the expression of form I.

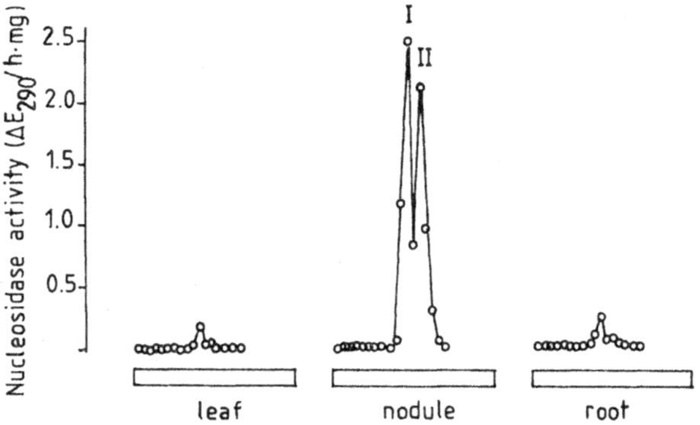

Figure 5. Native (7.5%) polyacrylamide gelelctrophoresis of G25 gelfiltrated extracts prepared from leaves, root nodules and roots from plants four weeks after infection with B. japonicum 110. Specific activities of purine nucleosidase in the extracts were: 3.1; 41.5 and 3.5 nmol/min mg protein respectively. Tube gels were sliced (2.5 mm) and each slice incubated with uridine to determine activity. Activities normalized (actual incubations were 9 hours for leaf and root, 2 hours for nodule.

Conclusion

The presented data indicate oxygen as one of the signals regulating expression of nodule-specific proteins: purine nucleosidase (form II) and uricase, both enzymes involved in ureide formation.

Acknowledgements

We wish to thank Dr. H. Hennecke for supplying *B. japonicum* strains. K.L. was supported by a Scholar stipendium from the Carlsberg Foundation.

References

Atkins CA, Ritchie A, Rowe PB, McCairns E and Saur D (1982) Plant Physiol. 70, 55-60.
Bergmann H, Preddie E and Verma DPS (1983) EMBO J. 2, 2333-2339.
Boland MJ and Schubert KR (1982) Arch. Biochem. Biophys. 213, 486-491.
Christensen TMIE and Jochimsen BU (1983) Plant Physiol. 72, 56-59.
Cullimore JV, Lara M, Lea PJ and Miflin BJ (1983) Planta 157, 245-253.
Hahn M and Hennecke H (1984) Mol. Gen. Genet. 193, 46-52.
Lara M, Cullimore JV, Lea PJ, Miflin BJ, Johnston AWB and Lamb JW (1983) Planta 157, 254-258.
Larsen K and Jochimsen BU (1986) EMBO J. 5, 15-19.
Lucas K, Boland MJ and Schubert KR (1983) Arch. Biochem. Biophys. 226, 190-197.
Schubert KR (1981) Plant Physiol. 68, 1115-1122.
Tajima S and Yamamoto Y (1975) Plant Cell Physiol. 16, 271-282.
Triplett EW, Blevins DG and Randall DD (1980) Plant Physiol. 65, 1203-1206.
Triplett EW, Blevins DG and Randall DD (1982) Arch. Biochem. Biophys. 219, 39-46.
Werner D, Mörschel E, Stripf R and Winchenbach B (1980) Planta 147, 320-329.

PROBING CELL WALL STRUCTURE IN THE SOYBEAN ROOT NODULE

S.-S.T. HUA[1], K.L. MILLER[1], V.J. VREELAND[2], and W.M. LAETSCH[2]

INTRODUCTION:

The infection of soybean (Glycine max) by Bradyrhizobium japonicum has been observed to occur through root hairs. The region of the root just below the smallest emergent root hairs and just above the zone of rapid root elongation is most susceptible to nodulation by rhizobia (1). An early response of the host plant is the marked curling of the root hairs, accompanied by the induction of meristematic activity in the cortex of the root adjacent to the point of infection. The infection thread is initiated at, or near, the growing tip of a root hair following the entrapment of rhizobia in the crook of the curled root hair. The stimulation of plant cell division induces active cell wall synthesis, which in turn facilitates the growth of the infection thread. Thus, the interaction of rhizobia with the apparatus of plant cell wall synthesis may be very important in nodule initiation and development.

Studies of the ultrastructure of the legume-Rhizobium symbiotic process have revealed that the interactions between the bacterium and host plant are maintained at the level of the cell surface. Determinants on the surface of the Rhizobium and on the infection thread wall may play important key roles during nodulation (1,2,3,4). The early stages of rhizobial infection are highly complex and multiphasic. The surface structures of both symbionts may become modified as a result of rhizobia-plant interactions. We are developing molecular probes for polysaccharides to study the molecular mechanisms by which the bacteria specifically attach, invade and initiate nodulation. These molecular probes include monoclonal antibodies to rhizobial capsular polysaccharide (CPS) and lipopolysaccharide (LPS), plus a fluorescein-conjugated pectate probe.

Pectins, which are heterogenous polysaccharides, predominate among the cell wall carbohydrates of legume roots (5,6). Variable aspects of the pectate component include the length of polygalacturonate sequences, the amount of esterification and the distribution of the esterified sugar in blocks or at random. These factors significantly affect the physical and biochemical properties of the pectate. The pectin-extensin interaction is another important part of cell wall growth and structure.

[1]Western Regional Research Center, ARS, USDA, 800 Buchanan Street, Berkeley, CA 94710, USA

[2]Department of Botany, University of California, Berkeley, CA 94720, USA

Vreeland et al. have recently prepared short alginate and pectate chains conjugated to fluorescein and used them to label plant tissue sections to examine polysaccharide sequences and their degree of esterification in the cell wall (7). The pectate probe consists of pectate blocks averaging 30 sugars in length and is based on the cooperative, calcium-mediated dimerization and aggregation of pectate. We have used this fluorescent pectate probe to label the cell wall of soybean nodules.

RESULTS AND DISCUSSION:

The fluorescent pectate probe labelled the plant cell walls of one- and three-week old soybean nodule sections in the presence of calcium chloride. Incubation of nodule sections with EDTA or pectinase significantly reduced the binding of the pectate probe to the cell wall, while treatment with proteinase K had no effect. These results demonstrate that the fluorescent probe is binding specifically to polygalacturonan (pectate) and not to protein. Postfixation of nodule sections with triazidotrinitrobenzene (TTB) resulted in more uniform labelling (Fig. 1). The pectate within the soybean nodule cell wall was highly esterified. Both chemical and pectinesterase (8) treatments of the nodule sections to remove the methyl esters increased the binding of fluorescent-pectate probe much more so in the infected cell zone than in the surrounding nodule cortex (Miller et al., manuscript submitted). This result implies that the cells in the central infected zone might have pectate with relatively long methyl esterified blocks while the cortical cells might have pectate with shorter, esterified blocks. Since cells in both the cortex and central infected zone are actively dividing under rhizobial influence, the differences in pectate esterification could reflect either a modification or differences in pectate synthesis.

The pectate probe concept could be extended to developing novel carbohydrate-carbohydrate hybridization probes (7,9). The like-like polymer associations form the basis for gelation and biological interactions (9). By labeling fragments of rhizobial polysaccharides with fluorescein (10), and using the proper incubation conditions, specific binding to plant cell wall polysaccharides may be observed. If

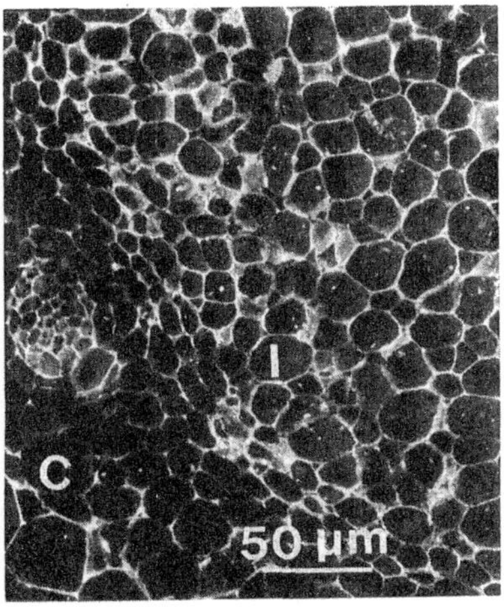

Figure 1. Section of a three-week-old soybean nodule labelled with fluorescent pectate. Nodules were fixed in 4% formalin in 70% ethanol, dehydrated in 95% and 100% ethanol sequentially and then embedded in glycol methacrylate. Sections 1-2 µm thick were cut, chemically deesterified and postfixed with TTB prior to incubation with probe. (C) nodule cortex, (I) infected cell zone.

so, this system may form the basis of an additional probe for examining the interaction of rhizobial polysaccharides with plant cell wall polymers.

Collodial gold-labeled monoclonal antibodies to various rhizobial and plant polysaccharides, in conjunction with transmission electron microscopy, will provide a third powerful probe for revealing the biochemical composition and details of changes in surface structures. These techniques, together with transposon Tn5 mutagenesis to produce genetically well defined infection mutants (3) blocked in their nodulation ability, will be powerful aids in defining the functional link between cell wall composition and rhizobial infection.

REFERENCES:

1. Bauer, WD: Infection of legumes by rhizobia. Ann. Rev. Plant Physiol. 32, 407-449, 1981.
2. Robertson JG, Wells B, Brewin NJ, Wood E, Knight CD, Downie JA: The legume-*Rhizobium* symbiosis: a cell surface interaction. J. Cell Sci. Suppl. 2. 317-331, 1985.
3. Rolfe BB, Shine J: *Rhizobium*-*Leguminosae* symbiosis: the bacterial point of view. In: Genes involved in microbe-plant interactions, pp. 95-128, Verma, DPS, Hohn, T. eds. Springer-Verlag, Vienna, Austria, 1984.
4. Verma DPS, Nadler K: Legume-*Rhizobium* symbiosis: host's point of view. In: Genes involved in microbe-plant interactions, pp. 57-93, Verma, DPS, Hohn, T. eds. Springer-Verlag, Vienna, Austria, 1984.
5. McNeil M, Darvill AM, Fry SC, Albersheim P: Structure and function of the primary cell walls of plants. Ann. Rev. Plant Physiol. 53, 625-663, 1984.
6. Selvendran RR: Developments in the chemistry and biochemistry of pectic and hemicellulose polymers. J. Cell Sci. Suppl. 2, 51-88, 1985.
7. Vreeland V, Morse S, Robichaux R, Laetsch WM: Pectate content and esteri-
fication in guard cell walls. J. Cell Biol. 101, 341a, 1985.
8. Pilnik W, Rombouts FM: Pectic enzymes. In: Polysaccharides in food, pp. 109-126, Blanchard J M V, Mitchell J M, (eds), Butterworths, London, 1978.
9. Rees DA: Polysaccharide Shapes., John Wiley and Sons, New York, 1977.
10. Dazzo FB, Hollingsworth JR, Sherwood JE, Abe M, Hrabak EA, Gardiol AE, Pankratz JT, Smith KB, and Yang H.: Receptor site on clover and alfalfa roots for *Rhizobuim*. Appl. Environ. Microbiol. 33: 132-136, 1985.

MONOCLONAL ANTIBODIES TO COMPONENTS OF RHIZOBIUM-INDUCED PEA NODULES

Desmond J. Bradley, Elizabeth A. Wood, Geoffrey W. Butcher[1], Giovanni Galfre[1], Nicholas J. Brewin.
John Innes Institute, Colney Lane, Norwich, NR4 7UH, U.K.
[1]AFRC Institute of Animal Physiology, Babraham, Cambridge, U.K.

Monoclonal Antibodies(McAb) are being used as a tool to identify and characterise molecular components of *Rhizobium*-induced pea nodules.

Fractionation of isolated root nodules and ultrastructural analysis of nodule thin-sections by Electron Microscopy(EM), together define the key interfaces involved in the Rhizobium-plant interaction, in particular, the infection-thread, the bacterial cell wall and the peribacteroid membrane.

Animals immunized with different nodule fractions were sacrificed and McAb derived. Screening by ELISA and Western Blot gave a number of interesting classes of McAb. The antigens were further characterised by immunogold EM localisation.

Range of McAb isolated

CLASS	ANTIGENS RECOGNISED	LOCALISATION	USES
1	LPS (ref.1)	bacterial cell wall	bacteroid/free liver comparison
2	LPS	infection thread only	infection process
	peptidoglycan component	bacteroid only	cell wall analysis
3	plant glycoprotein family (ref.2)	peribacteroid membrane(PBM)	PBM/bacteroid interactions (ref.3)
4	single bacterial protein	bacteria and nodule cytoplasm	characterising transferred antigens
	family of bacterial proteins	"	"

References:
1. Brewin, N.J. et al. (1986) J. Gen. Microbiol. 132:1959-1968.
2. Brewin, N.J. et al. (1985) EMBO J. 4:605-611.
3. Bradley, D.J. et al. (1986) J. Cell Sci., in the press.

LOCALIZATION OF THE GLUTAMINE SYNTHETASE POLYPEPTIDES IN *Phaseolus* ROOT NODULES

M. LARA, J.L. ORTEGA AND B. VALDERRAMA. CENTRO DE INVESTIGACION SOBRE FIJACION DE NITROGENO. UNIVERSIDAD NACIONAL AUTONOMA DE MEXICO. APDO. POSTAL 565-A. CUERNAVACA, MOR. MEXICO.

INTRODUCTION

Recently it has been established that *Phaseolus vulgaris* nodules produce a specific, physically separable glutamine synthetase (GS) during nodule development (called GSn-1) (1,2). This enzyme is probably the major route of ammonia assimilation during nitrogen fixation. The mature nodule contains two separable forms termed GSn-1 and GSn-2. In a previous work we describe that the GS form is composed mainly by one polypeptide denominated   which is specific of the nodule tissue, and that the GSn-2 form is composed by one polypeptide called , also present in the root tissue (3).

The GSn-1 form which appears simultaneously with the nitrogenase activity during nodule development, and is not present in Fix⁻ nodules, has been proposed as the GS form responsible for the ammonia assimilation in the nodule tissue (1). In order to elucidate if one or both of the nodule GS forms participate in the ammonia assimilation during nitrogen fixation, we analyzed the intracellular localization of these isoenzymes.

PROCEDURE

Materials and methods

Plant material. *Phaseolus vulgaris* L. cv. negro jamapa was inoculated with the wild type *Rhizobium phaseoli* strain CIAT-899 and grown as previously described (3). Nodules were harvested after 3 weeks and used immediately for fractionation.

Nodule fractionation. Peribacteroidal membranes (PBM), nodule soluble proteins and bacteroids were isolated as described by Fortin et al (4).

Electrophoresis. SDS-PAGE and two dimensional gel electrophoresis were done as described by Laemli (5) and O'Farrell (6) respectively.

The GS polypeptides were identified from the gels by immunoblot analysis using rabbit antiserum raised against nodule GS. Proteins were stained with Coomasie blue R.

RESULTS AND DISCUSSION

The localization of the GS was done by polyacrylamide electrophoresis of the total protein and immunoblot detection of the GS from the PBM, bacteroids and nodule soluble protein. In figure 1 is presented the protein profile of these fractions (A) and the immunoblot detection of GS (B). As is shown in figure 1-B lane 1 GS is present in the PBM and seems to be one of the principal proteins of the PBM fractions. (Fig. 1-A lane 1). In the soluble fraction which includes the cytoplasmic GS from the uninfected and the infected cells, the enzyme is present also in high amount. We are trying to determine the percentage of PBM recovery in the purification procedure and define the real proportion of the GS which is located in these membranes. We discard the possibility that the presence of GS in the PBM is an artefact or an unspecific association since leghemoglobin is not found in this fraction. The immunoblot analysis did not reveal any cross react material in the bacteroid fraction (Fig. 1-B lane 2). Cullimore et all report that the anti-GS serum raised against nodule GS did not cross react with the Rhizobium GS (7).

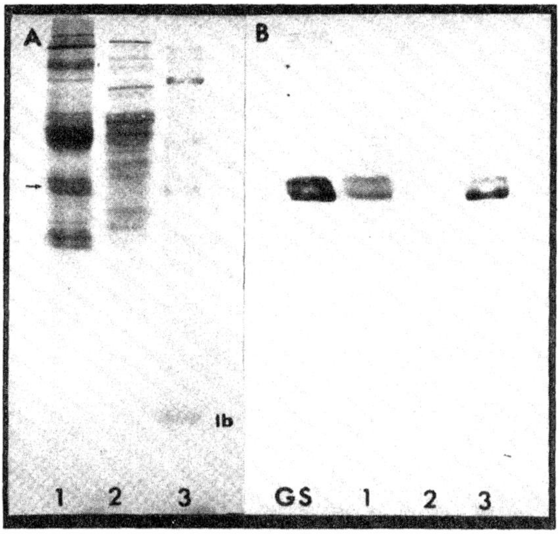

FIGURE 1. A) SDS-Polyacrilamide gel electrophoresis of nodule fractions: (1) PBM, (2) bacteroidal fraction, (3) nodule soluble proteins. Gel was stained with Coomasie blue after electrophoresis. lb; leghemoglobin arrow indicates GS position. B) Immunoblot of the different nodule fractions using anti-GS serum raised against nodule GS: lanes (1), (2) and (3) as in panel A. (GS) 3μg of purified nodule GS.

In order to know the polypeptide composition of the GS present in the PBM, two-dimensional gel electrophoresis of this fraction were performed and the GS monomers revealed by the immunoblot analysis. The result shows that the GS from PBM is composed by the β and the γ polypeptides (Fig. 2), with a higher proportion of the last one resembling the composition of the GSn-1 form (3). The presence of both polypeptides arise the question if the two nodule GS forms (GSn-1 and GSn-2) or if only one form of GS composed by the two type of monomers is associated to this fraction. At present is not possible to establish the significance of a specific nodule GS polypeptide.

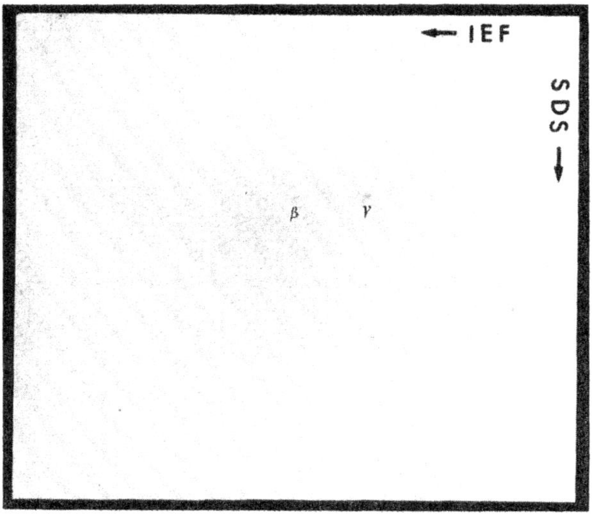

FIGURE 2. Immunoblot of the PBM after two-dimentional gel electrophoresis. GS polypeptides were immunodetected using rabbit antiserum against nodule GS.

The results presented here indicate that some proportion of the GS is associated to the PBM. The abundance and the strategic localization of this enzyme in the PBM makes this result significative suggesting that the ammonia assimilation could take place in this compartment of the infected cells.

REFERENCES

1. Lara M, JV Cullimore, PJ Lea, BJ Miflin, AWB Johnston and JW Lamb. 1983. Planta 157: 254-258.
2. Cullimore JV, M Lara, PJ Lea and BJ Miflin. 1983. Planta 157: 245-253.
3. Lara M, H Porta, J Padilla, J Folch and F Sánchez. 1984. Plant Physiol 76: 1019-1023.
4. Fortin MG, M Zelechowska and DPS Verma. 1985. EMBO J 4: 3041-3046.
5. Laemli UK. 1970. Nature 227: 680-685.
6. O'Farrell PH. 1975. J Biol Chem 250: 4007-4021.
7. Cullimore JV and BJ Miflin. 1984. J Exp Bot 35: 581-587.

CHANGES IN PROTEIN AND mRNA ACCUMULATION IN POTATO TUBERS TREATED WITH AN ELICITOR

H. GIROUX, C. MARINEAU AND N. BRISSON

1. INTRODUCTION

Potato tissues respond hypersensitively when inoculated with an incompatible race of the late blight fungus, Phytophtora infestans (Mont.) de Bary. In potato tubers the hypersensitive reaction is characterized by rapid but restricted cell death at the site of infection and changes in host metabolism, which usually culminates in the cessation of the growth and development of the invading microorganism. The metabolic changes observed in the host tissue are associated with lignification (1), localized browning of the tissue, accumulation of sesquiterpenoid phytoalexins (2) and inhibition of steroid glycoalcaloid synthesis (3). Cell free homogenates of the mycellium of P. infestans can also elicit the hypersensitive response in potato. The molecules with elicitor activity in such homogenates are two twenty carbon polyunsaturated fatty acids, arachidonic and eicosapentaenoic acids (4,5). Both of these fatty acids elicit all of the characteristics of the hypersensitive response described above and are present in all active fractions of the mycellium.

In order to understand the molecular basis of the hypersensitive response in potato we have analysed both the proteins and the messenger RNAs whose synthesis is specifically induced or repressed in tuber discs following treatment with the biotic elicitor arachidonic acid. We have observed changes in the level of at least three proteins by two-dimensional gel electrophoresis and shown that the amount of some mRNA sequences in polyribosomes from treated tissues varies significantly.

2. MATERIALS AND METHODS

2.1. Materials. Potato tubers (Solanum tuberosum L. cv. "Kennebec") were obtained from the research station Les Buissons, P.Qué. Tubers were stored at 4°C until 24 hours before use at which time they were brought to room temperature. Arachidonic acid (5,8,11,14-cis-eicosatetraenoic acid) was purchased from Sigma.

2.2. Elicitor treatment. Discs (2 mm thick and 2.5 cm in diameter) were cut from the medullary tissue of potato tubers, washed several times with water and aged for 6 hours at 19°C in the dark. They were then treated with arachidonic acid (75 µg). Control discs were treated with water or linoleic acid. The incubation was done in the dark at 19°C and stopped by freezing the discs in liquid nitrogen.

2.3. Protein extraction and analysis. Tuber discs were grounded in a mortar at 4°C and homogenized (2 ml/g) in TN buffer (10 mM Tris-Cl pH7.4, 1.5% W/V nonidet NP-40, 0.05% sodium bisulfite and 0.1 mM PMSF).

Starch and cell debris were removed by centrifugation and the samples prepared for analysis by two-dimensional gel electrophoresis (6). The first dimension gel contained 1.6% pH 5-8 and pH 3-10 ampholines. The second dimension was performed on a 14% SDS-polyacrylamide gel and the proteins were revealed by silver staining (7).

3. RESULTS AND DISCUSSION

Potato tuber discs treated with arachidonic acid elicitor show the characteristic browning reaction associated with the hypersensitive response about 24 hours following treatment. When the proteins from such tissues are analysed by two-dimensional gel electrophoresis and silver staining, more than 40% of the proteins that are detected correspond to the so-called patatins, which represent a family of heterogeneous storage proteins of MW 45 Kd (see Fig. 1-A and B). This, and another abundant class of proteins of MW 14 Kd, makes it difficult to detect less abundant proteins that could accumulate in tuber tissue following treatment with arachidonic acid. Representative patterns showing accumulated proteins after 60 hours of treatment with the elicitor or with water are shown in figure 1. Comparison of the two gel patterns reveals three spots (arrows) that correspond to proteins whose accumulation differs significantly and reproducibly in untreated (Fig. 1-A) and treated (Fig. 1-B) tissues. The level of one protein is reduced after treatment while the level of the two other proteins is increased.

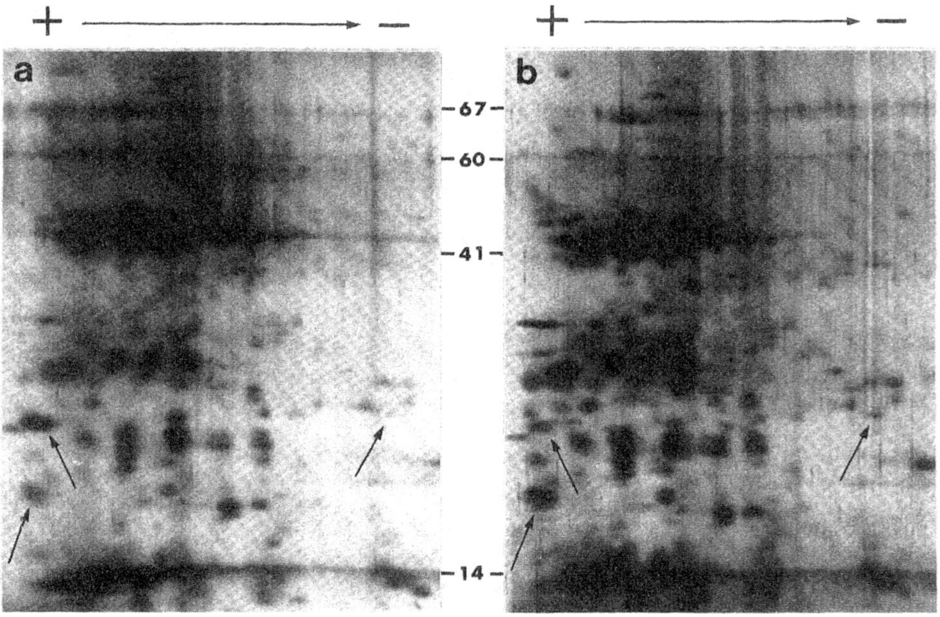

Figure 1. Proteins isolated from tuber discs were separated by two-dimensional gel electrophoresis and stained with silver. A) Control tissue. B) Tissue treated for 60 hours with the elicitor.

To confirm and analyse further these changes, an antiserum was prepared against the soluble proteins extracted from tissue that had been treated for 60 hours with the elicitor. Antibodies reacting with proteins from both treated and untreated tissues were eliminated by adsorption on a sepharose column on which proteins from untreated tissue has been covalently bound. The unadsorbed antibodies reacted specifically with the two proteins whose abundance is increased following elicitor treatment (Fig.1), thus confirming the results obtained by the silver staining technique.

The accumulation or disappearance of proteins in elicitor treated tissues could result from variations in the amount of specific messenger RNAs present in the cells. Polysomal mRNAs from control and arachidonic acid treated tissues were isolated at various time following treatment (1 to 120 hours) and translated in vitro. The analysis of the translation products by two-dimensional gel electrophoresis revealed that changes in the polysomal mRNA population were already taking place 24 hours after treatment with the elicitor, which corresponds to the onset of phytoalexin accumulation in arachidonic acid treated tuber discs (5). With longer time of treatment more changes could be detected between the two mRNA populations. After 72 hours at least 13 spots differed in intensity when the gel pattern corresponding to control tissue was compared to the pattern corresponding to treated tissue. Three spots were highly reduced in intensity while the intensity of ten spots was increased.

4.CONCLUSION

These results indicate that the hypersensitive response elicited in potato tuber discs by treatment with arachidonic acid is accompanied by changes in the population of mRNAs present in polysomes. These changes are also seen at the level of protein accumulation, although to a lesser extent. However, it is not clear yet whether these modifications in the pattern of gene expression are controlled by transcriptional or post-transcriptional events. We have recently isolated cDNA clones corresponding to mRNA more abundant in polysomes from treated tissues than from untreated tissues, and this should help answer this question.

ACKNOWLEDGEMENTS

We wish to thank Ms. M. Beauchemin for technical help, Mr. D.P. Matton for useful discussions, Dr. G. Banville for generously providing certified potato tubers and Ms. S. Beauchemin for typing this manuscript. This work was supported by a research grant from the NSERC, Canada. N.B. was supported by a NSERC University Research Fellowship.

REFERENCES

1.Friend J: Lignification in infected tissue. In: Friend J, Threlfall D.R. (eds.) Biochemical Aspects of Plant-Parasite Relationships, Academic Press, London, 1976, pp291-303.
2.Kuc J: Phytoalexins from Solanaceae. In: Bailey J., Mansfield J. (eds.) Phytoalexins, Blackie and Son, Glasgow, 1982, pp81-105.
3.Tjamos EC, Kuc J: Inhibition of steroid glycoalkaloid accumulation by arachidonic and eicosapentaenoic acids in potato. Science 217:542-544, 1982.

4. Bostock RM, Kuc J, Laine RA: Eicosapentaenoic and arachidonic acids from Phytophtora infestans elicit fungitoxic sesquiterpenes in the potato. Science 212:67-69, 1981.
5. Preisig CL, Kuc J: Arachidonic acid-related elicitors of the hypersensitive response in potato and enhancement of their activities by glucans from Phytophtora infestans (Mont.) de Bary. Arch. Biochem. Biophys. 236:379-389, 1985.
6. O'Farrell PH: High resolution two-dimensional electrophoresis of proteins. J. Biol. Chem. 250:4007-4021, 1975.
7. Merril CR, Goldman D, Van Kueren ML: Gel protein stains: Silver staining. Methods in Enzymology 104:441-447, 1984.

Section IV

MOLECULAR GENETICS OF *RHIZOBIUM*

ORGANIZATION OF THE Rhizobium phaseoli GENOME.

RAFAEL PALACIOS, MARGARITA FLORES, SUSANA BROM, ESPERANZA MARTINEZ, VICTOR GONZALEZ, SILVIA FRENK, CARMEN QUINTO, MIGUEL ANGEL CEVALLOS, LORENZO SEGOVIA, DAVID ROMERO, ALEJANDRO GARCIARRUBIO, DANIEL PIÑERO AND GUILLERMO DAVILA.
CENTRO DE INVESTIGACION SOBRE FIJACION DE NITROGENO, UNIVERSIDAD NACIONAL AUTONOMA DE MEXICO. AP. POSTAL 565-A, CUERNAVACA, MORELOS. MEXICO.

In the last few years we have directed our research efforts towards the understanding of the nature of Rhizobium phaseoli and its genome, in particular those features that could be relevant for its symbiotic properties. At the present time we are addresing some questions that we consider relevant to obtain an overall view of the organization of the Rhizobium genome that might contribute to better possibilities of manipulation. In the first part of this paper, we will comment on such questions and summarize results obtained. In the second part we will discuss some perspectives.

HOW MANY GENOTYPES CAN CONFER THE Rhizobium phaseoli PHENOTYPE?

We have analyzed Rhizobium strains that were isolated from Phaseolus vulgaris nodules in nature (Martínez et al, 1985). Two different types of strains were found. The most common type (R. phaseoli type I) comprises about 95% of the isolates. These strains have a limited host range of infection. We have been able to demonstrate nodulation only with P.vulgaris and P. coccineus, and they all share the characteristic of presenting reiteration of nitrogen fixation (nif) genes. On the other hand, R.phaseoli type II strains have a broad host range of infection, being able to establisl an effective symbiosis with legumes that belong to remote taxonomic groups such as Phaseolus vulgaris and Leucaena esculenta. Type II strains do not present reiteration of nif genes. It is interesting that the symbiotic plasmids of the two types of R. phaseoli strains share little homology among them (unpublished observation) suggesting different evolutionary lines for such genomes.

The nodulation of Phaseolus vulgaris by a wide range of strains from tropical legumes has been reported (Lange, 1961), however, there were no data on nitrogen fixation. In order to define the extent of strains that might present an R. phaseoli phenotype, we isolated Rhizobia from nodules of different legumes. Several strains had the capacity to effectively nodulate P. vulgaris (Martínez et al, 1985). Rhizobium strains from Dalea leporina, Crotalaria pumila and Macroptilum gibbosifolium, besides nodulating its original host, also nodulated Phaseolus and Leucaena effectively. Such strains do not present nif reiteration. They might represent members of the R. phaseoli type II group.

It is important to recall that R. phaseoli has been defined as an heterogeneous group on the basis of protein patterns (Roberts et al, 1980), plasmid profiles (Martínez, Palacios, 1982) antibiotic resistance (Beynon, Josey, 1980) and numeric taxonomy (Catteau et al, 1984). We are in the process of establishing evolutionary lines by comparing sequence homology of different regions of the genome.

HOW MUCH BACTERIAL GENETIC INFORMATION IS NEEDED TO ESTABLISH AN EFFECTIVE SYMBIOSIS WITH Phaseolus vulgaris?

Genetic information in Rhizobium is distributed among different replicons including the chromosome and several plasmids. In fast growing Rhizobium strains from different cross inoculation groups, nif genes as well as several genes that participate in nodulation, have been localized in a particular plasmid, the sym plasmid (Prakash et al, 1981; Hombrecker et al 1981; Rosenberg et al, 1981). To gain insight into the question addressed above, we have transferred symbiotic plasmids from Rhizobium phaseoli strains of both types I and II into an Agrobacterium tumefaciens strain that has been cured of its native plasmids, strain GMI9023 (Rosenberg, Hughart, 1984). Each Rhizobium strain was randomly mutagenized with Tn5 mob (Simon, 1984) and different Km^r derivatives were cross-mated with Agrobacterium tumefaciens GMI9023 in triparental matings using pJB3 (Brewin et al, 1980)as helper plasmid for mobilization. The mixture of Agrobacterium transconjugants that received the Km^r marker was used to inoculate Phaseolus vulgaris seedlings. Bacteria obtained from nodules were purified, checked for Agrobacterium markers and assayed again for nodulation with P. vulgaris. Finally, bacteria from nodules were characterized by analyzing plasmid profiles, nif genes organization, host range of nodulation and capacity to reduce acetylene.

We obtained Agrobacterium transconjugants able to nodulate P. vulgaris from two type I strains and two type II strains that were analyzed. In all cases, the host range of infection of the parental Rhizobium was maintained. Thus, derivatives from type II R. phaseoli were able to nodulate both Phaseolus and Leucaena while derivatives from type I only nodulated Phaseolus. Some Agrobacterium transconjugants were able to reduce acetylene in P. vulgaris. Of particular interest are transconjugants derived from type II strains CIAT899 and UMR1026. Transconjugants that received only the corresponding sym plasmid consistently reduced acetylene in different experiments, in the order of 10% as compared to the original Rhizobium strain. The capacity to reduce acetylene was slightly increased in transconjugants containing other plasmids in addition to the symbiotic one. Thus, it is possible to obtain a complete, although less efficient symbiosis, with Rhizobium plasmids in an Agrobacterium chromosomal background (Martínez et al, Brom et al, unpublished results).

An approach is now being followed that consists in complementing different genomic backgrounds for complete symbiosis with cosmid clones from different Rhizobium strains (Quinto et al, this volume).

WHICH IS THE ORGANIZATION OF SYMBIOTIC GENETIC INFORMATION IN R. phaseoli?

The most interesting result that we have found while addressing this question is the reiteration of nitrogen fixation gene sequences in type I Rhizobium phaseoli (Quinto et al, 1982; Quinto et al, 1985; Martínez et al, 1985). In our model strain CFN42 three different regions of the symbiotic plasmid contain nif genes. Two of the regions (a and b) share a 5 kb homology and contain the three structural genes for nitrogenase. The other region (c) contains only nif H gene. We have evidence that the three regions are functional.

The reiteration of nitrogen fixation gene sequences has also been observed in other organisms such as Rhodopseudomonas (Scolnick, Haselkorn, 1984), Anabaena (Rice et al, 1982), Calothrix (Kallas et al, 1983), Clostridium (Chen et al, 1985) and strains of Rhizobium, including R. fredii, a fast growing Rhizobium that establishes symbiosis with some soybean cultivars (Prakash, Atherly, 1984); the broad host range strain ANU240 (Morrison et al, 1983), and strains originally isolated from different species of Phaseolus and of Pachyrhyzus erosus (Martínez et al, 1985).

HOW GENERAL IS THE REITERATION OF DNA SEQUENCES IN THE Rhizobium GENOME?

To address this question we analyzed the genomes of two strains of R. phaseoli type I one strain of R. meliloti and one strain of Agrobacterium tumefaciens (Flores et al, unpublished experiments). Experiments similar to those previously reported by Sapienza and Doolittle for Halobacterium halobium and Halobacterium volcanii (1982) were performed. DNA from each strain was isolated and digested with the restriction endonuclease EcoR1. From each digest aliquots were used to obtain gene libraries in the EcoR1 site of pBR329 and to prepare Southern blots. Recombinant plasmids were purified from randomly selected clones and used as probes to hybridize against Southern blots of the corresponding digest. In this experiments probed bands in addition to the insert are evidence of DNA reiteration.

In each strain several recombinant plasmids probed bands in addition to the insert, indicating that DNA reiteration is a common feature in Rhizobium and Agrobacterium. Family size was usually low, from 2 to 5 elements, but some recombinant plasmids probed more than 10 bands. Intensity of hybridization suggested that in most cases reiterated segments corresponded to a short portion of the insert while in some cases reiteration might be several kilobases long. Analysis of strain CFN42 indicated that DNA reiteration is not confined to plasmids or chromosome but is a property of both types of replicons.

In addition to nif genes there is evidence on the nature of some repeated elements in Rhizobium of different cross inoculation groups. A DNA fragment that includes the promoter for nitrogenase structural genes was found to be reiterated several fold in R. meliloti (Better et al, 1983). This characteristic seems to be common to different cross-inoculation groups of Rhizobium. Recent evidence suggests that in R. meliloti (Long et al, 1985; Honma et al, 1985) and in R. fredii (Applebaum et al, 1985) an early nodulation gene, nodD, is present in several copies. Insertion sequences have also been described in some Rhizobium strains (Hennecke et al, 1985).

Besides the actual nature and biological significance of different repeated elements, they might be sites for homologous recombination leading to genomic rearrangements, as was first shown in the case of Halobacterium (Sapienza et al, 1982).

WHICH IS THE EXTENT AND FREQUENCY OF REARRANGEMENTS IN THE Rhizobium GENOME?

In search for frequent genomic rearrangements we have analyzed direct descendants of a single bacterial cell. A single colony was grown for a short period of time, resuspended in the presence of detergent and plated at high dilution. Several single colonies were randomly selected and used to prepare Southern blots from DNA digests. Such blots were hybridized against recombinant plasmids that probed repeated DNA families in the corresponding strain. Differences in the probed patterns are indication of genomic rearrangements. We have actually detected rearrangements that occur at high frequency for some reiterated DNA families (Flores et al, unpublished results). We are in the process of defining the frequency and extent of such rearrangements.

Soberón Chávez et al (1986) have found that several R. phaseoli type I strains loss their capacity to nodulate at high frequency due to a complex rearrangement of the sym plasmid. Thus, genomic rearrangements seem to be relevant in regard to the symbiotic properties of Rhizobium strains.

PERSPECTIVES

Our knowledge of the molecular genetics of the Rhizobium - plant symbiotic interaction is rapidly increasing due to the effort of several laboratories throughout the world. Rhizobium strains from different cross-inoculation groups have been selected as models to define the organization and expression of symbiotic genetic information. It is our view that in addition to the understanding of the genetic elements directly involved in the symbiotic process, it is important to understand the general organization of the genome in which such elements are immersed.

It seems that the potentiality of the genome of a given strain of Rhizobium is larger than that revealed by a single cell at certain time. Such potentiality might increase by two processes that we must eventually understand: internal plasticity and genetic exchange with other organisms in nature (see scheme). If the extent and frequency of these processes can be defined, we could obtain an actual view of the potentiality of the genome.

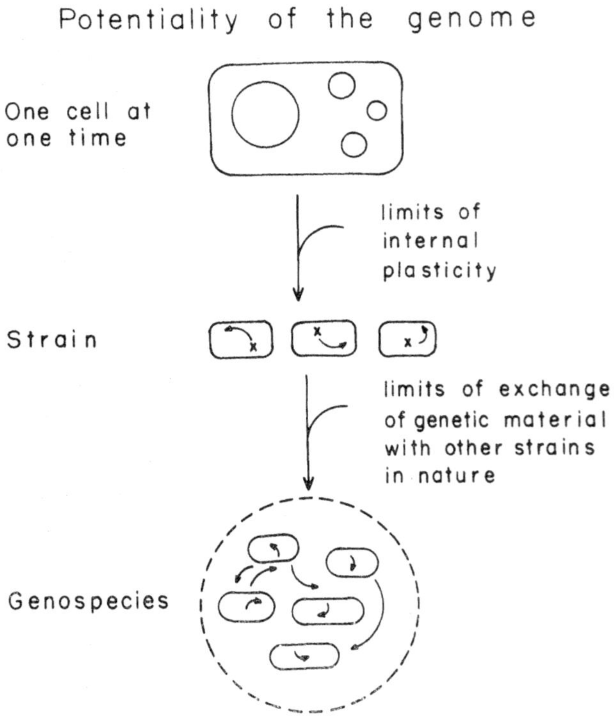

Instead of the concept of species, in the case of bacteria the term genospecies has been suggested (Reanney, 1976) for a group of organism that exchange genetic information among themselves. Accordingly, we should substitute our concept of the Rhizobium species for the broader one of genospecies and define their evolutionary origins. Such evolutionary lines might or might not be related with phenotypic features such as host range of infection.

To work with an expanded view of the genome might be important to further understand the variability and adaptability of soil bacteria populations to environmental conditions and to their interaction with their specific plant hosts.

ACKNOWLEDGEMENTS.

This work was performed with the technical assitance of Rosa María Ocampo and Virginia Quinto. Partial financial support for this research was provided by the U.S. National Academy of Science/National Research Council by means of a grant from the U.S. Agency for International Development, and by Grants from the Consejo Nacional de Ciencia y Tecnología and Fondo de Estudios e Investigaciones Ricardo J. Zevada.

REFERENCES

Applebaum E, Chartrain N, Thompson D, Johansen K, O'Connell M, and McLoughlin M (1985) In Evans HJ, Bottomley PJ and Newton WE eds, Nitrogen Fixation Research Progress, pp 101-107, Martinus Nijhoff, Dordrecht.
Better M, Lewis B, Corbin D, Ditta G and Helinski D (1983) Cell 35, 479-485.
Beynon JL, Josey DP (1980) J. Gen. Microbiol. 118, 437-442.
Brewin NJ, Beringer JE and Johnston AWB (1980) J. Gen. Microbiol. 120, 413-420.
Catteau M, Khanaka H, Segrand MD, Gillaume J (1984) In Veeger C and Newton WE eds, Advances in Nitrogen Fixation Research p. 330. Martinus Nijhoff/Junk, The Hague.
Chen KC, Chen JS and Johnson JL (1985) In Evans HJ, Bottomley PJ and Newton WE eds, Nitrogen Fixatin Research Progress, p 512, Martinus Nijhoff, Dordrecht.
Hennecke H, Alvarez-Moralez A, Betancourt-Alvarez M, Ebeling S, Filser M, Fisher HM, Gubler M, Hahn M, Kaluza K, Lamb JW, Meyer L, Regensburger B, Studer D, Weber J (1985) In Evans HJ, Bottomlet PJ and Newton WE eds, Nitrogen Fixation Research Progress, pp 157-163, Martinus Nijhoff, Dordrecht
Hombrecker G, Brewin NJ and Johnston AWB (1981) Mol. Gen. Genet. 182, 133-136.
Honma MA, Smith CA, Hirsh AM, Lang-Unnash N and Ausubel FM (1985) In Evans HJ, Bottomley PJ, Newton WE eds, Nitrogen Fixation Research Progress, p 120, Martinus Nijhoff, Dordrecht.
Kallas T, Rebiere MC, Rippka R and Tandeau de Marsac N (1983) J. Bacteriol. 155, 427-431.
Lange RT (1961) J. Gen. Microbiol. 26, 351-359.
Long SR, Egelhoff T, Fisher R, Jacobs T and Mulligan J (1985) In Evans HJ, Bottomley PJ and Newton WE eds, Nitrogen Fixation Research Progress, pp 87-93, Martinus Nijhoff, Dordrecht.
Martínez E, Palacios R (1984) In Veeger C and Newton WE eds, Advances in Nitrogen Fixation Research p 60, Martinus Nijhoff/Junk, The Hague.
Martínez E, Pardo MA, Palacios R and Cevallos M (1985) J. Gen. Microbiol. 131, 1779-1786.
Morrison NA, Hau CY, Trinick MJ, Shine J and Rolfe B (1983) J. Bacteriol. 153, 527-531.
Prakash RK and Atherly AF (1984) J. Bacteriol. 160, 785-787.

Prakash RK, Schilperoort RA and Nuti MP (1981) J. Bacteriol. 145, 1129-1136.
Quinto C, De la Vega H, Flores M, Fernández L, Ballado T, Soberón G and Palacios R (1982) Nature 299, 724-726.
Quinto C, De la Vega H, Flores M, Leemans J, Cevallos M, Pardo MA, Azpiroz R, Girard M, Calva E and Palacios R (1985) Proc. Natl. Acad. Sci. U.S.A. 82, 1170-1174.
Reanney DC (1976) Microbiol. Revs. 40, 552-590.
Rice D, Mazur BJ and Haselkorn R (1982) J. Biol. Chem. 257, 13157-13163.
Robert G, Leps WT, Silver LE and Brill WJ (1980) Appl. Environ. Microbiol. 39, 414-422.
Rosenberg C, Boistard P, Denarié J and Casse-Delbart F (1981) Mol. Gen. Genet. 184, 326-333.
Rosenberg C and Hughert T (1984) Mol. Gen. Genet. 196, 533-536.
Sapienza C and Doolittle WF (1982) Nature 295, 384-389.
Sapienza C, Rose MR and Doolittle WF (1982) Nature 299, 182-185.
Scolnick PA and Haselkorn R (1984) Nature 307, 289-292.
Simon R (1982) Mol. Gen. Genet. 196, 413-420.
Soberón Chávez G, Nájera R, Olivera H and Segovia L (1986) J. Bacteriol. in press.

RIFAMPIN RESISTANCE AND NODULATING COMPETITIVENESS IN RHIZOBIUM MELILOTI

D. MARK LEWIS, EDEN S.P. BROMFIELD, LESLIE R. BARRAN
Plant Research Center, Agriculture Canada, Ottawa, Canada K1A 0C6

INTRODUCTION

Resistance to high concentrations of rifampin has often been used to mark strains of *Rhizobium* spp. However, spontaneous mutation to rifampin may affect their nodulation and nitrogen fixation properties; for example, rifampin resistance has been correlated with the ability of slow-growing *Rhizobium* spp. to express nitrogenase in culture (1,2) and the formation of ineffective nodules in *R. leguminosarum* and *Rhizobium* nodulating *Lotus* spp. (3,4).

Recently, we reported (5) that a rifampin resistant mutant of *R. meliloti* strain IZ450 was significantly less competitive for nodulation of *Medicago sativa* than its wildtype parent. The mutant was further shown to possess an altered RNA polymerase insensitive to the action of rifampin. In order to determine whether these effects are generally associated with rifampin resistance in *R. meliloti* we examined the nodulating competitiveness of additional mutants from five strains and determined the sensitivity of their RNA polymerase to rifampin.

MATERIALS AND METHODS

R. meliloti strains used were IZ450, SU47, 102F70, Balsac and NRG 185; the last two strains are used in Canadian commercial inoculants. Spontaneous rifampin resistant mutants were obtained by plating wildtype rhizobia on minimal medium supplemented with mannitol and 200 ug/ml of rifampin. Ten separate mutants from each of the strains were isolated. All rifampin resistant mutants were competed against their respective wildtype parent strains using *Medicago sativa* cv. Saranac grown for six weeks in Leonard jars. Nodule typing was as previously reported (5) and RNA polymerase was purified according to Gross et al. (6).

RESULTS AND DISCUSSION

Symbiotic effectiveness (shoot dry weight) and phage sensitivity patterns using 16 typing phages showed no significant differences between the wildtype *R. meliloti* strains and their respective rifampin resistant mutants. The rifampin resistant mutants were less (P 0.02) competitive than their respective wildtype parent strains. RNA polymerase isolated from the wildtype strains were rifampin sensitive while that isolated from rifampin resistant mutants were rifampin insensitive.

These data suggest tht rifampin resistance in *R. meliloti* is generally associated with a significant loss of nodulating competitiveness and an altered RNA polymerase insensitive to the action of rifampin. Instances of ineffectiveness among rifampin resistant mutants of *R. leguminosarum* and *Rhizobium* nodulating *Lotus* spp. (3,4) contrast with the finding that rifampin resistant mutants of five strains of *R. meliloti* are symbiotically effective and probably reflects species differences. Rifampin resistance has been commonly used to mark *R. meliloti* strains for ecological studies. Our data suggest that rifampin resistant strains are unsuitable for such studies.

REFERENCES

1. Werner, D. 1978. Z. Naturforsch 33: 859-862.
2. Pankhurst, C.E., Scott, D.B. and C.W. Ronson. 1982. FEMS Microbiol. Lett. 15: 137-139.
3. Pain, A.N. 1979. J. Appl. Bacteriol. 47: 53-64.
4. Pankhurst, C.E. 1977. Can. J. Microbiol. 23: 1026-1033.
5. Bromfield, E.S.P., Lewis, D.M. and L.R. Barran. 1985. J. Bacteriol. 164: 410-413.
6. Gross, C., Engbaek, F., Flammang, T. and R. Burgess. 1976. J. Bacteriol. 128: 382-389.

A METHOD FOR ISOLATING COMPETITION DEFECTIVE MUTANTS IN RHIZOBIUM

THOMAS J. McLOUGHLIN, ANN OWENS MERLO, ERIC JOHANSEN
Agrigenetics Advanced Science Company
5649 East Buckeye Road
Madison, WI 53716 USA

1. INTRODUCTION

Competition between Rhizobium strains for nodulation of the legume host plant is of immense agronomic importance as it may determine whether or not nodules are occupied by an effective strain. In this study we have used a novel plant assay to isolate Tn5-induced mutants of *Rhizobium fredii* whose sole known defect is reduced competitiveness in mixed inoculations with a tester strain. Characterization of these mutants may provide a clue to the basis of the competitiveness of the parent USDA257 (1).

2. RESULTS

TABLE 1. Competition Studies With Comp$^-$ Mutants.

Strains[b] A:B	Input Ratio	% Nodules Formed By A	B	A+B	Plant Color	Significance[a]
257spc-2:EA213	1:1	54	45	2	Green	NS
	10:1	93	7	0	Green	NS
	1:10	14	86	0	Yellow	NS
TML90:EA213	1:1	8	91	1	Yellow	***
	10:1	37	62	1	Yellow	***
	1:10	0	100	0	Yellow	***
TML54:EA213	1:1	0	100	0	Yellow	***
	10:1	0	96	4	Yellow	***
	1:10	0	100	0	Yellow	***

[a] A chi-square analysis was used to test the deviation of the results from the expected ratio (1:1, 10:1 and 1:10) for the single-strain:single-strain nodules (df = 1); NS - not significant; *** - $P \geq 0.005$

[b] 257spc-2 is a spontaneous mutant of USDA257 resistant to 250 μg/ml spectinomycin; TML90 and TML54 are Tn5-induced mutants of 257spc-2 defective in competition; and EA213 is a Fix$^-$ mutant of 191str-1 (2) with Tn5 in nifD (3). EA213 is sensitive to spectinomycin and resistant to high levels of streptomycin (1000 μg/ml). The 257-2 derivatives are sensitive to high levels of streptomycin.

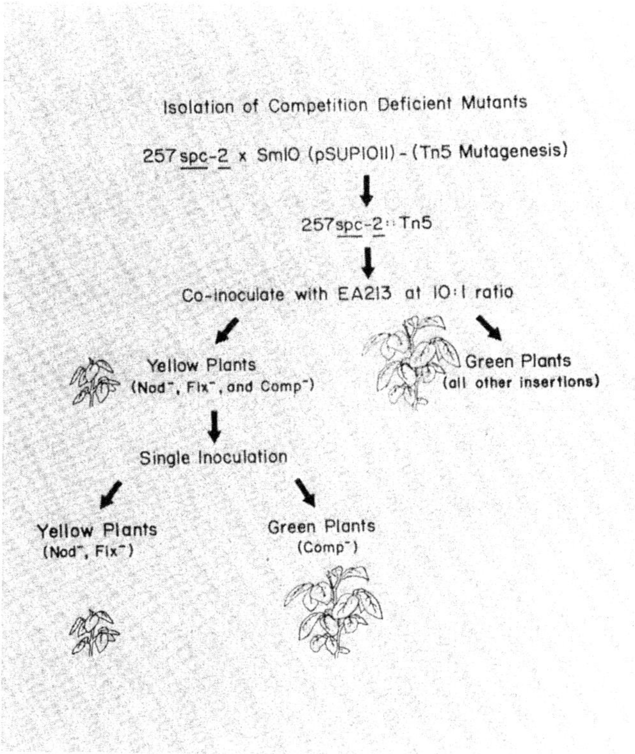

FIGURE 1. Transposon TN5 was introduced into R. fredii strain USDA257spc2 using Sm10/pSup1011 as described by Simon et al. (4). Competition studies were carried out in a modified Leonard jar assembly (E. Appelbaum, Pers. Comm.) containing vermiculite and a nitrogen-free solution (5).

Competition assays were carried out using a ratio of approximately 10:1 (USDA257spc-2::Tn5 to EA213). Plant color was scored visually after approximately 4 weeks of growth. Nodule occupancy was determined by transferring 80 surface sterilized nodules per treatment to TyAgar (6) with the appropriate antibiotics. Yellow plants resulted when the Fix^- strain EA213 formed the majority of the nodules due to the inability of the 257spc-2::Tn5 derivative to compete for nodulation. Yellow plants also arose when the 257spc-2 derivative was itself defective in nodulation or nitrogen fixation. When inoculated alone, $Comp^-$ mutants produce green plants while Nod^- and Fix^- mutants produce yellow plants.

3. CONCLUSIONS

1. The screening procedure described allows the isolation of competition-defective mutants on the basis of plant color. Yellow plants obtained in this assay could result from a mutation in a number of different genes (e.g. nif, nod, fix, or competition genes). When inoculated alone, Fix⁻ and Nod⁻ mutants produce yellow plants, whereas Comp⁻ mutants result in a green plant when inoculated alone.
2. With this assay, competition-defective and symbiotic mutants are screened simultaneously. Indeed, 3 mutants defective in nitrogen fixation were also identified during this study. In addition, one mutant was found to be delayed in nodulation.
3. With this novel screening assay, we have for the first time at our disposal a method to isolate competition mutants in Rhizobium species and 3 mutants whose sole known defect is reduced competitiveness. By characterizing genes that are involved in competition, we may soon understand why one strain is more competitive than another in forming nodules.

4. ACKNOWLEDGEMENTS

We would like to thank J. Pertzborn and S. Alt for excellent technical assistance and E. Appelbaum and Champa Sengupta-Gopalan for helpful discussions.

5. REFERENCES

1. McLoughlin, T. J., S. Alt, P. A. Owens, and C. Fetherston. 1986. Can. J. Microbiol. 32:183-186.
2. Appelbaum, E. R., Johansen, E., and Chartrain, N. 1986. Mol. Gen. Genet. 201:454-461.
3. Appelbaum, E., N. Chartrain, D. Thompson, K. Idler, E. Johansen, M. O'Connell, and T. McLoughlin. 1985. P. 101-107. In H. J. Evans, P. J. Bottomley, W. E. Newton, (eds.), Nitrogen Fixation Research Progress. Martinus Nijhoff.
4. Simon, R., U. Priefer, and A. Puhler. 1983. BioTechnology 1:784-791.
5. Cutting, J. A. and H. M. Schulman. 1969. Biochim. Biophys. Acta 192:486-493.
6. Beringer, J. 1974. J. Gen. Microbiol. 84:188-198.

GENETIC DETERMINANTS OF NODULATION IN pRle 1001*a*: *nodD*

A. SQUARTINI, P.J.J. HOOYKAAS*, M.P. NUTI
Istituto di Chimica e Industrie agrarie, Università di Padova, Italy
* Dept. of Biochemistry, University of Leiden, The Netherlands

1. INTRODUCTION

Previous studies had shown that symbiotic functions are encoded by, in *Rhizobium leguminosarum* strain 1001, on a large plasmid [1,2], subsequently called pRle1001a. This 240 Kb plasmid was further characterized and a circular restriction map is available [3] along with the homology regions with *Agrobacterium* and other rhizobia [4].

The aim of this study was to extend our knowledge of the nodulation region in this fast-growing *Rhizobium.* A combined molecular and genetical analysis was made possible by the availability of cloned nod genes from other fast-growing rhizobia of the same species [5].

2. METHODS

2.1. Microbiological techniques and DNA manipulation. Rhizobial strains were grown as described elsewhere [6,7]. *E. coli* strains were grown on LB medium supplemented with the relevant antibiotic. A BamHI gene bank of pSym1 was constructed using pUC18. Hybridization experiments were performed using routine methods. Plasmid DNA was isolated according to the methods of Birnboim and Doly [8] and Prakash et al. [1].

2.2. DNA sequencing and computer analyses. The DNA from selected clones was sequenced using the dideoxy chain termination method [9]. DNA sequences were analyzed using the Microgenie computer program (Beckman).

3. RESULTS AND DISCUSSION

A 6.6 Kb EcoRI fragment from pRL1JI has been shown to contain part of the nod region [5]. Using this fragment as a probe, the BamHI gene bank in pUC18 was screened. Homologous nod sequences were identified having sizes 3.7, 2.6 and 2.4. These correspond to a region extending from *nodH* (H. Spaink and R. Okker, pers. comm.) to *nodC* in pRL1JI. Further efforts were undertaken to elucidate the molecular structure of the cloned nod genes. Putative *nodD* containing regions were identified by comparing the conservation of restriction sites between pSym1 and pRL1JI. The DNA sequence of the cloned *nodD* region is shown in Fig. 1. *NodD* is represented by an open reading frame containing 966 bp. The distance from the *nodA* translational start on the complementary strand is tentatively assigned to 256 bp(*), though a second possible start is located 355 bp(**) upstream with respect to the *nodD* start. Homology between *nodD* of p Sym1 and *nodD* of pRL1JI [10] is estimated to be 90.2%, and 71.8% when compared with *nodD* of *R. meliloti* 1021 [11]. Consensus sequences reported as nod promoters [10,12] were found between *nodA* and *nodD*, consisting of six highly conserved units. It appears that the region sequenced in this study contains the *nodA* promoter and that there is extensive homology with published sequences of *nodD*. However, it is interesting to note that termination occurs 60 bp downstream of *nodD* in pRL1JI, possibly implying that both proteins have diverged to some extent.

```
CCAGCTTCGC CCAGTCTCGA ACGGCTGGC ATTGAAAGCG TGCCCGGTCG GCCCATAGGG  60
TCTTGCAAAA AAACCTCAGC GAGTTCCGCG TGGTCTGAGG CTTCCAGCTC ATTTTCCCAG 120
CATATTTTCC ATCGCACTTG AGAAGACATG CAAAAGCTCC AACTTGTTTC CTTTCCAACC 180
CTGCCACGAT CAAAGGTGCG CGTGATTGTT GCGCATACTA TTGGATTGCC GTTTGACAAT 240
CGAGGCGGTT AAGAGAATTA AGAATTCAAA TCGGGCCCCT GGCCAATTTC GTTTTTTCGT 300
TCGCGGCCTT GATATTGATC AAGTTCCGGC TTTTCTATAA GCGATCCGAA AGATTGGTAA 360
AATTGATTGT TTGGATGGCA ATCATCTATG GAATGGATAT AAACTT    ATG  CGT  TTT   415
                                                      Met  Arg  Phe
                                                      *    *    *

AAA  GGC  CTA  GAT  CTT  AAT  CTT  CTT  GTA  GCG  CTC  GAC  CGT  CTA  ATG    460
Lys  Gly  Leu  Asp  Leu  Asn  Leu  Leu  Val  Ala  Leu  Asp  Arg  Leu  Met
*    *    *    *    *    *    *    *    *    *    *    *         *    *
ACC  GAG  CGC  AAG  CTG  ACA  GCA  GCG  GCA  CGA  GCC  ATC  AAC  CTC  AGT    505
Thr  Glu  Arg  Lys  Leu  Thr  Ala  Ala  Ala  Arg  Ala  Ile  Asn  Leu  Ser
*    *    *    *    *    *    *    *    *    *         *    *    *    *
CAA  CCG  GCG  ATG  AGC  GCT  GCC  ATC  TCT  AGG  TGG  CGC  GAC  TAT  TTC    550
Gln  Pro  Ala  Met  Ser  Ala  Ala  Ile  Ser  Arg  Trp  Arg  Asp  Tyr  Phe
*    *    *    *    *    *    *    *    *    *         *    *    *    *
CGC  GAC  GAC  CTT  TTT  ATC  ATC  CAG  AGA  CGG  GAG  CTA  AAT  CCG  ACC    595
Arg  Asp  Asp  Leu  Phe  Ile  Ile  Gln  Arg  Arg  Glu  Leu  Asn  Pro  Thr
*    *    *    *    *    *         *    *    *    *    *         *    *
CCG  GCT  GCA  GAG  CCA  CTT  GCC  CCC  GTC  GTG  CGC  GAG  GCC  CTG  CTG    640
Pro  Ala  Ala  Glu  Pro  Leu  Ala  Pro  Val  Val  Arg  Glu  Ala  Leu  Leu
*         *    *         *    *    *         *    *    *    *    *    *
CAT  ATT  CAG  CTT  TCC  GTC  ATC  GCA  TGG  GAT  CCA  ATA  AAC  CTG  CGG    685
His  Ile  Gln  Leu  Ser  Val  Ile  Ala  Trp  Asp  Pro  Ile  Asn  Leu  Arg
*    *    *    *    *    *    *    *    *    *    *    *    *         *
AAG  TCC  GAC  CGC  CGA  TTC  AGA  ATT  ATC  CTT  TCA  GAT  TTC  ATG  GCC    730
Lys  Ser  Asp  Arg  Arg  Phe  Arg  Ile  Ile  Leu  Ser  Asp  Phe  Met  Ala
     *    *    *    *    *    *    *    *    *    *    *    *    *    *
TTG  GTC  TTC  TTC  GAA  AAA  ATC  ATA  GTG  CGC  TTA  GCT  CGG  GAG  GCG    775
Leu  Val  Phe  Phe  Glu  Lys  Ile  Ile  Val  Arg  Leu  Ala  Arg  Glu  Ala
*    *    *    *         *         *         *    *    *    *    *    *
CCA  GGG  GTC  AGC  TTC  AAG  TTG  CTG  CCA  CTT  GAC  GAC  GAT  CCC  GAG    820
Pro  Gly  Val  Ser  Phe  Lys  Leu  Leu  Pro  Leu  Asp  Asp  Asp  Pro  Glu
*    *    *    *    *         *    *    *    *    *    *    *    *    *
GAG  CTT  CTC  CGC  CGT  GGG  GAT  GTT  GAT  TTT  CTG  ATC  CTA  CCC  GAT    865
Glu  Leu  Leu  Arg  Arg  Gly  Asp  Val  Asp  Phe  Leu  Ile  Leu  Pro  Asp
*    *    *    *    *    *    *    *    *    *    *    *    *    *    *
CTA  TTC  ATG  TCT  GGC  GCC  CAT  CGG  AAG  GCA  AGG  CTT  TTC  GAA  GAG    910
Leu  Phe  Met  Ser  Gly  Ala  His  Arg  Lys  Ala  Arg  Leu  Phe  Glu  Glu
*    *    *    *    *    *    *         *    *    *    *    *    *    *
AGA  CTG  GTG  TGC  GTC  GGC  TGC  TCC  ACG  AAC  GAG  CAG  TTG  CAA  GGG    955
Arg  Leu  Val  Cys  Val  Gly  Cys  Ser  Thr  Asn  Glu  Gln  Leu  Gln  Gly
*    *    *    *    *    *    *         *    *    *    *    *    *    *
AAG  CTC  TTC  CTG  GAG  CAA  TAT  ATT  GCC  ATG  GGA  CAT  GTT  GCG  GCT   1000
Lys  Leu  Phe  Leu  Glu  Gln  Tyr  Ile  Ala  Met  Gly  His  Val  Ala  Ala
     *         *    *    *    *              *    *    *    *    *    *
AAG  TTC  GGA  CGT  GGT  CTT  AAG  CCT  TCC  GTT  GAG  CAA  TGG  TTA  TTG   1045
Lys  Phe  Gly  Arg  Gly  Leu  Lys  Pro  Ser  Val  Glu  Gln  Trp  Leu  Leu
*    *    *    *    *    *    *    *    *    *    *    *    *    *    *
```

CTG	CAG	CAA	GGT	CTT	AAG	AGG	CGT	ATT	GAA	CTC	GTC	GTC	CCG	GGG	1090
Leu	Gln	Gln	Gly	Leu	Lys	Arg	Arg	Ile	Glu	Leu	Val	Val	Pro	Gly	
	*	*	*	*	*	*	*	*	*	*	*	*	*	*	
TTT	AAC	TTG	ATC	CCG	CCG	TTG	CTG	TCA	GGC	ATT	AAT	CGA	ATA	GCA	1135
Phe	Asn	Leu	Ile	Pro	Pro	Leu	Leu	Ser	Gly	Ile	Asn	Arg	Ile	Ala	
*	*	*	*	*	*	*	*	*	*		*	*	*	*	
ACC	ATC	CCC	CTG	CGG	CTG	GTC	AAA	CAT	TAC	GAA	CAA	ACT	ATC	CCC	1180
Thr	Ile	Pro	Leu	Arg	Leu	Val	Lys	His	Tyr	Glu	Gln	Thr	Ile	Pro	
*	*	*	*	*	*	*	*	*	*	*		*	*	*	
CTG	CGG	ATT	ATT	GAG	CAT	CCT	TTG	CCA	CTT	CTT	TCG	TTC	ACT	GAG	1225
Leu	Arg	Ile	Ile	Glu	His	Pro	Leu	Pro	Leu	Leu	Ser	Phe	Thr	Glu	
*	*	*	*	*	*	*	*	*	*		*	*	*	*	
GCT	GTC	CAA	TGG	CCG	GCT	CTT	CAC	AAC	TCT	GAT	CCT	GGA	AAC	ATA	1270
Ala	Val	Gln	Trp	Pro	Ala	Leu	His	Asn	Ser	Asp	Pro	Gly	Asn	Ile	
*	*	*	*	*	*	*	*	*		*	*	*	*	*	
TGG	ATG	CGC	GAG	ATT	ATG	ATC	CAA	GAG	GCT	TCG	CGC	CAT	TGG	AAT	1315
Trp	Met	Arg	Glu	Ile	Met	Ile	Gln	Glu	Ala	Ser	Arg	His	Trp	Asn	
*	*	*	*	*	*	*	*	*	*		*				
CCG	AGG	CCG	AAA	GTT	GTA	CGT	CTT	AAG	CGA	CCC	CGG	TCA	TTT	CAC	1360
Pro	Arg	Pro	Lys	Val	Val	Arg	Leu	Lys	Arg	Pro	Arg	Ser	Phe	His	

AGC CGC AGT AGT TAG ACGGGCGCTGATCATTGGACCTATGCATTCCGGTCCAACGA 1416
Ser ARg Ser Ser **END**

CGCATCACGTTACAGGCAGGCCC 1439

FIGURE 1. DNA sequence of *nod* D from pSym1. The conserved nucleotides between pSym1 and pRL1JI are underlined, those conserved between pSym1 and *R. meliloti* 1021 megaplasmic are overlined. Asterisks represent the conserved a.a. between pSym1 and pRL1JI. Sequences reported as consensus for nod promoters are boxed.

4. ACKNOWLEDGEMENTS

Research work supported by CNR, Italy. Special grant IPRA, paper n.937. Mr A. Giacomini is gratefully acknowledged for computer analyses and Dr S.Fanning for stimulating discussions.

5. REFERENCES

1. Prakash R.K., Schilperoort R.A., and Nuti M.P. (1981), J. Bacteriol. 145: 1129-1136.
2. Hooykaas P.J.J., Snijdewint F.G.M., and Schilperoort R.A. (1982), Plasmid 8: 73-82.
3. Prakash R.K., van Veen R.J.M., and Schilperoort R.A. (1982), Plasmid 7: 271-280.
4. Prakash R.K., Ph.D. Thesis (1981), University of Leiden .
5. Downie J.A., Hombrecher G., Ma Q.S., Knight C.D., and Wells B. (1983), Mol. Gen Genet. 190: 359-365.
6. Hooykaas P.J.J., Klapwijk P.M., Nuti M.P., Schilperoort R.A. and Rörsch A. (1977), J Gen. Microbiol. 98: 477-484.
7. Beringer J. (1974), J. Gen. Microbiol. 84: 188-198.
8. Birnboim M.C. and Doly J. (1979), Nucleic Acids Res. 7: 1513.
9. Sanger F., Coulson A.R., Barrell B.G., Smith A.J. and Roe B.A. (1980), J. Mol. Biol. 143 161-178.
10. Shearman C.A., Rossen L., Johnston A.W.B. and Downie J.A.(1986), EMBO J. 5: 647.
11. Egelhoff T.T., Fisher R.F., Jacobs T.W., Mulligan J.T. and Long S.R. (1985), DNA 4 241-248.
12. Rostas K., Kondorosi E., Horvath B., Simoncsits A. and Kondorosi A. (1986), Proc. Nat Acad. Sci. USA 83: 1757-1761.

SYMBIOTIC MUTANTS OF *RHIZOBIUM MELILOTI* WHICH PRODUCE NON-SUCCINYLATED EXOPOLYSACCHARIDE

JOHN A. LEIGH
(Department of Microbiology SC-42, University of Washington, Seattle, WA 98195)

JASON REED
GRAHAM C. WALKER
(Department of Biology, Massachusetts Institute of Technology, Cambridge, MA 02139)

We reported previously (1) that mutants (Exo$^-$) of *Rhizobium meliloti* SU47 which failed to secrete the acidic extracellular polysaccharide formed abnormal nodules on *Medicago sativa* that did not fix nitrogen and did not contain bacteroids. This was true of mutants belonging to six different genetic complementation groups, and established a strong correlation between the ability to secrete the polysaccharide and the ability to invade nodules. These mutants were isolated on the basis of their non-fluorescent colony phenotype on agar medium containing the fluorescent stain calcofluor. Three of the loci defined by these mutants (ExoA, ExoB, and ExoF) were shown to reside in a cluster on the megaplasmid pRmeSU47b (2).

We now report on a seventh category of mutant which produced exopolysaccharide differing from the normal polysaccharide in its pattern of acyl modification. These mutants shared the nodule invasion defect of the Exo$^-$ mutants. We called them "haloless", since we detected them as colonies on calcofluor-containing agar medium which were fluorescent but lacked zones of fluorescence in the agar surrounding the colony.

We isolated four independent transposon Tn5 insertion mutants which had the haloless phenotype. All four of these mutants could be complemented by a single cosmid clone isolated from a pLAFR1 clone bank of *R. meliloti*, indicating that the four mutations belonged to a single genetic locus. The cosmid also complemented the Exo$^-$ phenotype of the ExoA mutants, indicating that the new locus was part of the Exo cluster of pRmeSU47b. We also demonstrated transductional linkage between the haloless mutants and the ExoA mutants.

Wild type *R. meliloti* formed pink, cylindrical, nitrogen fixing nodules on *Medicago sativa*. In contrast, the haloless mutants formed round, white nodules which did not fix nitrogen. Crushing nodules formed by the haloless mutants released few if any bacteria, while crushing nodules formed by the wild type strains released turbid suspensions of bacteroids. Thus the haloless mutants resembled the previously described Exo$^-$ mutants in their failure to invade nodules.

We subcloned an EcoRI fragment from the complementing cosmid onto pSUP104. The resulting plasmid was able to complement both the haloless and the nodule invasion defects of the mutants.

Since the haloless mutants produced exopolysaccharide but shared a nodule invasion defect with mutants that did not produce exopolysaccharide, we decided to determine whether a structural defect in the polysaccharide might explain the nodule invasion defect. Proton NMR spectroscopy showed that while the wild type polysaccharide contained close to one unit each of pyruvate, acetate, and succinate per oligosaccharide subunit, the polysaccharide secreted by the haloless mutants contained no succinate.

Because the haloless mutants differ from wild type in their production of exopolysaccharide and contain mutations mapping close to the Exo cluster of pRmeSU47b, we refer to the new locus as ExoH. Our results establish succinylation as a structural feature of the polysaccharide which is necessary for its function in nodule invasion. It now seems virtually certain that normal exopolysaccharide is required for the invasion of alfalfa nodules by *R. meliloti*.

REFERENCES

1. Leigh, JA, Signer, ER, and Walker, GC (1985) Exopolysaccharide-deficient mutants of *Rhizobium meliloti* that form ineffective nodules. Proc. Natl. Acad. Sci. U.S.A. 82, 6231-6235.
2. Finan, TM, Kunkel, B, De Vos, GF, and Signer, ER (1986) Second symbiotic megaplasmid in *Rhizobium meliloti* carrying exopolysaccharide and thiamin synthesis genes. J. Bacteriol. 167, 66-72.

RHIZOBIUM MUTANTS DEFECTIVE IN LIPOPOLYSACCHARIDE AND INFECTION

K.D. NOEL, P. PACHORI, B. KULPACA, K.A. VANDENBOSCH[1], B.A. BRINK, AND J.R. CAVA
Marquette University, Milwaukee, WI, and [1]University of Wisconsin, Madison, WI, USA

One nodulation phenotype accounts for half of the isolated mutants of *Rhizobium phaseoli* CFN42 defective in symbiosis with bean. The phenotype is to elicit small, white nodules that are obviously not completely developed and lack nitrogen-fixing activity (Noel et al, 1984). One mutant class with this phenotype (Ndv$^-$) has been described previously (VandenBosch et al, 1985). These mutants induce early stages of nodule development without infection thread formation; in minimal agar culture they are not stained by Calcofluor, a β-1,4-glycan stain. Another major class is described below. Mutants of this second class elicit an abortive infection thread with abnormal morphology; they are defective in lipopolysaccharide (LPS).

Nodules induced by the mutants were sectioned for microscopy 9 days and 21 days after inoculation to assess early and mature development. Anatomy of early development closely resembled development induced by wild-type bacteria except for wide, globular infection threads having unusually thick walls with abundant accumulations of amorphous and fibrillar material. These infection threads ceased development prematurely, almost always within root hairs. After 21 days nodule structures elicited by these mutants were identical in anatomy, ultrastructure, and protein content to nodules induced by non-infective mutants (VandenBosch et al, 1985).

Cultured mutant strains of this class stained normally with Calcofluor. However, in tryptone-yeast extract agar they formed rough colonies and in broth culture they autoagglutinated.

Acrylamide gel electrophoresis of washed broth cultures of the wild-type solubilized in dodecyl sulfate (SDS) revealed two bands after periodic acid-Schiff staining for carbohydrate. The more abundant and more slowly migrating band, termed Saccharide I, was missing from all five mutants of this class, strains CE109, CE113, CE121, CE125, and CE126.

Saccharide I was extracted from washed wild-type cells by the hot phenol-water procedure for obtaining LPS (Westphal and Jann, 1965). It co-eluted with the ketodeoxyoctonate (KDO) peak on Sepharose 4B chromatography (Carlson et al, 1978) of the phenol-water extract. The less abundant, more rapidly migrating carbohydrate band on SDS gels of the wild-type (Saccharide II) and the only carbohydrate band of the mutants also co-eluted in this peak with Saccharide I and the

KDO content. To separate Saccharide I and II, the Sepharose 4B peak material was subjected to preparative scale SDS gel electrophoresis and the bands were excised. Each had substantial KDO content.

LPS molecules are known to be susceptible to mild acid hydrolysis of KDO linkages, releasing lipid A and a polysaccharide or oligosaccharides (Carlson, 1984). When purified Saccharide I was treated in this way, it was no longer detectable in SDS gel electrophoresis. The oligosaccharides released from mutant LPS lacked a higher molecular weight component released from Saccharide I by this treatment.

Co-transmission in conjugation experiments, reversion analysis, and complementation by cloned DNA have shown that the Ndv$^-$ phenotype and absence of Saccharide I are due to a single mutation in each mutant. Four of the mutations (in at least three genes) are clustered within a 30 kilobase stretch of chromosomal DNA. This region has been linked by conjugation with several auxotrophic markers and is located between cys-3 and met-4 of *R. phaseoli* CFN42. Mutations causing the non-infective, Calcofluor-negative phenotype (VandenBosch et al, 1985) are widely dispersed on either side of this region of LPS determinants.

Hence, it appears that at least two classes of *R. phaseoli* CFN42 surface polysaccharides and two sets of chromosomal loci are important for infection thread development. One set of determinants is necessary for initiation of the infection thread. The other (related to LPS biosynthesis) is important for the continuation of infection thread growth after it is initiated.

REFERENCES

1. Carlson RW: J. Bacteriol. 158:1012-1017 (1984).
2. Carlson RW, Sanders RE, Napoli C, and Albersheim P: Plant Physiol. 62:921-917 (1978).
3. VandenBosch KA, Noel KD, Kaneko Y, and Newcomb EH: J. Bacteriol. 162:950-959 (1985).
4. Westphal O and Jann K: Methods Carbohydr. Chem. 5:83-91 (1965).

ANALYSIS OF THREE RHIZOBIUM PHASEOLI GENES, psi, psr AND pss, WHICH AFFECT EXOPOLYSACCHARIDE SYNTHESIS AND SYMBIOTIC NITROGEN FIXATION AND/OR NODULATION

D. BORTHAKUR, J.W. LAMB* & A.W.B. JOHNSTON

John Innes Institute, Colney Lane, Norwich NR4 7UH, UK
*ETH-Zentrum, Zurich, Switzerland

In *R. phaseoli*, as in other species of fast-growing *Rhizobium spp*, genes required for nodulation and nitrogen fixation are on large 'symbiotic' plasmids (Lamb *et al.* 1982). Although such plasmids are not required for the synthesis of exopolysaccharide (EPS) we show here that pRP2JI, the symbiotic plasmid of one strain of *R. phaseoli* contains at least two genes that appear to regulate EPS synthesis.

1. *psi* (polysaccharide inhibition)

This is a gene, located close to *nod* genes which when cloned in multicopy vectors causes inhibition of EPS synthesis in *Rhizobium* and abolishes nodulation ability. *psi*::Tn*5* mutant strains induce non-fixing nodules on *Phaseolus* beans (Borthakur *et al.* 1985). It is possible that *psi* is normally expressed only in the nodule and serves to repress genes (e.g. those for EPS synthesis) which are normally expressed in free-living culture but not in the bacteroids.

1.1 *psr* (polysaccharide restriction)

psr was identified as a gene, located 13kb from *psi*, which, when cloned, conferred an EPS$^+$ phenotype to strains of *Rhizobium* containing multicopy *psi*. By constructing *psi*::*lacZ* fusions, it was shown that *psr* repressed transcription of *psi* and so *psr* may be a gene which acts to repress expression of genes that are expressed in the nodule but not in the free-living state.

1.2 *pss* (polysaccharide synthesis)

Mutations in the *pss* gene (which is not on the symbiotic plasmid) abolish the ability to synthesize EPS. In *R. phaseoli* such mutations had no effect on symbiotic nitrogen fixation on *Phaseolus* beans, but, in a near-isogenic strain in which a *R. phaseoli* symbiotic plasmid was replaced by a plasmid that specified the ability to nodulate peas, the *pss*::Tn*5* mutation abolished the ability to nodulate peas.

When the *pss* gene was cloned in a multicopy plasmid, it overcame the EPS$^-$ phenotype of strains harbouring multicopy *psi*. However, at a molecular level, the relationship between *psi* and *pss* has not been established.

It was found (Borthakur *et al.*, 1986) that a recombinant plasmid from the phytopathogen *Xanthomonas campestris* could correct both the EPS$^-$ and Nod$^-$ defects of strains of *pss*::Tn*5* *R. leguminosarum* mutant strains.

2. CONCLUSION

Studies on the three genes *psi*, *psr* and *pss* have shown that they each have subtle and complex effects on the synthesis of EPS and that their presence or absence, and/or their effects when cloned in multicopy plasmids have profound effects on symbiotic nitrogen fixation.

3. REFERENCES

Borthakur D, Downie JA, Johnston AWB, Lamb JW: Mol. Gen. Genet. **200**:278-

-282, 1985.
Borthakur D, Barber CE, Lamb JW, Daniels MJ, Downie JA, Johnston AWB: Mol. Gen. Genet. **203**:320-323, 1986.
Lamb JW, Hombrecher G, Johnston AWB: Mol. Gen. Genet. **186**:449-452, 1982.

INVOLVEMENT OF pSYM NODULATION GENES IN PRODUCTION OF SURFACE AND EXTRACELLULAR COMPONENTS OF RHIZOBIUM TRIFOLII WHICH INTERACT WITH WHITE CLOVER ROOT HAIRS

Frank B. Dazzo*, Rawle I. Hollingsworth, Saleela Philip, Kathryn B. Smith, Margaret A. Welsch, Janet Salzwedel, Pamela Morris, and Lorna McLaughlin
Department of Microbiology and Public Health, Michigan State University, East Lansing, Michigan 48824 USA

We are studying surface and extracellular molecules of Rhizobium trifolii which interact with white clover root hairs as a model of cell-cell communication in plant-bacterial interactions. To determine whether production of these components requires functional nodulation genes, we have examined a collection of mutant and recombinant strains of wild type R. trifolii 843. These have alterations in genes encoding 5 contiguous "nodulation regions" on a 14 kb HindIII fragment of pSym from the wild type strain. These strains were provided by Barry G. Rolfe, ANU, Canberra, Australia, and include: a heat-cured pSym-minus strain, mutants with a single Tn5 insertion in nodA, nodB, nodC, nodD, nodE, nodF, nodI, nodJ, or nodulation region IV; a deletion in nodulation region V, and recombinants of the pSym-minus strain containing all of the nod genes on the 14 kb HindIII fragment, nodABCDF only, or nodABCDIJ only (1). Using a variety of microscopic and biochemical techniques, we have found that some of these nod genes on the symbiotic plasmid are required for wild type expression of certain surface and extracellular components which interact with the white clover root hair.

Binding of a white clover lectin, trifoliin A, to R. trifolii (ex planta or in situ) requires the pSym plasmid. Mutations in either nod A, B, C, D, E, F, I, J, or nodulation region V significantly lower the level of trifoliin A binding to the bacterium in the clover root environment (hence, under isoflavone induction conditions). This effect on pSym nod genes is most pronounced for the polar pattern of trifoliin A binding to the bacterial cell. However, because nodABCIJ reside on one operon, mutations in some of these genes affecting trifoliin A binding may be due to pleotropic polar effects. Transfer of R. trifolii pSym nodulation regions III, IV, and V to a wild type R. leguminosarum strain results in a hybrid which now can infect white clover root hairs and bind trifoliin A extensively in the clover root environment. Of great interest to us is the analysis of a mutant strain having a Tn5 insertion in one of the pSym host specificity genes (nod E) of R. trifolii. This mutant has an altered host range: it no longer can infect white clover, it retains the ability to infect subterranean clover, and it has gained pea infection ability. In addition to a significant loss of ability to bind trifoliin A in the clover root environment, this nodE mutant has an altered ratio of noncarbohydrate substitutions (acetate, pyruvate, 3-hydroxybutyrate) in its capsular polysaccharide (CPS). We therefore conclude that certain pSym nodulation genes of R. trifolii encoding the molecular determinants of host range are required for production of the correct CPS chemistry (especially noncarbohydrate substituents) necessary for molecular recognition by trifoliin A. These results strongly support a role for the interaction between acidic heteropolysaccharides of R. trifolii and trifoliin A in infection of white clover root hairs (2).

By analyzing B. Rolfe's mutant strains, we have also discovered two other aspects of the interaction between R. trifolii and white clover root hairs which are affected by pSym nodulation genes. In the first case, we found that mutations in hac genes (nod D, A, B, and C) almost totally abolish the accumulation of extracellular microfibrils associated with the attached bacteria as a characteristic feature of Phase II attachment to clover root hairs (3). Secondly, we have isolated a novel aromatic compound (tentatively called Bacterial Factor - 1, or BF-1 for short) which is elaborated by the bacterium in defined, minimal medium (no flavone inducer), and can modulate root hair differentiation and development when applied in minute quantities ($\geq$10 ng) to white clover seedling roots. We were able to isolate BF-1 from cultures of the wild type 843 strain and from a recombinant containing the 14 kb pSym nodulation region but lacking the rest of its pSym plasmid, but not from the heat-cured pSym-minus mutant. Thus, the pSym nodulation genes are involved in BF-1 production/excretion by the bacterium. BF-1 induces profound growth responses of the root hairs, alterations in the root cortex, and significantly promotes shepherd crook and infection thread formation by R. trifolii. We believe that BF-1 affects biosynthesis of clover root hair walls and its action may be required for successful infection. Structural elucidation of BF-1 is in progress. It will be interesting to determine if BF-1 is the bioactive molecule produced by rhizobia which is responsible for the thick, short root response as described by B. Zaat and colleagues at Leiden University (4) and B. Rolfe and colleagues at ANU (5).

Literature Cited

1. Rolfe BG, Redmond JW, Bately M, Chen H, Djordjevic SP, Ridge RW, Bassam BJ, Sargent CL, Dazzo FB, Djordjevic MA: NATO Adv. Research Workshop on Plant-Microorganism Recognition (B. Lugtenberg, ed), 1986, in press.

2. Abe M, Sherwood JE, Hollingsworth RI, Dazzo FB: J. Bacteriol. 160, 517-520, 1984.

3. Dazzo FB, Truchet GE, Sherwood JE, Hrabak EM, Abe M, Pankratz HS: Appl. Environ. Microbiol. 48, 1140-1150, 1984.

4. van Brussel AAN, Zaat SAJ, Cantercremers HCJ, Wijffelman CA, Pees E, Lugtenberg BJJ: J. Bacteriol. 165, 517-522, 1986.

5. Cantercremers HCJ, van Brussel AAN, Plazinski J, Rolfe BG: J. Plant Physiol., 122, 25-40, 1986.

RHIZOBIUM EXOPOLYSACCHARIDES ARE ESSENTIAL FOR THE FORMATION OF NITROGEN FIXING NODULES IN THE RHIZOBIUM-LEGUME SYMBIOSIS

S.P. DJORDJEVIC, H.C. CHEN, J.X. GRAY, J.J. WEINMAN, M.A. DJORDJEVIC, J.W. REDMOND, M. BATLEY AND B.R. ROLFE

A characteristic of many wild-type Rhizobium strains is their capacity to produce large quantities of exopolysaccharides (EPS) and form mucoid (Muc$^+$ phenotype) colonies on various laboratry media. Rhizobium polysaccharides, particulary EPS and lipopolysaccharide (LPS), have been postulated to be involved in the early recognition steps between the plant and the bacteria including specific adhesion to the root hair surfaces (5) the determination of host specificity (9) and the ability to induce proper nodule development (1,10). Tn5 transposon mutagenesis of the broad host range fast-growing Rhizobium strain NGR234 has yielded 90 EPS mutants showing altered colony morphology. These mutants were classified on the basis of their symbiotic properties on 4 host legumes (Macroptilium atropurpureum, Desmodium intortum, Desmodium uncinatum, and Lablab purpureus) which form spherical (determinant) nodules and on Leucaena leucocephala, where cyclindrical (indeterminant) nodules are induced (2).

Muc$^-$ mutants of NGR234 with unconditionally rough colony morphologies (group 2 mutants) were still able to infect and induce nodules on most plant hosts (2). These infections, however, usually resulted in the production of non-nitrogen fixing nodules. Particularly interesting was the finding that most unconditionally rough (group 2) mutants produced calli-like structures on Leucaena plants instead of indeterminate nodules. An important feature of the calli was the presence of osmiophilic droplets which are often observed in plant hypersensitive reactions (3). These calli structures appear to be grossly deformed nodules which resulted from an early inhibition of infection as little or no evidence of infection thread formation was seen (3).

The first indications that EPS might be important for functional nodule development came from cell-mixing experiments involving a Muc$^+$

non-invasive strain of NGR234, and several group 2 mutants. The Sym plasmid cured derivative of NGR234, (ANU265) produces Muc^+ colonies, but is unable to initiate any detectable symbiotic response on any legume since many of the essential nodulation genes, as well as the nitrogen-fixation genes, have been shown to be deleted from this strain (8). When strain ANU265 was mixed in equal amounts with many of the group 2 mutants (produced by Tn5 mutagensis) and coinoculated onto *Leucaena* plants, nitrogen-fixing nodules were produced (8). Sections through these nodules showed a meristematic zone with well formed vascular bundles, and an extensive bacteroid zone containing non-swollen bacteria. Bacteria reisolated from these nodules retained their original colony morphology (Muc^+ for ANU265, and Muc^- for the mutant strains) antibiotic resistence markers (Sp^r for ANU265 and Rf^r, Sm^r and Km^r for the Muc^- mutants) and their original nodulation phenotypes on *Leucaena* plants, thus indicating that no genetic transfer had occurred between the two strains.

If the EPS produced by strain ANU265 is essential for effective nodulation of *Leucaena*, then the amount of the surface polysaccharide produced by ANU265 could affect cooperation with Muc^- mutants. A Muc^- Nod^- derivative of strain ANU265 was constructed using the plasmid R68.45 Km^s which has chromosome mobilization ability (3). Strain ANU265 was mated with the donor strain ANU2811(R68.45 Km^s) and kanamycin resistant transconjugants were found to arise at a frequency of 10^{-9}. Verification that Tn5 was mobilized into the ANU265 genome came from the following observations (a) all Km^r isolates contained a copy of R68.45 but did not carry a Sym plasmid; (b) the Km^r isolates contained a single copy of Tn5 which was located in the same sized DNA fragment as Muc^- mutant (donor) strain ANU2811, (c) all Km^r isolates retained a Nod^- phenotype and (d) the ANU265 transconjugant possessed a Muc^- phenotype. This Muc^- Nod^- kanamycin resistant derivative of strain ANU265 was named ANU266. The colony morphology and the amount of EPS produced ANU266 was similar to that produced by the donor strain ANU2811. Coinoculation of strain ANU266 with several Muc^-, group 2 mutants failed to initiate nitrogen-fixing nodules on *Leucaena*. Finally, the combination of two different Muc^- strains (that normally induce callus structures on *Leucaena*) failed to induce nitrogen-fixing nodules when coinoculated onto *Leucaena* plants indicating that the EPS produced by Muc^+ Nod^- strain ANU265 is important for the effective nodulation of *Leucaena* (3).

As part of the programme to determine the role of surface polysaccharides in the Rhizobium-legume symbiosis the structure of the EPS from strain ANU280 (Rf^r, Sm^r derivative of NGR234) and strain ANU265 (Sp^r Sym plasmid cured derivative of NGR234) was determined. Polysaccharides were isolated using a Amicon DC101 hollow-fibre filtration system fitted with a 0.1um filter, and subjected to exhaustive diafiltration with distilled water to remove salts and other nutrients. After removing the cells by centrification, the resulting crude EPS was purified by precipitation as the cetyltrimethylammonium (CTAB) salt. Fragments generated by graded acid hydrolysis of the EPS were fractionated by sequential ion exchange and gel chromotography, and their structures were assigned by ^{13}C-N.M.R. spectroscopy (Figure 1)(6,7). The ^{13}C-N.M.R. spectrum of the isolated oligosaccharide repeat-unit from strain ANU265 was indistinguishable from the spectrum of the oligosaccharide from the parent strain ANU280.

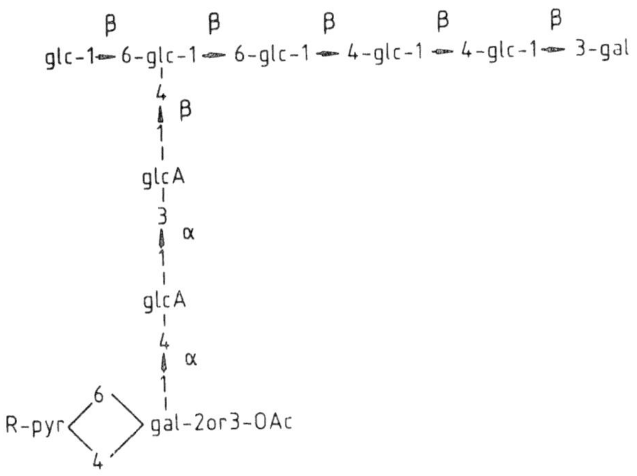

Figure 1 The structure of the acidic EPS produced by wild-type parent strain ANU280 Sm^r, Rf^r derivative of NGR234.

To determine if the helper effect of strain ANU265 could be substituted for by the addition of purified EPS from the parent strain ANU280 or from the Sym plasmid cured strain ANU265, purified EPS or oligosaccharide repeat-unit isolated from strain ANU280 was inoculated onto Leucaena plants together with one of the Muc$^-$ mutants. In all tested cases, the coinoculation of the EPS or the oligosaccharide repeat-unit enabled the Muc$^-$ mutants to induce nitrogen-fixing nodules on Leucaena plants, although some calli were still produced (8). The same behaviour was observed whether the EPS was obtained from ANU280 or ANU265. Only Muc$^-$ bacteria were detected when the contents of at least 30 pigmented Leucaena nodules were analysed and the Muc$^-$ bacteria invaribly retained their defective phenotype, original colony morphology and genetic markers.

Parallel experiments similar to those described above for Leucaena were performed on siratro plants. Strain ANU280 normally induces determinant nitrogen-fixing nodules on siratro whereas Muc$^-$ mutants appear to nodulate normally but fail to induce nitrogen-fixing nodules. The addition of EPS isolated from parent strain ANU280 together with several Muc$^-$ group 2 strains of NGR234 enabled these bacteria to induce determinant nitrogen-fixing nodules on siratro plants (8).

Several experiments suggested that the structure of the EPS was important for the success of the correction phenomena. EPS isolated from the unrelated R.trifolii strain ANU843 was chemically sequenced using a similar approach used to the structure of the NGR234 EPS. The addition of EPS isolated from R.trifolii strain ANU843 to together with Muc$^-$, group 2 mutants of NGR234 was unable to facilitate the induction of nitrogen-fixing nodules on Leucaena plants. Similarly, coinoculation of strain ANU845 (Muc$^+$, Sym plasmid cured derivative of strain ANU843) with Muc$^-$, group 2 mutants of NGR234 also failed to induce nitrogen-fixing nodules on Leucaena plants (8).

In order to determine the exact role the EPS plays in the Rhizobium-legume symbiosis, an understanding of the organization and regulation of genes responsible for the synthesis of surface polysaccharides is required. We have isolated R-prime plasmids carrying the group 2 mutant loci (4). These R-primes were isolated by using the kanamycin sensitive derivative of R68.45 (pMN2) to permit selection for the mobilization of the kanamycin resistence of transposon Tn5 from 28 Muc$^-$ group 2 mutants of strain NGR234 to an E.coli recA strain. A number of these the kanamycin-resistant transconjugants contained R-prime plasmids since the kanamycin resistence marker was shown to be 100% co-linked to the tetracycline resistence determinant located on the pMN2 plasmid. Futhermore, physical analyses showed that these R-prime plasmids contained overlapping segments of Rhizobium DNA sequences flanking the Tn5 insertion site indicating that the group 2 mutant loci were clustered. The intergrated DNA fragments inserted into pMN2 ranged in size from 57-84 Kb. The conjugal transfer of these R-prime derivatives to the original parent strain (ANU280) permitted the group 2 mutants to be subdivided into two classes. Of the 28 R-primes carrying group 2 chromosomal DNA, 17 were able to alter the mucoid phenotype of wild-type strain ANU280 from Muc^+ to Muc^- and were thus termed dominant R-primes. The remaining 11 R-prime derivatives did not repress EPS synthesis when transferred to ANU280. The Muc^- transconjugants of strain ANU280 carrying the dominant R-prime plasmids had a colony morphology and symbiotic phenotype of the Muc^- strain from which the R-prime was originally constructed (4).

The Tn5-containing fragments from several group 2 mutants have been cloned. ^{32}P labelled Rhizobium DNA fragments (isolated from the cloned Group 2 DNA fragments) have been used to analyse a 60kb area of DNA which has been shown by complementation analysis to code for all but one of the "group 2" genes. Several DNA fragments have been cloned onto broad host range plasmids to facilitate fine point mapping and complementation analysis. These wild-type DNA fragments together with the large segments of chromosomal DNA contained in the R-primes will enable us to gain a thorough understanding of the organization and regulation of chromosomal genes responsible for the synthesis of surface polysaccharides essential to the establishment of a successful nitrgen-fixing symbiosis.

Our studies with the surface polysaccharides of Rhizobium have thoroughly established that the EPS of strain NGR234 is essential for the ability of this strain to form nitrogen-fixing nodules on host legumes.

ACKNOWLEGEMENTS

S.P.D., H.C.C. and J.X.G. are recipients of ANU PhD scholarships; J.J.W. is a National Research Fellow; M.A.D. is an Australian Wool Board Fellow.

REFERENCES

1. Chakravorty et. al.; J. Molec. Appl. Genet. 1: 585-596, 1982.
2. Chen HC et. al.; J. Plant Physiol 120: 331-349, 1985.

3. Chen HC and Rolfe BG; J. Plant Physiol., in press.

4. Chen HC and Rolfe BG; (Manuscript in preparation).

5. Dazzo FB and Brill WJ; J. Bacteriol. 137: 1362-1373, 1978.

6. Djordjevic SP et. al.; Carbohydr. Res. 148: 87-99, 1986.

7. Djordjevic SP et. al.; J. Chromot. 354: 507-510, 1986.

8. Djordjevic SP et. al.; (Submitted).

9. Dudman WF: In Sutherland I (ed.), Surface carbohydrates of the prokaryote cell, New York, Academic Press, pp 357-414, 1978.

10. Leigh et. al.; Proc. Natl. Acad. Sci. (U.S.A.) 82: 6231-6235, 1983.

COINOCULATION WITH SYMBIOTICALLY DEFECTIVE MUTANTS OF *RHIZOBIUM MELILOTI*

S. KLEIN[1], A.M. HIRSCH[2], C.A. SMITH[2], and E.R. SIGNER[1]

[1]Dept. of Biology, Massachusetts Institute of Technology, Cambridge, MA 02139, and [2]Dept. of Biological Sciences, Wellesley College, Wellesley, MA 02181.

INTRODUCTION

The ability of critical symbiotic functions to be provided to one bacterium by another (i.e., in *trans*) during the course of nodule development has been examined with derivatives of the alfalfa symbiont *R. meliloti* SU47.

nod$^-$ AND *nif*$^-$ MUTANTS

Coinoculation with pairs of strains carrying wild or mutant alleles of *nod* (*A*,*B* or *C*) and *nif* (*A* or *H*) genes in all combinations generally gave Fix$^+$ nodules. For the combination *nod*$^+$*nif*$^-$ with *nod*$^-$*nif*$^+$, all plants were nodulated, and roughly half of them fixed nitrogen (Fix$^+$). Thus the *nod*$^+$ *nif*$^-$ helper can allow the *nod*$^-$*nif*$^+$ member to nodulate, implying that the *nod*$^+$ function (Nod$^+$) can act in *trans*. Both genotypes were recovered approximately equally from Fix$^+$ nodules, whereas Fix$^-$ nodules gave mostly *nod*$^+$*nif*$^-$ bacteria.

exo$^-$ MUTANTS

With *exo*$^-$ mutants, however, results depended on the relative orientation of the alleles involved. *Exo*$^-$ mutants, which are deficient in extracellular acidic heteropolysaccharide (EPS), form nodules with a complex phenotype, namely no infection threads (Inf$^-$), no intracellularly located bacteria, and no fixation of nitrogen (Fix$^-$; Finan et al., *Cell*, *40*, 869, 1985; Leigh et al., *PNAS* *82*, 6231, 1985). Coinoculated pairs included one *exo*$^+$ and one *exoB* member, with various combinations of *nod* and *nif* alleles as above. Nodules were scored for nitrogen fixation (by acetylene reduction, and by morphology of seedling tops); for occupancy by *exo*$^+$ and *exo*$^-$ (by calcofluor fluorescence, and by drug resistance of Tn*5* insert derivatives); and for morphology of intracellular bacteria(by transmission electron microscopy).

NITROGEN FIXATION AND NODULE OCCUPANCY

Nitrogen fixation and nodule occupancy are presented in Table 1. In the experimental coinoculations the *exo*$^+$ member was always *nif*$^+$, so that any fixation observed must have been carried out by the *exo*$^-$ helped in *trans* by the coinoculated *exo*$^+$.

Clearly, fixation occurred only when the helper *exo*$^+$ was also *nod*$^-$ (i.e., *nod*$^-$*exo*$^+$*nif*$^-$ with *nod*$^+$*exo*$^-$*nif*$^+$). As in the control (*nod*$^-$*exo*$^+$*nif*$^+$ with *nod*$^+$*exo*$^-$*nif*$^-$) all nodules contained both *exo*$^+$ and *exo*$^-$ bacteria in a ratio of approximately 3:1. By contrast, coinoculation with *exo*$^+$ that was also *nod*$^+$ (i.e., *nod*$^+$*exo*$^+$*nif*$^-$ with either *nod*$^+$*exo*$^-$*nif*$^+$or *nod*-*exo*$^-$*nif*$^+$) did not help the *exo*$^-$ member to fix nitrogen. Nevertheless, a minority of the Fix$^-$ nodules produced did contain both coinoculated genotypes, albeit fewer of the *exo*$^-$ (9:1). Bacteria in these nodules, however, were ultrastructurally abnormal.

TABLE 1.

Coinoculants		Phenotype*		Nodule Occupancy (Percent)			
exo$^+$	exo$^-$	nod	fix	Empty	exo$^+$	exo$^-$	exo$^+$:exo$^-$
Control:							
nod$^+$nif$^+$	nod$^+$nif$^+$	+	+				
nod$^+$nif$^+$	nod$^-$nif$^+$	+	+				
nod$^-$nif$^+$	nod$^+$nif$^-$	+	+	0	0	0	100 (3:1)
nod$^-$nif$^-$	nod$^+$nif$^-$	+	–				
nod$^-$nif$^-$	nod$^-$nif$^+$	–					
Experimental:							
nod$^+$nif$^-$	nod$^+$nif$^+$	+	–	0	75	0	25 (9:1)
nod$^-$nif$^-$	nod$^+$nif$^+$	+	+	0	0	0	100 (3:1)
nod$^+$nif$^-$	nod$^-$nif$^+$	+	–	10	85	0	5 (9:1)
nod$^-$nif$^-$	nod$^-$nif$^+$	–					

*Fix$^+$; plants green and tall, acetylene reduced
Fix$^-$; plants chlorotic and stunted, acetylene not reduced

MORPHOLOGY OF INTRACELLULAR BACTERIA

Nodules made by coinoculation showed differences in bacterial morphology between nod$^-$exo$^+$ and nod$^+$exo$^+$ helpers that corresponded to the differences in fixation. With nod$^-$exo$^+$ helper, all intracellular bacteria had the elongated appearance of normal bacteroids. With nod$^+$exo$^+$ helpers, however, two types of intracellular bacteria were found (see Figures), namely, elongated bacteroids and non-elongated forms. Often one of each type was enclosed within the same peribacteroid membrane, even though multiply enclosed bacteroids are rarely found in alfalfa nodules.

The non-elongated forms appear not to have differentiated completely. Given that these nodules are Fix$^-$, the results suggest that the elongated forms are normally-differentiated exo$^+$ that cannot fix because they are nif$^-$, whereas the non-elongated forms are exo$^-$ that cannot fix because they are not completely differentiated.

This suggests that exo$^-$ bacteria make abberant Fix$^-$ nodules not just for lack of intracellular location, but rather because exo$^+$ function is essential for proper differentiation to nitrogen-fixing bacteroids.

EXO$^+$ IN *trans* AND *cis*

The biologically active exo$^+$ function (Exo$^+$) could in principle be EPS, or its oligosaccharide repeat unit, or another glycosyl product. Whatever it may be, Table 1 shows that this Exo$^+$ is effective in *trans* when it comes from a nod$^-$ helper, but not from a nod$^+$ helper. Formally, this implies that Nod$^+$, which itself acts in *trans* (see above) can modify Exo$^+$ (e.g., by cleavage, or alteration of non-saccharide substituents, etc.). Modified Exo$^+$ would become ineffective for the step in question, but whether or not it would be essential in some other way is unclear.

This idea was tested by a triple coinoculation, of the nod$^+$exo$^-$nif$^+$ member with both nod$^+$exo$^+$nif$^-$ and nod$^-$exo$^+$nif$^-$ helpers. This was to show whether in *trans*, ineffective Exo$^+$ from the nod$^+$ helper would supersede effective Exo$^+$ from the nod$^-$ helper. Table 2 shows that this is in fact the case. This could mean that modified (ineffective) Exo$^+$ excludes unmodified (effective) Exo$^+$ from a critical target, or, that Nod$^+$ from the nod$^+$ helper modifies Exo$^+$ from the nod$^-$ helper as well.

TABLE 2.

	Coinoculants		Phenotype	
	exo$^+$	exo$^-$	nod	fix
Control:				
	nod$^+$nif$^-$	nod$^+$nif$^+$	+	-
	nod$^-$nif$^-$	nod$^+$nif$^+$	+	+
Experimental:				
	nod$^+$nif$^-$)) nod$^-$nif$^-$)	nod$^+$nif$^+$	+	-

In trans modification by Nod$^+$ makes Exo$^+$ ineffective, but in cis this is clearly not so, because Exo$^+$ from wild-type (nod$^+$exo$^+$) is obviously effective in cis. (Note that Exo$^+$ from nod$^-$ cannot be tested in cis). Although that does appear paradoxical, it could mean, for instance, that Exo$^+$ is originally made unmodified (and thus effective in cis), but is then modified by Nod$^+$ (and thus made ineffective in trans) before it has a chance to be translocated from helper to coinoculant exo$^-$.

CONCLUSIONS

1. Whatever its molecular nature, Exo$^+$ appears necessary, not only for infection threads and Fix$^+$ nodules, but also for normal bacteroid differentiation.

2. In coinoculation, however, whereas Exo$^+$ from a nod$^-$ helper is sufficient, Exo$^+$ from a nod$^+$ helper allows aborted differentiation at best; only a minority of exo$^-$ are intracellularly located, and these do not seem to fix. Thus in trans, Exo$^+$ from nod$^-$ is effective, whereas Exo$^+$ from nod$^+$ is apparently ineffective.

3. In trans, Exo$^+$ from nod$^+$ may also lead to enclosure of two bacteria within the same peribacteroid membrane, otherwise rare in alfalfa. Moreover, besides having different properties coming from nod$^+$ and nod$^-$ when in trans, when in cis (i.e., in wild-type) Exo$^+$ from nod$^+$ clearly is effective. Thus, physiologically, Exo$^+$ appears to have a complex (and possibly multiple?) role.

4. The results suggest that, at the molecular level, Exo$^+$ can evidently be modified by Nod$^+$. This is now being tested experimentally.

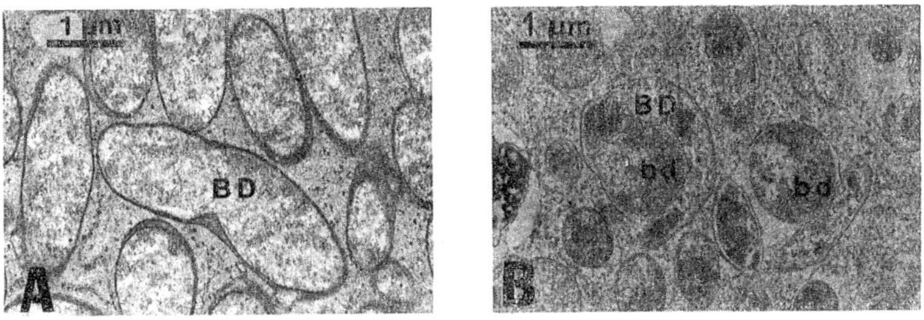

FIGURE 1. Coinoculatons. A. nod$^-$exo$^+$nif$^+$with nod$^+$exo$^+$nif$^-$ results in a Fix$^+$ nodule with elongated bacteroids. B. nod$^+$exo$^-$nif$^+$ with nod$^+$exo$^+$ nif$^-$ results in a Fix$^-$ nodule with two different types of bacteria.

SURFACE PROPERTIES OF *Rhizobium meliloti* ASSOCIATED WITH SYMBIOSIS

JOSEPH KIEBER, RALPH CLOVER, TURLOUGH M. FINAN* and ETHAN R. SIGNER
Dept. of Biology, Mass. Inst. of Technology, Cambridge MA 02139; *present address: Dept. of Biology, McMaster Univ., Hamilton, Ontario L8S 4K1

INTRODUCTION

Earlier, we described mutants of *R. meliloti* SU47 deficient in extracellular polysaccharide (EPS) that make aberrant Fix⁻ nodules[1,2]. Here, we report on other symbiotic mutants with altered lipopolysaccharide (LPS), and also on alterations in the membranes of bacteroids.

BACTERIAL STRAINS

SU47 was isolated by Vincent. Rm1021[4] is SU47 *str21*. EJ312[3] is SU47 *str3 rif100 sag1 sag2,3* (*sag*: surface antigen). EJ360 is EJ312 *sag4,5 sag6*.

During isolation[1] of EPS-deficient mutants from EJ312 by selection with phages ØM5, ØM9h1 and ØM10, mutants *fix20* and *fix21*, which are Fix⁻ but EPS⁺, were found coincidentally. EPS⁺ mutants were then isolated from SU47 by selection with phages ØM9, ØM10 and Ø14, and classified (Table 1) by resistance to phage and to sodium deoxycholate (DOC, diagnostic for LPS[6]). SU47 mutants of class 2 have the same phage and DOC phenotypes as EJ312 *fix20* and *fix21*. A genomic library[4] from Rm1021 in cosmid pLAFR1 was mated into EJ312 *fix21*, pooled transconjugants were inoculated on alfalfa, and bacteria were recovered from rare Fix⁺ nodules. Complementing clone pIA, isolated in this way, restores wild phage and DOC phenotypes to all mutants.

MEMBRANE PROTEINS

Bacteria or bacteroids (in 10 mM HEPES, 25% sucrose, pH 7.5) are passed twice through a French press, cleared at 7000 rpm, and pelleted (150,000 g, 1 hr). For bacteroids, nodules on 3 wk old perlite-grown alfalfa seedlings, ground and filtered through miracloth, are spun in a sucrose step gradient (34/45/60%, 150,000 g, 1 hr); peribacteroid membrane (PBM) enclosed bacteroids at the 45/60% interface are then shocked twice in 10mM HEPES and passed through a 26-gauge needle. Outer and inner membranes are separated on a sucrose step gradient (30-55% in 5% steps, 150,000 g, 12 hr). Bacteria and bacteroids have similar outer membrane 2-keto-3-deoxyoctanoic acid as well as inner membrane succinate and lactate dehydrogenases, but inner membrane NADH oxidase, present in bacteria, is undetectable in bacteroids.

SDS-PAGE membrane protein patterns are different for bacteria and bacteroids in several conditions. The three strongest bands in membranes but not cytoplasm of bacteroids (Fig. 1) are not present in bacterial membranes (not shown). Molecular weights (~35 KD, ~55 KD, ~60 KD) are those of the *nifH*, *D* and *K* gene products, and Western blotting confirms that these are indeed the nitrogenase proteins (Fig. 1). Presence in the membrane fraction could be due to non-specific absorption, or to aggregation during aerobic isolation. However, the bands are not released by 3M NaCl, which suggests tight binding. Moreover, membrane association of nitrogenase has been noted ultrastructurally in Bradyrhizobium japonicum bacteroids[6].

MONOCLONAL ANTIBODY (MAb) ANALYSIS

Identical results are obtained by agglutination and by peroxidase ELISA. Six MAb isolated earlier[3] have been characterized by sensitivity to proteases, boiling and periodate, and by competition with crude LPS[7]. Antigens (Ag) 1, 2, 4 and 5 appear to be LPS, and Ag6 to be protein.

For R. leguminosarum, a MAb to the bacterial surface is also bound by the PBM[8]. For wild-type R. meliloti SU47 as well, two MAb are bound by membrane-enclosed bacteroids (Table 1). Thus association of bacteroid epitopes with the PBM may be generally true for rhizobia.

For wild-type SU47, both bacteria and bacteroids bind all six MAb (Table 1). For mutant EJ360, bacteria no longer bind any of the MAb due to mutation. However, EJ360 bacteroids do bind four of the MAb (Table 1). Thus differentiation has exposed four surface epitopes. Antigenic changes in bacteroids have also been reported for R. leguminosarum[9].

The phage-resistant mutants, selected for altered surface, are also changed in binding of anti-LPS MAb (Table 1). Moreover, while crude LPS[7] from wild-type blocks MAb binding by bacteria, LPS from mutants does not. (EJ312 LPS does not block due to prior mutations; failure of SU47 LPS to block Ab5 is not understood.) Thus the mutant alterations seem to be in LPS.

SYMBIOTIC PHENOTYPES

The original LPS mutants, EJ312 fix20 and fix21, make small, white, Fix$^-$ nodules with aborted infection threads[10]. In this way they resemble LPS mutants of R. phaseoli[11], and like them are chromosomal (data not shown). However, similar Class 2 Tn5 inserts, isolated in complementing plasmid pIA and then transduced into different backgrounds, have a conditional symbiotic phenotype. In SU47 these inserts are Fix$^+$, whereas in EJ312 the same inserts are Nod^d Fix$^-$ (nodules at 5 wk instead of 2 wk, no fixation at 8 wk; inserts may be less leaky than the spontaneous mutations of fix20 and fix21).

SUMMARY

1. Alterations in membrane proteins are correlated with bacteroid differentiation. In particular, as may be true for B. japonicum[6], the nitrogenase proteins are associated with the bacteroid membrane.

2. For wild-type, some bacterial epitopes are found not only in bacteroids but in the PBM as well; for mutant EJ360, bacteroids expose four epitopes not displayed in bacteria. Thus, as in R. leguminosarum[8,9], differentiation to bacteroids alters the bacterial envelope.

3. Mutants with altered LPS have been isolated, and the phenotype of Class 2 (at least) suggests LPS is involved in infection thread elongation, as in R. phaseoli[11]. Dependence of that phenotype on genetic background indicates that for R. meliloti the role of LPS in symbiosis is complex.

TABLE 1.

	growth of phage ØM								binding of Ab						blocking by LPS of Ab					
	5	7	9	10	12	14	col[#]	DOC[%]	1	2	3	4	5	6	1	2	3	4	5	6
EJ312	+	+	+	+	+	+	s	R	+	+	+	+	+	+	-	-	-	+	+	-
fix21	-	-	-	-	+	-	r	S	+	-	+	-	-	+	-	-	-	-	-	-
SU47	+	+	+	+	+	+	s	R	+	+	+	+	+	+	+	+	-	+	-	-
#cl 1	+	+	±	±	-	±	m	R	+	+	+	+	+	+	+	+	-	+	+	-
2	-	-	-	-	+	-	r	S	+	-	+	-	-	+	+	-	-	-	-	-
3	+	-	+	-	+	±	s	R	+	+	+	-	-	+	+	+	-	-	-	-
SU47 bacterial membrane									+	+	+	+	+	+						
bacteroid membrane									+	+	+	+	+	+						
peribacteroid membrane									±	-	-	-	-	+						
EJ360 bacterial membrane									-	-	-	-	-	-						
bacteroid membrane									±	-	-	+	+	±						

#cl, class. [#]col, colonial morphology: s, smooth; m, mucoid; r, rough.
%DOC: sodium deoxycholate 500 ug/ml in LB agar: R, resistant; S, sensitive.

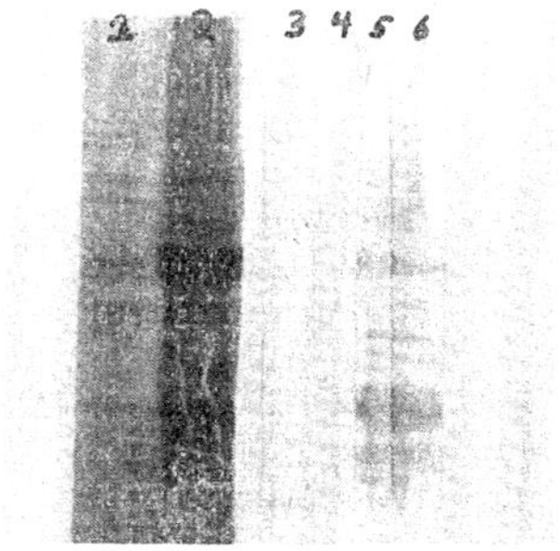

FIGURE 1. Nitrogenase in bacteroid membranes. Rabbit anti-*Klebsiella pneumoniae* nitrogenase antiserum was the generous gift of Dr. K. Howard.

1 SDS-PAGE, bacteroid cytoplasm
2 SDS-PAGE, bacteroid membrane
3 Western blot, bacteroid cytoplasm
4 duplicate of lane 3
5 Western blot, bacteroid membrane
6 duplicate of lane 5

REFERENCES

1. TM Finan et al 1985, Cell 40:869.
2. JA Leigh ER Signer & GC Walker 1985, PNAS 82:6231.
3. E Johansen TM Finan ML Gefter & ER Signer 1984, J Bacteriol 60:454.
4. SR Long WT Buikema & FM Ausubel 1982, Nature 298:485.
5. KE Sanderson T Macalister & JW Costerton 1978, Can J Microbiol 20:1135.
6. F Bergersen 1984, in Adv Nitr Fix Res ed Veeger & Newton, Nijhoff, p171.
7. O Westphal & K Jann 1965, in Meth Carb Chem ed Whistler, Acad Pr, V:83.
8. DJ Bradley et al 1986, J Cell Sci in press.
9. NJ Brewin et al 1986, J Gen Microbiol in press.
10. JF Hanks et al, this volume.
11. KD Noel et al, this volume.

DEGRADATIVE ENZYMES IN *Rhizobium meliloti*

MARY F. LOPEZ and ETHAN R. SIGNER
Dept. of Biology, Mass. Inst. of Technology, Cambridge MA 02139

INTRODUCTION

Ultrastructural evidence[1,2] clearly shows that rhizobial infection involves degradation of the root hair wall at the infection thread origin. This could be due to plant activities[3], and/or to the low and variable activities reported for rhizobia[4-7], toward cellulose, hemicelluloses or pectic substances. We have found two such activities in *R. meliloti*.

DEGRADATIVE ACTIVITIES

Bacteria were grown in minimal M9 medium + 0.1% yeast extract + 0.5% inositol. Degradation was assayed by reducing sugar production[8] for polysaccharides and by thiobarbituric acid test[9] for polygalacturonate.

Rm1021 culture supernatants degrade substrates in the order gum arabic (GA) > xylan (XY) > Rm1021 exopolysaccharide (EPS) > celluloses > poly-galacturonate (PGA) (Table 1). XY (a hemicellulose) has a β(1->4)-D-xylose backbone with galacturonosyl and arabinosyl side chains, while GA (an exudate from the woody legume *Acacia*) is a highly branched L-arabinan with galactosyl, rhamnosyl and glucuronosyl units[10]. Activities toward GA and XY are inducible (Table 2), with GA the best inducer for both; inositol alone, reported to induce cellulase in *R. trifolii*[7], has little effect.

GA and XY activities appear to be due to different enzymes. With XY or CMC + PGA as inducers, GA activity is highest in late log but XY activity is highest in stationary phase. In assays GA activity has a lag to 2 hr and then increases to 24 hr, while XY activity is linear to 2 hr and then plateaus to 24 hr. Temperature optima are 37° for GA, but 30° for XY. Both activities are stable (room temperature to 4 da, freezing to 1 mo).

PARTIAL PURIFICATION

Rm1021 bacteria have been centrifuged from culture supernatant, sonicated (2 min in 30 sec bursts), and centrifuged again to give intracellular and particulate (cell wall and membrane) fractions. GA and XY activities are found in both fractions, but only at over 10-fold lower specific activity than in culture supernatants.

Supernatant from a culture induced with CMC+PGA, made 2M in urea, was applied to a DEAE Trisacryl column, which was eluted with a linear gradient of 0.01M Tris - 0.1M Tris + 0.5M NaCl (in 2M urea, pH 7.5). Whereas XY activity was not recovered from the column, for GA the most active fraction (#13 out of 60) had an increase in specific activity of nearly 5-fold.

ACTIVITIES IN SYMBIOTIC MUTANTS

Assays have been done on culture supernatants and bacterial sonicates of mutants affected in entry into the plant host. These include nod mutants[11], which form no nodules at all, and exo mutants[12,13,14], which are deficient or altered in EPS and (except for exoD[12]) form aberrant Fix$^-$ nodules by invading intercellularly rather than through root hairs.

nod mutant activities to GA and XY, and exo mutant activities to GA, are comparable to wild-type (Table 3). However, nearly all exo mutants have no activity to XY in culture supernatants (the sole exception being exoD, also the only Fix$^+$). Moreover, whereas exoB, exoE (deletion[15] of exoB, F and A) and exoH mutants have no XY activity in sonicates, mutants in exoA, exoC and exoF do have XY activity in sonicates. That XY activity is therefore intracellular only, rather than also extracellular as in wild-type.

SUMMARY

1. With celluloses or polygalacturonate as substrate, R. meliloti has only low degradative activity. Much higher activities are found with gum arabic (GA) or xylan (XY). These activities are inducible, particularly by GA, and both are found at highest specific activity in culture supernatants.

2. The properties of the GA and XY activities suggest they are due to different enzymes, and partial purification gives GA without XY activity.

3. Both nod and exo mutants have wild-type GA activity, and nod mutants have wild-type XY activity. However, exo mutants generally have no extracellular XY activity; mutants in exoA, C and F do have intracellular XY activity, but mutants in exoB and exoH do not. One interpretation is that EPS is required for secretion or stability of xylan-degrading enzyme(s).

4. These results suggest that extracellular xylan-degrading activity may be required for root hair penetration, presumably by degradation of hemicellulose. The role of gum arabic-degrading activity is not yet clear.

TABLE 1.

substrate*	OD/ml/24hr x100	nmol red sug equiv/ml/24hr
carboxymethylcellulose (CMC)	4.2 (1)	1.8 (0.4)
microcrystalline cellulose (MCC)	3.0 (1)	1.3 (0.4)
Rm1021 exopolysaccharide (EPS)	6.5 (2)	2.9 (0.9)
xylan (XY)	13.8 (2)	6.2 (0.9)
gum arabic (GA)	30.1 (7)	20.7 (0.6)
arabinogalactan	NDA	NDA
mannan	NDA	NDA
#polygalacturonate (PGA)	2.0 (0)	

*Rm1021 EPS prepared by ethanol precipitation, all others from Sigma
#assayed by thiobarbituric acid test
inducer: CMC + PGA; NDA, no detectable activity.

TABLE 2.

inducer	XY (nmol red sug equiv/ml/24hr)	GA (nmol red sug equiv/ml/24hr)
none	NDA	0.7 (0.05)
CMC	NDA	2.0 (0.6)
PGA	7.2 (0.04)	3.4 (0)
CMC + PGA	6.5 (1.1)	18.2 (1)
XY	4.9 (0.4)	23.0 (0)
GA	19.1 (1.3)	22.4 (2)

TABLE 3.

strain	fluorescence with calcofluor (EPS)	symbiotic phenotype	XY activity supernatant	XY activity sonicate	GA activity supernatant
wild (1021)	+	Nod^+Fix^+	+	+	+
nodA (S1A3)	+	Nod^-	+	+	+
nodC (S170)	+	Nod^-	+	+	+
nodC (S8A2)	+	Nod^-	+	+	+
exoA (7031)	-	Nod^+Fix^-	-	+	+
exoB (355)	-	Nod^+Fix^-	-	-	+
exoC (7025)	-	Nod^+Fix^-	-	+	+
exoD (7053)	±	Nod^+Fix^+	+	+	+
exoE (7022)	-	Nod^+Fix-	-	-	+
exoF (7055)	-	Nod^+Fix-	-	+	+
exoH (7154)	+(no halo)	Nod^+Fix-	-	-	+

REFERENCES

1. D Callaham & JG Torrey 1981, Can J Bot 59:1647.
2. RW Ridge & B Rolfe 1985, Appl Env Micro 50:217.
3. G Fahraeus & H Ljunggren 1959, Nature 184:1578.
4. DH Hubbel VM Morales & M Umali-Garcia 1978, Appl Env Micro 35:210.
5. DPS Verma V Zogbi & AK Bal 1978, Pl Sci Lett 13:137.
6. E Martinez-Molina VM Morales & DH Hubbell 1979, Appl Env Micro 38:1186.
7. VM Morales E Martinez-Molina & DH Hubbell 1984, Pl Soil 80:407.
8. N Nelson 1944, J Biol Chem 195:19.
9. A Weissbach & J Hurwitz 1959, J Biol Chem 234:205.
10. A Darvill et al 1980, in The Biochemistry of Plants ed PK Stumpf & EG Conn, Academic Press NY, 1:91.
11. TT Egelhoff & SR Long 1985. J Bacteriol 164:591.
12. JA Leigh ER Signer & GC Walker 1985, PNAS 82:6231.
13. TM Finan et al 1985, Cell 40:869.
14. GC Walker et al, this volume.
15. TM Finan et al, this volume.

IDENTIFICATION OF HOST SPECIFICITY DNA REGIONS DETERMINING THE BROAD HOST RANGE NODULATION OF *RHIZOBIUM* STRAIN NGR234.

MURALI NAYUDU, GREG L. BENDER, BRANT J. BASSAM, MARTHA SINCLAIR AND BARRY G. ROLFE.

1. INTRODUCTION

Rhizobium strain NGR234 has generated much interest becuase this fast-growing strain is able to form nitrogen-fixing (effective) nodules with a wide range of legumes including *Macroptilium atropurpureum* (siratro), *Desmodium intortum* (desmodium), *Vigna unguiculata* (Cowpea), *Lablab purpureus* (lablab), *Leucaena leucocephala* and ineffectively nodulate *Glycine max* (soybean) and the non-legume *Parasponia andersonii*. A large symbiotic plasmid (approx. 470kb) in strain NGR234 (Morrison *et al.*, 1984; Pankhurst *et al.*, 1983) has been identified.

An alternative approach of *in vivo* genetic engineering technique using an "R68.45" like plasmid has been used to construct over 100 hybrid plasmids (R-primes) carrying various segments of the symbiotic (Sym) plasmid of this strain NGR234. The construction of these 'mini Sym' plasmids enables a more detailed study to be made of genes contributing to broad host range in strain NGR234.

2. RESULTS

R primes were isolated by selection for mobilization of Kanamycin resistance of Tn5 from strain ANU1245 (Tn5: Sym insertion mutant of NGR234, nod+Fix-). The R-primes were identified by showing that the Kanamycin resistance of Tn5 was genetically linked to the R-plasmid antibiotic resistance markers; that they had large DNA insertions in them and they had the capacity to nodulate siratro in the *Rhizobium* strain ANU265 (NGR234 Sym$^-$). Three R-primes which conferred inefficient (pMN23) and efficient (pMN31, pMN49) nodulation of siratro were selected for further study. It was shown by Southern hybridization that they carried contiguous overlapping segments of the NGR234 Sym plasmid: pMN23 (180kb); pMN31 (220kb); pMN49 (330kb). Further hybridization studies using gene specific probes showed that pMN23 contained *nod D* and siratro host specific nodulation genes (described later) while pMN31 and 49 contained *nod A B C D*, region II and siratro *hsn* genes on different restriction enzyme fragments. Both the copies of the structural genes for nitrogen fixation (*nif H D* and *K*) present in this strain (J. Badenoch-Jones, *pers. comm.*) were shown to be on all three R-primes.

Only the largest of these R-primes (pMN49) could determine the complete broad host range of NGR234 in strain ANU265. We report here for the first time the ability of strain NGR234 to inefficiently nodulate the stems and roots of the tropical legume *Sesbania rostrata*. One interesting result was that the R-prime construct ANU265 (pMN49) was able to nodulate the non-legume *Parasponia andersonii*, and the stems and roots of the tropical legume *Sesbania rostrata* more efficiently and reliably than the parent strain NGR234.

4. REFERENCES

Morrison, N.A., Cen, Y.H., Chen, H.C., Plazinski, J., Ridge, R. and Rolfe, B.G. (1984) Mobilization of a Sym plasmid from a fast-growing cowpea *Rhizobium* strain. *J. Bacteriol.* 160: 483-487.

Pankhurst, C.E., Broughton, W.J., Bachem, C., Kondorosi, E. and Kondorosi A. (1983) Identification of nitrogen fixation and nodulation genes on a large plasmid from a broad host range *Rhizobium* sp. In: Puhler A. (ed.) *Molecular Genetics of the Bacterial-Plant Interaction*, Springer Verlag, Berlin, Heidelberg, New York, pp. 169-176.

5. Table 1: Nodulation pattern of different R-primes in *Rhizobium* strain ANU265 (Sym^-)[a]

Rhizobium Strain	Siratro	Desmodium Lablab, *leucaena leucocephala*, cowpea	soybean	*Sesbania rostrata* roots/stems	*Parasponia andersonii* plate assay
ANU265	-	-	-	-	-
ANU265(pMN23)	±	-[b]	-[c]	-[c]	-
ANU265(pMN31)	+	+	-[d]	+	-[e]
ANU265(pMN49)	+	+	+	+	+
NGR234	+	+	+	±	-[f]

[a]The symbols in this table represent the following:

+ reliable nodulation on all replicates
- no nodulation observed
± mostly 'callus-like' structures with occasional pseudonodule (<10%plants)

[b]'callus-like' structures observed rarely (<10% of plants)

[c]'callus-like' structures observed on all plants

[d]'gross gall like' structure observed or all plants

[e]rare nodule observed (<10% of plants) in Leonard jar assay.

[f]reliable nodulation in Lenoard jar assay

The host range genes were shown to be Sym plasmid determined by the ability of pMN49 to express the same broad host range in *Agrobacterium tumefaciens* strain A136. The host range of the R-primes was examined in ANU265 and the results summarised in Table 1. The R-prime plasmids were shown to contain different regions of distinctive host specific nodulation (hsn) for tropical legume infection and for the nodulation of the non-legume *Parasponia*. Soybean nodulation, however, required additional regions that were not essential for nodulation of other tropical legumes.

A 6.7 kb *Hin*dIII fragment from the Sym-plasmid of strain NGR234 was found to code a *nodD* gene flanked by two loci which were required for siratro host range. Transfer of the 6.7 kb fragment from NGR234 to *R. trifolii* strain ANU843 conferred extended host range ability to this strain on siratro plants but not to other plants normally nodulated by strain NGR234. Tn5 mutagenesis of the 6.7 kb fragment showed that insertions located into loci flanking the *nodD* gene abolished the extended host range phenotype. Complementation and DNA hybiridization data showed that the *nodD* gene of strain NGR234 was functionally similar to that in *R. trifolii*. Transfer of the 6.7 kb *Hin*dIII to *R. trifolii* derivatives containing Tn5 insertions into either *nodA*, *B* or *C or other* R. *trifolii nod* genes failed to confer siratro nodulation to these recipients indicating these *R. trifolii* genes are essential for extension of siratro host range. DNA sequence analysis of this 6.7 kb *Hin*dIII region showed that the *nod D* gene shares 60% DNA sequence homology with *nod D* of *R. trifolii* and 65% homology with *nod D* of *R. meliloti* and *Bradyrhizobium* sp. ANU289. Using *lac* operon transcriptional fusions with a mini-Mu-lac bacteriophage transposon Mu*d*I1734 on this 6.7 kb *Hin*dIII fragment we have identified promotors which are induced by specific plant signal(s).

3. CONCLUSION

The symbiotic genes of strain NGR234 have been shown to have a complex organization with the genes (two copies of *nif H, D* and *K; nod A, B, C* and D; regionII ; host specific nodulation ,*hsn*) being located over half of the 470 kb Sym plasmid. Furthermore, the results show there are distinctive host specific nodulation regions for different host plants, i.e., there appears to be different sets of *hsn* genes required for different plants.

The isolation of these broad host range genes on a transmissible R-prime vector will make them amenable to genetic manipulation in understanding the nature of the broad host range phenotype.

NIF, FIX AND NOD GENE CLUSTERS IN BRADYRHIZOBIUM JAPONICUM, AND NIFA-MEDIATED CONTROL OF SYMBIOTIC NITROGEN FIXATION

H. HENNECKE, H.-M. FISCHER, S. EBELING, M. GUBLER, B. THÖNY, M. GÖTTFERT, J. LAMB, M. HAHN, T. RAMSEIER, B. REGENSBURGER, A. ALVAREZ-MORALES AND D. STUDER

MIKROBIOLOGISCHES INSTITUT, EIDGENÖSSISCHE TECHNISCHE HOCHSCHULE, ETH-ZENTRUM, UNIVERSITÄTSTRASSE 2, CH-8092 ZÜRICH, SWITZERLAND

1. INTRODUCTION

Bradyrhizobium japonicum is the slow-growing microsymbiont of soybean (Glycine max L. Merr.). Research done by our group is concerned with studies on nodulation (nod) genes and on the organization and regulation of symbiotic nitrogen fixation genes (nif and fix). The 'nif' terminology refers to genes that are homologous to corresponding genes in Klebsiella pneumoniae, whereas fix genes have first been found in rhizobia. B.japonicum (strain 110) has the advantageous trait of being able to derepress its nitrogen fixation genes in free-living, microaerobic culture; the phenotype of mutants unable to do so is called Nif$^-$. In symbiosis the nitrogen fixing phenotype is called "Fix". In this article we summarize our recent data on the identification of new symbiotic genes, and present findings on the control of symbiotic genes by the NifA protein and by oxygen.

2. ORGANIZATION OF NIF AND FIX GENES IN CLUSTER I

The nif and fix genes in B.japonicum are clustered. So far, a total of nine genes were found in the so-called cluster I (Fig. 1). Further genes, including the "common" nod genes, are grouped in cluster II (see following paragraph). In addition to clusters I and II we have identified at least ten new regions harboring essential genes for effective symbiosis. Most of these have been found as a result of random Tn5 mutagenesis (1); five of these new regions have been cloned from two separate cosmid libraries of B.japonicum genomic DNA in E.coli, and have been mapped extensively. They are not, however, described in this report except for the comparative phenotypic analysis of one selected novel regulatory mutation in one region (mutant 3160). None of those five newly cloned regions is closely linked to clusters I and II or to each other.

The presence of all genes in cluster I has been proven by mutational analysis, interspecies hybridization, and by complete or partial sequencing. The genes are transcribed in the same direction (from left to right in Fig. 1). NifD and nifK are coding for the α and β subunits of the nitrogenase MoFe protein (2,3). NifE and nifN have been located immediately downstream of nifDK. They were found to hybridize to the corresponding K.pneumoniae genes (Ebeling et al., submitted). Interestingly, nifE expression is controlled by the nifD promoter. This was shown with help of nifE'-'lacZ fusion which expressed β-galactosidase activity only when the nifD promoter was present, but not when it was deleted. Nevertheless, part of the mRNA population initiated at the nifD promoter appears to terminate immediately after the end of nifK, as was shown by an S1 mapping experiment to determine the 3' end of nifDK message (K. Kaluza, unpublished). Thus, it is probable that a sig-

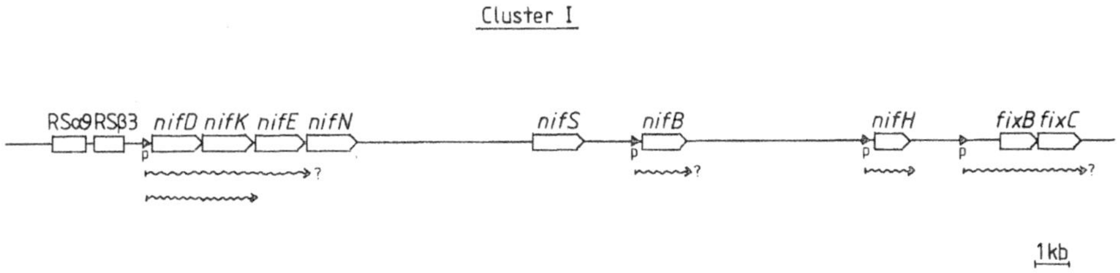

FIGURE 1. Organization of B.japonicum symbiotic genes in cluster I. Wavy lines indicate transcripts initiated from identified promoters (p). The question marks at the end of transcripts denote that transcription termination sites are not known. For references and further details see text.

nificantly lower amount of mRNA is read through into nifE, thereby expressing less nifE gene product as compared to the MoFe protein subunits. It is not known whether nifDKE transcription extends into nifN. A Tn5 insertion located between nifE and nifN was found to be completely Fix$^+$. This would speak against a putative nifDKEN operon but one has to be cautious because of the notorious outreading promoter activity of Tn5 whenever it is located in intergenic regions (4). Approximately 6 kb downstream of nifN we found a nifS-homologous gene. Tn5 mutations as well as a small deletion in the nifS region still resulted in 30 % residual Fix activity in the corresponding mutant strains (Ebeling et al., submitted). The nifS promoter has not yet been identified. NifB was found previously (5) and has now been confirmed by a nifB$^-$ insertion mutant that was Nif$^-$Fix$^-$. The nifB promoter has been mapped by Chelm and coworkers (6). The four genes described before (nifE, N, S, and B) are known in K.pneumoniae to be involved in the synthesis of the FeMo cofactor and its incorporation into nitrogenase (7). Hence, it is not surprising to see that these genes are conserved in B.japonicum. The nifH gene (coding for the nitrogenase Fe protein) has been described previously (8). About 2.5 kb downstream of nifH lies the fixBC operon. These genes were first identified in Rhizobium meliloti (9,10,11) in which they are organized as a fixABC operon while the B.japonicum fixA gene is located in a separate cluster (see below). The fixBC promoter was mapped ca. 1 kb upstream of the start of fixB. This may be indicative of the presence of another potential gene in front of fixB for which, however, we have not found experimental evidence. FixB$^-$, C$^-$ and fixA$^-$ mutants are Fix$^-$ in symbiosis and Nif$^-$ in free-living, microaerobic culture (12). Thus, these genes appear to be of direct functional importance for nitrogenase activity. By hybridization, homologues to these genes were found in all rhizobia tested (12,13) as well as in the aerobic or microaerobic diazotrophs Azotobacter vinelandii (12) and Azospirillum brasilense (14), but not in the anaerobically N_2-fixing K.pneumoniae. On the basis of these findings and other considerations (12) we favor the hypothesis that the products of fixB, C and fixA could possibly function as electron donor proteins to nitrogenase in aerobic/microaerobic N_2 fixing bacteria.

3. GENES ORGANIZED IN CLUSTER II

Fig. 2 depicts what we know about the genetic organization of cluster II. As indicated before, the fixA gene is located there, and is read from its own promoter (5). Immediately upstream of fixA we detected an operon carrying a promoter-proximal 1 kb long open reading frame (ORF) plus the promoter-distal nifA gene. NifA codes for a regulatory protein that positively controls nifH and nifDKE expression and probably other nif and fix genes (15). The pleiotropic phenotype of nifA⁻ mutants will be described in a separate paragraph (see below). An insertion in ORF, provided that it is not polar on nifA, has no distinct phenotype; hence, ORF does not appear to be strictly required under N_2 fixing conditions (15).

Relatively closely linked to the nifA and fixA genes are the common nodulation genes (Fig. 2). They consist of nodABC (16), nodI and nodJ (only determined by interspecies hybridization), and nodD which is transcribed in the opposite direction. NodA⁻ and C⁻ mutants were constructed and found to be clearly Nod⁻, whereas nodD⁻ mutants are at most only marginally delayed in nodulation (depending on the position and orientation of a nptII cartridge inserted into nodD). This result would argue for the presence of a second functional copy of nodD in the B.japonicum genome, similarly to some other fast-growing rhizobia (17,18), but this second, hypothetical nodD copy has not (yet) been identified. A so-called nod-box sequence, which is likely to be involved in the control of nodABC expression (19-21) is located 730 bp upstream of the start of nodA. This distance is larger than in fast-growing rhizobia, but similar to the situation in Bradyrhizobium sp. Parasponia in which a novel gene, nodK, has been postulated upstream of nodA (22). Whether or not B.japonicum harbors a nodK-like gene is currently under investigation.

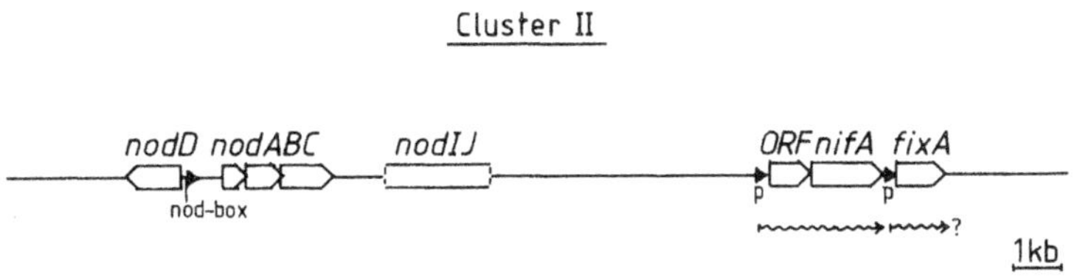

FIGURE 2. Organization of B.japonicum symbiotic genes in cluster II. For references and further details see text. Additional explanations are in the legend to Fig. 1.

4. SYMBIOTIC GENES ARE LOCATED IN AN UNUSUAL GENOMIC REGION CARRYING NUMEROUS, DIFFERENT REPEATED SEQUENCES

In contrast to the fast-growing rhizobia, the B.japonicum nif, fix and nod genes appear to be located on the chromosome rather than on a symbiotic plasmid. Nevertheless, most of these genes are clustered (as shown before), and they are located in a highly specialized area of the genome which does not carry essential genes for growth (23). In addition, the area in particular around cluster I is characterized by the presence of many copies of different repeated sequences (RS). So far, we have characterized at least five different types of RSs around cluster I, called RSα, -β, -γ, -δ and -ε. In the total genome, these are present in 12, 6, 12, 10 and 4 copies, respectively, of which 6, 3, 7, 9 and 3, respectively, are confined to the region around cluster I. Their relative positions to each other has been deduced by an analysis of large deletions. Moreover, extensive chromosome walking has been started which has led to the cloning of more than 250 kb of DNA. Apart from a confirmation of the positions of many RS copies, these clonings have also resulted in the identification of new DNA regions involved in symbiosis. For example, 67 kb and 84 kb downstream of the fixBC operon of cluster I we have located two further nod-box sequences. A deletion of these regions makes the corresponding strain less competitive than the wild-type, and it will be of interest to see whether this phenotype is associated with genes flanked by the nod-boxes. Within cluster II we have also located at least one prominent RS, but linkage between clusters I and II has not yet been established.

5. REGULATION OF SYMBIOTIC NITROGEN FIXATION GENES

5.1. Structure of *nif* and *fix* promoters

The target sites for transcriptional control are the promoters. In B. japonicum, the following promoters have been mapped and identified: nifDKE (2), nifH (8), nifB (6), glnAII (6), fixA (5), fixBC and nifA (M. Gubler and B. Thöny, unpublished). All have the typical nif(ntr)-consensus sequence found in many diazotrophs: NTGGYRYR-N_4-TTGCT. In order to be activated by the NifA protein (24) the presence of an upstream activator sequence (UAS) is required: TGT-N_4-T-N_5-ACA. Two such sequences are found upstream of the nifDKE and nifH promoters (25). Experiments to activate the fixA'-, fixBC'- and nifA'-'lacZ fusions by NifA have failed, and this may be due to the absence of UAS in the lacZ-fusion constructs. Nevertheless, it appears as if at least fixA and fixBC expression are nifA-dependent: in nifA$^-$ strains we could not detect fixA and fixBC mRNA.

5.2. The pleiotropic phenotype of *nifA* mutants

The nifA gene does not only regulate nif and fix promoters but also seems to control earlier events in bacterial development and/or persistence and in the formation of determinate soybean root nodules (15). This became evident from the phenotypic analysis of nifA$^-$ mutants. The pleiotropic nature of nifA$^-$ mutants could be seen at several levels: (i) nodulation frequency was increased, and nodules were distributed all over the root system; (ii) nodule weight was drastically reduced: (iii) the nodules were initially white inside but then exhibited dark-brown zones of necrotic appearance; (iv) bacteroids were released from the infection thread, but were then apparently heavily degraded; concomitantly, plant cells started to degrade. The interpretation of all these observations is that nifA controls one or more essential bacterial genes that enable the bacteria to proliferate, to overcome the plant defense reactions and to persist as endosymbiotic bacteroids.

There is an additional level of complexity: nifA controls the synthesis of a number of proteins that are subject to regulation by oxygen, i.e. in nifA$^-$ mutants many proteins are missing the synthesis of which is also repressed when a microaerobic wild-type culture is shifted to aerobiosis (1,15). After random Tn5 mutagenesis another Nif$^-$Fix$^-$ mutant, 3160, was obtained with a similar, but not identical, regulatory phenotype. In this mutant, too, the majority of the nifA- or oxygen-controlled proteins are missing in microaerobic culture. These results lead to two conclusions: (i) at some level, the regulatory circuitry mediated by oxygen and by nifA must be tied together; (ii) nifA may not be the only regulatory gene involved in control of symbiotic nitrogen fixation. As the example with mutant 3160 shows, there may be one or more additional genes that either exert their function in parallel to nifA, or they are superimposed to, or dependent on, nifA.

5.3. Possible connection between *nifA*- and oxygen-mediated control

There are several possibilities how control of symbiotic gene expression by oxygen may work: (i) as was first proposed for K.pneumoniae (26) a NifL-type repressor protein may inactivate the nifA product in response to oxygen; (ii) alternatively, the B.japonicum NifA protein itself could be sensitive to oxygen; (iii) expression of the nifA operon may be controlled by oxygen which would require the presence of a superimposed regulatory system. We have done some attempts to examine the first two possibilities.

It was tested whether the ORF upstream of nifA could code for a putative nifL-like protein, since its position within the operon, and the fact that non-polar mutations are Fix^+, are reminiscent of the K.pneumoniae nifLA operon. However, it turned out that synthesis of the "microaerobic" proteins in a B.japonicum ORF^-nifA^+ strain was still repressible by oxygen. Furthermore, the ORF sequence does not share homology to K.pneumoniae nifL (provided by M. Drummond). We then tested in an E.coli background whether activation of a nifD'-'lacZ fusion by the NifA protein was sensitive to oxygen. The surprising result was (Table 1) that a plasmid that expressed the B.japonicum NifA protein constitutively (pRJ7551) conferred sensitivity to O_2 while another plasmid that expressed the K.pneumoniae NifA protein (pKP7533) rendered β-Gal expression insensitive to O_2. We have not yet tested whether the Kp and Bj nifA mRNAs are of different stabilities in the presence of O_2, nor have we investigated whether there is a specific influence of the E.coli background on Bj nifA, but the most plausible (and most attractive) interpretation of this result would be that the Bj NifA protein itself is sensitive to oxygen. Further experiments are underway to prove (or disprove) this idea. If the NifA protein is indeed oxygen-sensitive, at least some aspects of symbiotic nif/fix regulation could be simpler than expected. One might conceive a model in which the cell could afford to express nifA constitutively at low levels the product of which would then become active when the appropriate level of microaerobiosis is reached during the symbiotic infection process. It must be made clear at this point that this is only one of several speculative models. Before going on with speculations several pertinent questions need to be answered: (i) Is (also) nifA expression controlled by O_2? (ii) What is the role of the ORF protein, if it has any? (iii) What is the role of additional regulatory genes, such as the one uncovered by mutant 3160? Our future research will certainly be directed towards a solution of some of these key problems.

TABLE 1. Activation of a B.japonicum nifD'-'lacZ fusion (pRJ1025) in E.coli strain MC1061 by K.pneumoniae and B.japonicum NifA proteins.

	nifA expression plasmid			β-Gal activity (Miller units)		Ratio $-O_2/+O_2$
	Plasmid No.	(Vector)	Promoter	in oxygen-deprived cells	in aerobically grown cells	
	---	---	---	67	124	0.54
Kp nifA	pKP7533	(pBR329)	P_{cat}	1258	1197	1.05
Bj nifA	pRJ7551	(pBR329)	P_{cat}	4048	260	15.6

REFERENCES

1. Regensburger B et al. (1986) Arch.Microbiol. 144, 355-366.
2. Kaluza K, Hennecke H (1984) Mol.Gen.Genet. 196, 35-42.
3. Thöny B et al. (1985) Mol.Gen.Genet. 198, 441-448.
4. Berg DE et al. (1980) J.Bacteriol. 142, 439-446.
5. Fuhrmann M et al. (1985) Mol.Gen.Genet. 199, 315-322.
6. Chelm BK et al. (1985) In Evans HJ et al., eds, Nitrogen Fixation Research Progress, p.217, Martinus Nijhoff Publishers, Dordrecht.
7. Orme-Johnson WH (1985) Annu.Rev.Biophys.Chem. 14, 419-459.
8. Fuhrmann M, Hennecke H (1984) J.Bacteriol. 158, 1005-1011.
9. Ruvkun GB et al. (1982) Cell 29, 551-559.
10. Corbin D et al. (1983) Proc.Natl.Acad.Sci.USA 80, 3005-3009.
11. Pühler A et al. (1984) In Veeger C, Newton WE, eds, Advances in Nitrogen Fixation Research, pp.609-619, Nijhoff/Junk, The Hague.
12. Gubler M, Hennecke H (1986) FEBS Lett. 200, 186-192.
13. Donald RGK et al. (1986) J.Bacteriol. 165, 72-81.
14. Fogher C et al. (1985) FEMS Microbiol.Lett. 30, 245-249.
15. Fischer H-M et al. (1986) EMBO J. 5, 1165-1173.
16. Lamb JW, Hennecke H (1986) Mol.Gen.Genet. 202, 512-517.
17. Appelbaum E et al. (1985) In Evans HJ et al., eds, Nitrogen Fixation Research Progress, pp.101-107, Martinus Nijhoff Publishers, Dordrecht.
18. Göttfert M et al. (1986) J.Mol.Biol., in press.
19. Mulligan JT, Long SR (1985) Proc.Natl.Acad.Sci.USA 82, 6609-6613.
20. Rostas K et al. (1986) Proc.Natl.Acad.Sci.USA 83, 1757-1761.
21. Schofield PR, Watson JM (1986) Nucl.Acids Res. 14, 2891-2903.
22. Scott KF (1986) Nucl.Acids Res. 14, 2905-2919.
23. Kaluza K et al. (1985) J.Bacteriol. 162, 535-542.
24. Alvarez-Morales A, Hennecke H (1985) Mol.Gen.Genet. 199, 306-314.
25. Alvarez-Morales A et al. (1986) Nucl.Acids Res. 14, 4207-4227.
26. Buchanan-Wollaston V, Cannon F (1984) In Veeger C, Newton WE, eds, Advances in Nitrogen Fixation Research, p.732, Nijhoff/Junk Publishers, The Hague.

MOLECULAR GENETICS OF NODULATION OF SOYBEAN BY BRADYRHIZOBIUM JAPONICUM

G. STACEY, A. J. NIEUWKOOP, Z. BANFALVI, J.-S. SO, N. DESHMANE, M. G. SCHELL, AND D. GERHOLD. Department of Microbiology and Graduate Program of Ecology, The University of Tennessee, Knoxville, Tennessee, U.S.A.

The Gram-negative, aerobic, soil bacteria of the root nodule group comprise two taxonomically distinct genera: the fast-growing *Rhizobium* species and the slow-growing *Bradyrhizobium* species (1). All species of rhizobia have the ability to infect leguminous plants and establish a nitrogen fixing symbiosis. The genetics of this process has been intensely studied in several fast-growing *Rhizobium*. The initial interactions of plant and symbiont that lead to establishment of the symbiosis requires at least two sets of genes. One set (*nod*ABCDIJ), the so-called "common" nodulation genes due to their sequence homology between species, encodes functions necessary for the early events of nodulation (2, 3, 4). A second set of genes, *nod*EFGH, imposes on the plant-symbiont interaction a degree of specificity; these genes determine the host range of the particular rhizobia (5, 6). Induction of the above nodulation genes takes place in response to the presence of the plant. Induction appears to be due to the *Rhizobium* sensing plant produced flavones (7, 8). Induction also requires that a functional *nod*D gene be present (7). In addition, Rostas et al. (9) have identified a 47 bp sequence upstream of all the known nodulation genes of *R. meliloti* that is essential for *nod* gene induction. This sequence has been termed the Nod Box.

A general picture of the genetics of nodulation in fast-growing *Rhizobium* is emerging. Recently, several authors (10, 11, 12) have described mutants of rhizobia defective in polysaccharide production. These mutants are defective in nodulation. These genes may represent a third set of nodulation genes found in rhizobia. It is unclear how many genes may be required for complete nodulating ability.

The view of the genetics of nodulation by *Bradyrhizobium* species is very incomplete when compared to that of fast-growing *Rhizobium*. Recently, four groups reported the cloning of the common nodulation genes from three different species of *Bradyrhizobium* (13, 14, 15, 16). This is an important step to understanding the genetics of *Bradyrhizobium*, which nodulate some of the most agronomically important plants in the world. In this report, we outline our progress in elucidating the genetics of nodulation in *Bradyrhizobium japonicum*, the symbiont of soybean. We have found many similarities between *B. japonicum* and the better studied *Rhizobium* species. However, some differences are also apparent. Most interesting, we have located a locus which is essential for *B. japonicum* to nodulation siratro (*Macroptilium atropurpurium*), an alternate host.

The Common Nodulation Genes

Previous work by our laboratory identified a 40 Kb DNA region of *B.*

japonicum a portion of which showed homology to the nodABCD gene region of R. meliloti (13). A segment of this region also conferred root hair curling ability (hac) to a Hac$^-$ mutant of R. fredii (13). We have now characterized this region further. We have utilized gene-specific M13 clones of the R. meliloti genes to localize the nodABCD genes by hybridization (Figure 1). The orientation of these genes is similar to that of Rhizobium species.

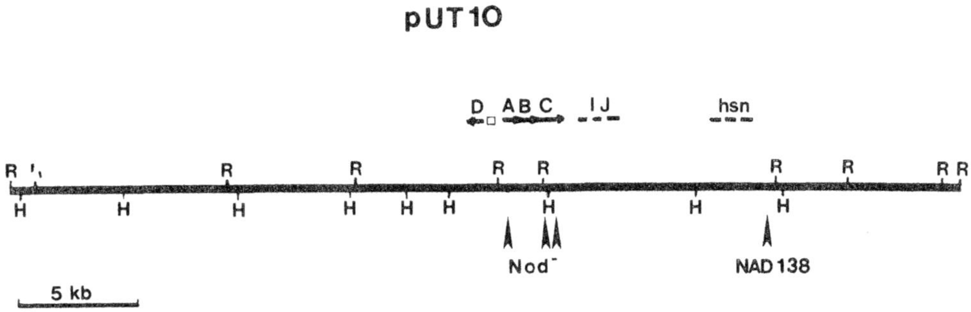

Figure 1. Restriction map of B. japonicum DNA showing the location of the nodABCDIJ, nod box, and siratro hsn locus. Arrows refer to Tn5 insertions.

A synthetic oligonucleotide (25-mer, dATAAAAACAATCGATTTTACCATC) of the R. meliloti Nod Box sequence was used to locate a comparable region in B. japonicum. Hybridization of this probe to B. japonicum EcoRI digested genomic DNA gives two strongly hybridizing bands (5.8 and 4.0 Kb). The 5.8 Kb fragment is contained in the common nodulation region (Figure 1). The 4.0 Kb band is unlinked to this region.

The nodIJ genes from R. leguminosarum were subcloned into pBR329 from the plasmid pIJ1089 (17). This subclone was then used to localize the nodIJ genes by hybridization (Figure 1). The orientation of these genes with respect to the nodABCD genes is comparable to that found in Rhizobium species.

To confirm the nodABCD and Nod Box homologous regions of B. japonicum, this region of DNA was subjected to DNA nucleotide sequence analysis. A partial sequence was obtained in order to confirm the presence and orientation of these genes. When the amino acid sequence of this region was compared to the R. meliloti common nod region (2, 18), the B. japonicum DNA was found to have 60, 82, 47, and 55% homology with nodABC and nodD, respectively. The homology of the B. japonicum sequence was much greater to the nodABCD genes of B. parasponiae (19); 82, 86, 78, and 90% homology between the nodABC and nodD genes, respectively. The Nod Box DNA sequence obtained from B. japonicum was highly homologous to similar sequences from B. parasponiae (19) and R. meliloti (9).

The functionality of the identified nodulation regions was tested by site-directed Tn5 mutagenesis. As expected, Tn5 insertions in the nodABC genes resulted in a Nod$^-$ phenotype. A Tn5 insertion in the nodD region resulted in a slight delay (2-4 days) in nodule formation. This result, however, may be due to transcription of an effective *nodD* product from a promoter on the *Tn5*. The functionality of the *nodD* gene is of particular interest because of its postulated regulatory role (7). A 3.9 Kb *Hind* III fragment containing the *nodD* gene was transferred to a NodD⁻ mutant of *R. meliloti*. This mutant nodulates alfalfa with a 9 day delay when compared to wild type. The nodulation defect by the mutant was largely complemented by the *B. japonicum nodD* gene with nodulation occurring with only a 2 day delay when compared to wild type. Therefore, the identified *B. japonicum nodD* appears to be functionally equivalent to the *R. meliloti nodD* gene.

Host Range Locus Linked to the Common Nod Genes

In addition to the *nodABCDIJ* genes, the gene region isolated appears to encode other symbiotically important genes as indicated by site-directed mutagenesis (Figure 1). For example, a region essential for nitrogen fixation was identified by mutagenesis. This region has been shown previously by hybridization to have homology to the *nifA* and *fixA* genes of *R. meliloti* (16). Most interesting, however, is a region between the identified *nodABCD* and *nifA/fixA* genes that shows, by hybridization, homology to the *hsn* genes isolated from *Rhizobium* sp. MPIK3030 (Figure 1, 21). These *hsn* genes were isolated from strain MPIK3030 by their ability to confer to *R. meliloti* the potential to nodulate siratro (21). *B. japonicum* can also nodulate siratro as an alternative host to soybean. We constructed a *Tn5* insertion in the region of *B. japonicum* showing homology to the siratro *hsn* genes (NAD138, Figure 1). This mutant does not nodulate siratro but still nodulates soybeans with little or no delay. This region appears to encode the siratro *hsn* functions of *B. japonicum*.

Other Nodulation Genes of B. japonicum

In addition to the genetic studies described above, we sought to identify other loci in *B. japonicum* essential for nodule formation. Conceivably, such loci could encode host specificity, polysaccharide formation, or some other unknown function. In order to identify new *nod* genes, we first isolated a number of random *Tn5* mutants of *B. japonicum* strains USDA110 and AN16. Mutants defective in nodulation were isolated from both strains. In the case of strain USDA110, a number of Nod⁻ mutants were isolated that were later found to be histidine auxotrophs. Similar mutants have previously been isolated by Sadowsky et al. (22). However, both in the case of USDA110 and AN16, non-auxotrophic Nod⁻ mutants were obtained.

Hybridization of the previously cloned DNA (containing the *nodABCDIJ* and *nifA-fixA* genes) to genomic DNA of the Nod⁻ mutants indicated that the *Tn5* insertion was not located in these regions. The DNA containing the *Tn5* insertions of three Nod⁻ mutants of *B. japonicum* USDA110 were cloned. The cloned DNA was then used to mutate wild type *B. japonicum* by conjugation and reciprocal exchange of the *Tn5* region into the genome. The resulting mutant homogenotes exhibited a Nod⁻ phenotype. This result indicates that the *Tn5* is indeed the cause of the observed phenotype. The three *Tn5* regions cloned map to two apparently unlinked regions. These regions also appear unlinked to the previously cloned DNA. These new *nod* gene regions show no homology, by hybridization, to the nodABCD, nodEFGH genes, or the Nod Box sequence of R. meliloti, nor to the nodIJ genes of R. leguminosarum. Work is in progress to further characterize these genes.

Summary

Our work on the genetics of nodulation in B. japonicum has progressed to the point where genes comparable to the nodABCDIJ genes of Rhizobium species have been identified. The orientation of these genes is comparable to that found in Rhizobium species. The presence of a functional nodD gene and a conserved Nod Box sequence suggests that the regulation of the nodABCDIJ genes in B. japonicum will be similar to that in Rhizobium species. We have now constructed lacZ fusions in these genes to test this prediction.

The study of the genetics of host range in Bradyrhizobium species may be a fruitful area of research in that, in general, a broader range of hosts are infected when compared to Rhizobium species. We have identified a locus essential for the ability of B. japonicum to nodulate siratro. Mutations in this region only slightly affect nodulation of soybean suggesting that B. japonicum may have distinct sets of hsn genes for each host nodulated.

Our work is just beginning on other nodulation loci identified by random Tn5 mutagenesis. The regions cloned, thus far, appear to be unique. Of course, the hope is that one or more of these regions will encode the soybean host specific functions.

ACKNOWLEDGEMENTS

The studies reported here have been supported in part by U.S. Public Health grant 1-R01-GM 33494-01A1 from N.I.H. and grant 04-CRCR-1-1419 from the U.S. Department of Agriculture.

REFERENCES

1. Jordan DC: Int J Syst Bacteriol 32, 136-139, 1982.

2. Jacobs TW, Egelhoff TT, Long SR: J Bacteriol 162, 469-476, 1985.

3. Torok I, Kondorosi E, Stepkowski E, Postfai J, Kondorosi A: Nucl Acids Res 12, 9509-9514, 1984.

4. Rossen L, Johnston AWB, Downie JA: Nucl Acids Res 12, 9497-9502, 1984.

5. Kondorosi E, Banfalvi Z, Kondorosi A: Mol Gen Genet 193, 445-452, 1984.

6. Horvath B, Kondorosi E, John M, Schmidt J, Torok I, Gyorgypal Z, Barabas I, Wieneke U, Schell J, and Kondorosi A: Cell (in press), 1986.

7. Mulligan JT, Long SR: Proc Natl Acad Sci USA 82, 6609-6613, 1985.

8. Innes RW, Kuempel PL, Plazinski J, Canter-Cremers H, Rolfe BG, Djordjevic: Mol Gen Genet 201, 426-432, 1985.

9. Rostas K, Kondorosi E, Horvath B, Simoncsits A, Kondorosi A: Proc Natl Acad Sci USA 83, 1757-1761, 1985.

10. Leigh JA, Signer ER, Walker GC: Proc Natl Acad Sci 82, 6231-6235, 1985.

11. Vandenbosch KA, Noel KD, Kaneko Y, Newcomb EH: J Bacteriol 162, 950-959, 1985.

12. Hynes MF, Simon R, Muller P, Niehaus K, Laes M, Puhler A: Mol Gen Genet 202, 356-362, 1986.

13. Russell P, Schell MG, Nelson KK, Halverson LJ, Sirotkin KM, Stacey G: J Bacteriol 164, 1301-1308, 1985.

14. Noti JD, Dudas B, Szalay AA: Proc Natl Acad Sci USA 82, 7379-7383, 1985.

15. Marvel DJ, Kuldau G, Hirsch A, Richards E, Torrey JG, Ausubel FM: Proc Natl Acad Sci USA 82, 5841-5845, 1985.

16. Lamb JW, Hennecke H: Mol Gen Genet 202, 512-517, 1986.

17. Downie JA, Ma Q-S, Knight CD, Hombrecher G, Johnston AWB: EMBO J 2, 947-952, 1983.

18. Egelhoff TT, Fischer RF, Jacobs TW, Mulligan JT, Long SR: DNA 4, 241-248, 1985.

19. Scott KF: Nucl Acids Res 14, 2905-2910, 1986.

20. Bachem CWB, Banfalvi Z, Kondorosi E, Schell J, Kondorosi A: Mol Gen Genet 203, 42-48, 1986.

21. Sadowsky MJ, Rostas K, Sista PR, Bussey H, Verma DPS: Arch Microbiol (in press), 1986.

CHARACTERIZATION OF GENES ESSENTIAL FOR SYMBIOTIC NITROGEN FIXATION FROM *BRADYRHIZOBIUM JAPONICUM* STRAIN I110

John D. Noti, Allen C. Yun, Otto Folkerts, Istvan Torok*, and Aladar A. Szalay
Boyce Thompson Institute for Plant Research, Cornell University, Ithaca, New York 14853 and *Institute of Biochemistry, Hungarian Academy of Sciences, H-6701, Szeged, Hungary

SUMMARY

A total of 96 independent Tn*5* insertions within a 39 kilobase pair (kb) segment of chromosomal DNA containing the three structural genes for nitrogenase, (*nifH*, *nifD*, and *nifK*) from *Bradyrhizobium japonicum* strain I110 were obtained in *Escherichia coli* and transferred to the wild-type strain by marker exchange. Individual transconjugants containing a Tn*5* insertion were inoculated onto *Glycine max* cv. Wilkin (soybean) and analyzed for their effect on symbiotic nitrogen fixation. In addition to the three structural genes, genes essential for nitrogen fixation (*fix* genes) were located in three separate regions: 1) 9 kb upstream of the *nifDK* operon (*fix* region I); 2) 1.5 kb downstream of the *nifDK* operon (*fix* region II); 3) 4.5 kb upstream of *nifH* (*fix* region III). All of the *fix*::Tn*5* insertion strains formed nodules which contained low or undetectable levels of nitrogenase activity. Bacteroids isolated from these nodules had approximately the same levels of the *nifDK* and *nifH* transcripts as those detectable from nodules formed by the wild-type strain. Western blot analysis of bacteroid proteins from nodules formed by the *fix*::Tn*5* mutants or the wild-type strain showed the presence of similar levels of the nitrogenase protein subunits.

DNA fragments containing either the *nifD* or *nifH* promoter and 5'-structural gene sequences from *Bradyrhizobium japonicum* strain I110 were fused in-frame to the *lacZ* gene. Stable integration of these *nif* promoter-*lacZ* fusions by homologous double reciprocal crossover into a symbiotically nonessential region of the *B. japonicum* chromosome provided an easy assay for the effects of potential *nif* regulatory mutants. The level of -galactosidase activity expressed from these two *nif* promoter-*lacZ* fusions was assayed in bacteroids of *B. japonicum* I110 wild-type and Fix mutants generated by transposon Tn*5* mutagenesis. No *nif* positive regulatory mutants were identified from among this array of Fix mutants. This result indicates that there are no genes in these regions involved in the regulation of nitrogenase structural gene expression.

Fix region III was characterized further by DNA sequence analysis and was shown to contain three open reading frames (ORF). One ORF corresponds to the *nifB* gene. The coding sequence of the *nifB* gene consists of 1494 nucleotides and is preceded by putative promoter (5'ACGG-8bp-TTGCT 3') and upstream activator (5'TGT-4bp-T-5bp-ACA 3') sequences. The second ORF is upstream of *nifB* and consists of a coding sequence 1356 nucleotides long that is preceded by a putative promoter sequence (5'CTGG-9bp-TTGCT 3'). The third ORF is 831 nucleotides long and immediately follows *nifB*.

INTRODUCTION

We have cloned a 39 kb region of DNA from *Bradyrhizobium japonicum* strain I110 that contains all three structural genes for nitrogenase (5). In order to determine whether genes essential for symbiotic nitrogen fixation other than *nifH*, *nifD*, and *nifK* were present, site-directed Tn*5* mutagenesis was performed. A total of 96 independent Tn*5* insertions distributed throughout this 39 kb region of DNA were transferred to the wild-type strain I110 by marker exchange. The effect of each Tn*5* insertion on symbiotic nitrogen fixation was determined by inoculation of the strains onto soybean.

RESULTS

Localization and characterization of *fix* genes

Fix genes were found to be separated into three regions that we refer to as *fix* regions I, II and III. These regions were located 9 kb upstream (*fix* region I) and 1.5 kb downstream (*fix* region II) of the

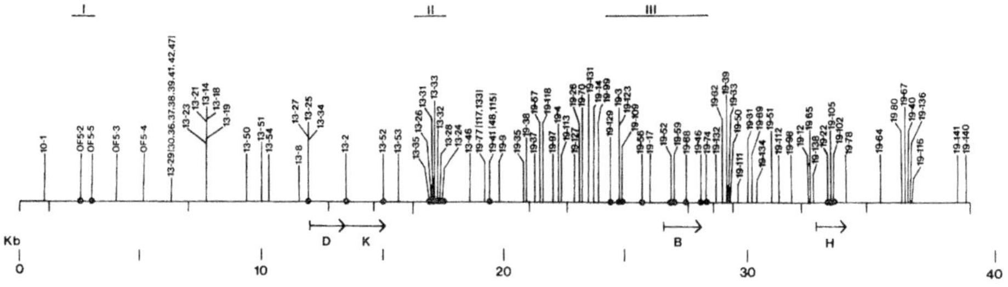

Fig. 1. Location of Tn*5* insertions within *B. japonicum* I110 DNA in pJN110-22. Tn*5* insertions (closed circles) within three regions (*fix* regions I, II, III) resulted in Fix$^-$ or low Fix phenotype. All other Tn*5* insertions resulted in a Fix$^+$ phenotype. Several independently generated Tn*5* insertions were in the same location (branching vertical lines). *Eco*RI sites are shown below the location of the Tn*5* insertions.

nifDK operon and 4.5 kb upstream (*fix* region III) of the *nifH* operon. Twenty Tn*5* insertions distributed throughout *fix* regions I, II and III resulted in a complete or partial loss of nitrogenase activity.

Effect of Tn*5* insertions on the transcription of the *nif* genes. In order to determine whether the Tn*5* insertions in *fix* regions I, II and III prevented the transcription of the *nifDK* or *nifH* operons, bacteroid RNA was isolated from nodules formed by strains carrying the Tn*5* insertions. The effect of these insertions on the levels of the *nifDK* and *nifH* transcripts was determined by northern blot analysis. None of the Tn*5* insertions outside of the nitrogenase structural genes prevented the transcription of the nifDK and nifH operons although there was some variability in the amounts of these transcripts relative to those found in nodules formed by the wild-type strain.

Effect of Tn5 insertions on the translation of the nif genes. In order to determine if the Fix$^-$ phenotypes were caused by a failure of the nifDK and nifH transcripts to be translated, bacteroid protein preparations were assayed for the presence of the nitrogenase subunits. The individual subunits of nitrogenase were detected in preparations of bacteroid proteins after incubation with antisera made against component I (the products of nifD and nifK) and component II (the product of nifH) from R. leguminosarum. The results of this experiment showed not only that the three polypeptides of nitrogenase were synthesized, but also that the levels of components I and II in the Fix$^-$ strains were approximately the same as those found in the wild-type strain.

Translational fusions of nif DNA to lacZ

Quantitation of nif mRNA levels in bacteroids is difficult and time-consuming. As an alternative to monitoring the expression of the nifDK and nifH operons, translational fusions of the two nif promoters to the lacZ gene of Escherichia coli were constructed. A single copy of either the nifD promoter-lacZ or nifH promoter-lacZ fusion was stably integrated into a symbiotically nonessential region of the B. japonicum chromosome. In order to determine whether any of the fix genes identified in this study were required for the activation of the nifDK or nifH operons, we examined the effect of inactivation of these genes by Tn5 mutagenesis on nif-promoter dependent expression of β-galactosidase (7).

Construction of nif promoter-lacZ translational fusions. Because the DNA sequences of the nifD and nifH genes and their 5'-flanking regions were known (2,4), the appropriate fragments could be isolated to create fusions to lacZ in the correct translational frame. The 0.81 kb ClaI fragment containing the nifD promoter and 330 bp of the N-terminal sequence was made blunt with the Klenow fragment of DNA polymerase I and the resulting fragment was ligated to a polylinker fused to the 5' end of lacZ. The 0.75 kb SmaI-XhoI fragment containing the nifH promoter and 219 bp of the N-terminal sequence was also fused to lacZ by blunt-end ligation. DNA sequence analysis verified both nif promoter-lacZ fusions to be in correct translational frame.

For the stable integration of the nif promoter-lacZ fusions into the B. japonicum chromosome an insertional vector must be available that contains a symbiotically nonessential region of the chromosome within which the fusion can be inserted. This nonessential region must also contain a unique restriction site (that can be used to clone the nif-lac fusions) flanked by two large contiguous DNA regions. In addition, this vector must contain a mob region (mobilization) to enable it to be transferred from E. coli to B. japonicum. Integration of nif promoter-lacZ fusions occurring by a double reciprocal crossover into the B. japonicum chromosome can be distinguished on the basis of their hybridization pattern (Fig. 2).

Analysis of β-galactosidase activity. The levels of β-galactosidase activity expressed from the nifD or nifH promoter in Fix$^-$::Tn5 mutants of fix regions I, II, and III were determined. Nodules were harvested from soybean plants thirty days after inoculation with B. japonicum I110 wild type and Fix$^-$ mutants (containing nif promoter-lacZ fusions) to assay for nitrogenase, β-hydroxybutyrate dehydrogenase and β-galactosidase activities.

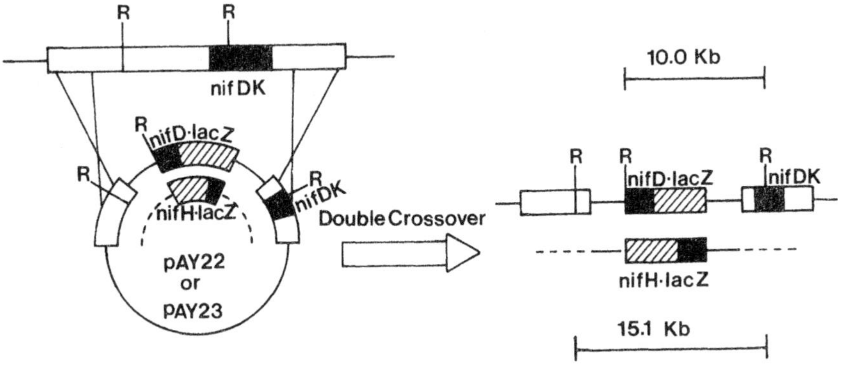

Fig. 2. Integration of nif promoter-lacZ fusions by a double crossover. Dashed lines represent identity to the DNA regions shown directly above. Double crossover events were identified by Southern blot analysis of chromosomal DNA with nifD as probe. Only the pertinent EcoRI sites (R) are shown.

Total nodule and shoot dry weights correlated well with nitrogenase activity as measured by the acetylene reduction assay. High values of acetylene reduction, and total nodule and shoot dry weights were observed for wild type strain I110 (with or without nif promoter-lacZ fusions). In contrast, low values of acetylene reduction and total nodule and shoot dry weights were observed with all Fix$^-$ mutants containing nif promoter-lacZ fusions. The level of β-galactosidase activity, however, was essentially the same in nodules formed by either the wild-type strain or strains carrying Tn5 insertions in the three fix regions. β-galactosidase activity was only observed in bacteroids of strains containing nif promoter-lacZ fusions. As a control for measuring bacteroid-specific enzyme activity, β-hydroxybutyrate dehydrogenase activity was assayed and found to be present at approximately the same levels in all wild type and Fix$^-$ mutant strains.

Identification of fix genes

Fix region II was found to hybridize to the nifE and nifN genes of Klebsiella pneumoniae (6). The arrangement of these genes in B. japonicum is nifD, nifK, nifE, and nifN. One Tn5 insertion that was localized between the nifDK operon and fix region II had no effect on nitrogen fixation. This suggests that fix region II is not part of the nifDK operon. However, we have found that in Bradyrhizobium sp. (Vigna) strain IRc78, the nifE gene is downstream and part of the nifDK operon (3). By analogy, the same organization may be present in B. japonicum. DNA sequence analysis of the region is currently in progress.

Fix region III spans at least 4 kb of DNA. One Tn5 insertion (19 17) within fix region III had no effect on nitrogen fixation and, thus, indicates that these fix genes make up more than one operon. The 4 kb region between two Fix$^+$ Tn5 insertions that bound fix region III was sequenced and shown to contain the nifB gene (Fig. 3) by comparative

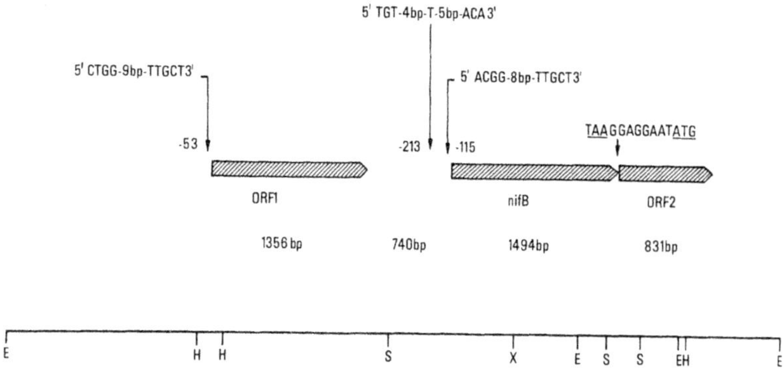

Fig. 3. Summary of the DNA sequence analysis of fix region III. The location of the promoter and activator sequences is indicated relative to the translational initiation codon (+1). The 7 bp region between the stop codon (TAA) for nifB and the start codon (ATG) for ORF2 is shown. (E) = EcoRI; (H) = HindIII; (S) = SalI; (X) = XhoI.

sequence analysis with the R. meliloti and K. pneumoniae nifB sequences (1, F.M. Ausubel and W.J. Buikema, personal communication). A single unambiguous open reading frame (ORF) 1494 nucleotides long preceded by a putative promoter (5'ACGG-8bp-TTGCT 3') at position -115 was found. Additionally, the upstream activator sequence (5'TGT-4bp-T-5bp-ACA 3') was found at position -213. The first ATG of the ORF was chosen as the translational initiation codon because it is immediately preceded by a putative Shine-Dalgarno sequence. On the basis of the positions of the putative promoter sequence and the ORF, the direction of transcription of nifB is the same as that of the nifDK and nifH operons.

The DNA sequence of nifB of B. japonicum is 54% and 47% homologous, respectively, to the nifB genes of R. meliloti and K. pneumoniae. Similarly, 51% and 45% of the amino acids of the R. meliloti and K. pneumoniae genes, respectively, are conserved. Further, those cysteine residues that are conserved among the nifB genes of R. meliloti, K. pneumoniae and R. leguminosarum are also conserved in B. japonicum.

In addition to the ORF corresponding to the nifB gene, two other ORFs were found within fix region III. The nifB coding sequence is immediately followed by an ORF 831 nucleotides long. The nifB gene and this ORF are separated by an AG-rich region seven nucleotides long (Fig. 3) that could serve as a possible ribosome binding site. This finding suggested that nifB and the small ORF are part of the same transcriptional unit. In K. pneumoniae the nifQ gene lies downstream of nifB and is part of the same operon (F.M. Ausubel and W.J. Buikema, personal communication). A comparison of this ORF with nifQ however, did not reveal any significant sequence homology. The relevance of this ORF to nitrogen fixation is unknown.

The second ORF is upstream of the nifB gene. The ORF is 1356 nucleotides long and is preceded at position -53 (upstream from the putative initiation codon) by a putative promoter sequence (5' CTGG-9bp-TTGCT 3'). This ORF is separated from the putative start codon of nifB by 740 nucleotides. However, the putative promoter that we have identified is atypical of the nif promoters identified to date in that 9 bp, rather than 8 bp, separate the conserved bases. This ORF therefore could conceivably be longer than indicated. Hennecke et al. (this volume) have shown that nifS (6) lies upstream of nifB.

CONCLUDING COMMENTS

Although no nif regulatory gene has been identified, further efforts to obtain nif regulatory mutants using the wild-type I110 strains that contain nif promoter-lacZ fusions can be envisioned. With the development of an ex plant method of induction of nitrogenase from B. japonicum on defined media, the screening of mutants generated from random Tn5 mutagenesis can be used to obtain white rhizobial colonies on X-gal plates that are unable to activate the expression of β-galactosidase.

REFERENCES

1. Ausubel, F.M, W.J. Buikema, C.D. Earl, J.A. Klingensmith, B. Nixon, and W.W. Szeto. 1985. In Nitrogen Fixation Research Progress. ed. Evans, H.J., P.J. Bottomly and W.E. Newton. (Martinus Nijhoff Publ., Dordrecht, The Netherlands) pp. 165-172.
2. Fuhrmann, M. and H. Hennecke. 1984. Mol. Gen. Genet. 187:419-425.
3. Jagadish, M.N., A.C. Yun, J.D. Noti, O. Folkerts, and A.A. Szalay. 1985. In Advances in Molecular Genetics of the Bacteria-Plant Interaction, ed. Szalay, A.A. and R.P. Legocki. (Cornell Univ. Publ., Ithaca, NY) pp. 27-31.
4. Kaluza, K. and H. Hennecke. 1984. Mol. Gen. Genet. 196:35-42.
5. Noti, J.D., O. Folkerts, A.N. Turken, and A.A. Szalay. 1986. J. Bacteriol. in press.
6. Roberts, G.P. and W.J. Brill. 1981. Ann. Rev. Microbiol. 35:207-235.
7. Yun, A.C., J.D. Noti, and A.A. Szalay. 1986. J. Bacteriol. in press.

NODULATION GENES OF THE STEM NODULATING SESBANIA ROSTRATA SYMBIONT, STRAIN ORS571

M. HOLSTERS[1], G. VAN DEN EEDE[1], K. GOETHALS[1], M. VAN MONTAGU[1], B. DREYFUS[2]

[1] Laboratorium voor Genetica, Rijksuniversiteit Gent, B-9000 Gent (Belgium); [2] Laboratoire de Biologie des Sols, O.R.S.T.O.M., Dakar (Sénégal)

1. INTRODUCTION

Strain ORS571 nodulates both stems and roots of the tropical leguminous plant Sesbania rostrata (1). Stem nodules are induced at the level of dormant root primordia via a crack entry mechanism (2) whereas root nodulation involves root hair curling (3). Among symbiotic N_2 fixing bacteria strain ORS571 is unique because it can fix N_2 in the free-living state and grow at the expense of N_2 as nitrogen source (4). Taxonomically the strain is more related to the slow growing bradyrhizobia than to the fast growing rhizobia (5). It will represent a new genus, more closely related to the genus Xanthobacter than to the genus Bradyrhizobium (Dreyfus et al., in preparation). Here we describe the identification and cloning of ORS571 genes essential for root and stem nodulation.

2. RANDOM TRANSPOSON MUTAGENESIS OF STRAIN ORS571

In order to identify ORS571 nodulation genes we carried out a Tn5 mutagenesis using the pSUP2021 vector system (6). The Tn5-encoded resistance to streptomycin is expressed in ORS571. After mating between E. coli SM10 (pSUP2021) and ORS571 streptomycin was used both to eliminate the E. coli donor and to select, in combination with kanamycin, for Tn5 harbouring ORS571 derivatives. 2000 $Sm^R Km^R$ ORS571 transconjugants were purified and screened. Both true Tn5 transposition and vector cointegration was observed in approximately equal proportion. Auxotrophic mutants appeared with a frequency of 1%. Among the non-auxotrophic ORS571 derivatives 2 Nod^-, 4 Nif^- and 7 Fix^- mutants were found. The Nif^- mutants had lost both free-living and symbiotic N_2 fixation capacity. They were phenotypically identical to the Nif^- mutants isolated by Donald et al. (7), some of which were mapped in the ORS571 genes homologous to the Klebsiella pneumoniae nifHDK genes that code for the subunits of the nitrogenase complex. The Fix^- mutants fixed N_2 normally in the free-living state. The nodules they induced did contain bacteria but were ineffective, without measurable acetylene reduction activity. These Fix^- mutations therefore affect functions that may be involved in bacteroid differentiation and regulation of symbiotic N_2 fixation. The two Nod^- mutant strains, ORS571-1 and ORS571-2 had lost both root and stem nodulation capacity. With strain ORS571-1 root hair curling was no longer observed. Strain ORS571-2 induced a slight swelling at the junction of the lateral roots to the main root of S. rostrata and root hair curling and infection thread initiation still occurred.

3. PHYSICAL CHARACTERIZATION OF TWO NOD LOCI

By selecting for the Tn5 encoded resistance to kanamycin, the mutated DNA regions of strains ORS571-1 and ORS571-2 were cloned in E. coli plasmid vectors. From a pLAFR1 gene library of ORS571 DNA two clones, pRG70 and pRG20 were isolated which fully restored the nodulation capacity of strains ORS571-1 and ORS571-2 respectively. From the comparison of the physical maps of the Tn5 containing clones with the map of the complementing clones it was concluded that the Nod⁻ 1 and 2 mutations were caused by Tn5 insertion. The two mutations are located in separate loci of the genome. The restriction map of a 20 kb DNA region of Nod locus 1 showed no overlap with the map of a 40 kb region containing Nod locus 2. We also determined a restriction map of a 50 kb DNA region containing Nif locus 1 and saw no linkage to either Nod locus. We never could demonstrate the presence of large plasmids in ORS571, therefor we assume that these symbiotic loci are located on the chromosome, an organization which may reflect the closer relatedness of ORS571 to the bradyrhizobia than to the fast growing rhizobia.

4. HOMOLOGY BETWEEN NOD LOCUS 1 AND THE NODC GENE OF R. MELILOTI

Southern blotted EcoR1 digests of the clones pRG70 and pRG20 were hybridized to the ^{32}P-labeled 3.5 kb EcoR1/BamH1 insert of the plasmid pEK12 (8) which contains the common nodABC genes of R. meliloti. Under conditions of low stringency the probe hybridized weakly to a 12.7 kb EcoR1 fragment of pRG70 (pRG801, Fig. 1). The homology was narrowed down to a 3 kb EcoR1/BamH1 subfragment, several digests of which were Southern blotted and hybridized to a nodC specific R. meliloti probe [the 1.8 kb EcoR1 fragment from pJS209 (9)] and to a nodAB specific probe [the EcoR1 insert of pJS204 (8)]. With the latter probe no convincing results were obtained. The nodC probe hybridized to a 400 bp sequence 1.5 kb away from the Nod⁻ 1 mutation in Nod locus 1 (Fig. 1). By site specific mutagenesis we recently isolated a Tn 5 insertion located in the region of nodC homology. This mutation abolished both stem and root nodulation capacity and could be complemented, although not to the full wild type phenotype, by pGMI149, a clone which contains the common nod genes of R. meliloti 2011 (10). The Nod⁻ 1 Tn5 insertion outside the nodC homology region was not complemented by pGMI149. At present we are carrying out an extensive site-specific mutagenesis of the two Nod loci and we are comparing the effect of mutations on the interaction of ORS571 with stems and roots of S. rostrata.

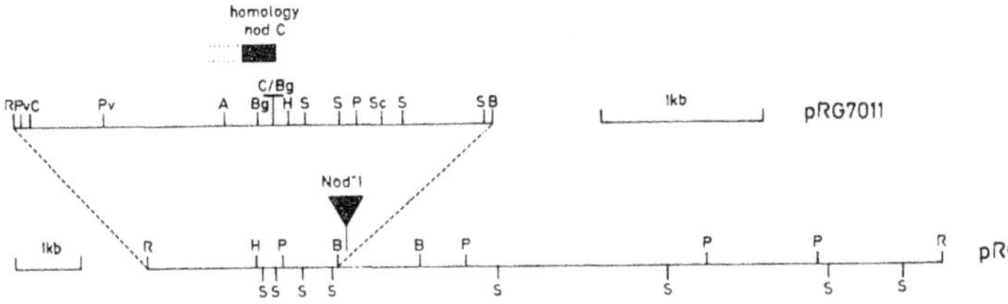

FIGURE 1. Location of a Nod⁻ 1 Tn5 insertion and of nodC homology in a 12.7 kb EcoR1 fragment of Nod locus 1. A, ApaI; B, BamHI; Bg, BglII; C, ClaI; H, HindIII; P, PstI; Pv, PvuII; R, EcoRI; S, SalI; Sc, SacI.

REFERENCES

1. Dreyfus B, Dommergues YR : Nitrogen fixing nodules induced by Rhizo-bium on the stem of the tropical legume Sesbania rostrata. FEMS Microbiol Lett 10:313-317, 1981
2. Tsien HC, Dreyfus BL, Schmidt EL : Initial stages in the morphogenesis of nitrogen-fixing stem nodules of Sesbania rostrata. J Bacteriol 156:888-897, 1983.

3. Olsson, JE, Rolfe BG : Stem and root nodulation of the tropical legume Sesbania rostrata by Rhizobium strains ORS571 and WE7. J Plant Physiol 121:199-210, 1985.
4. Dreyfus B, Elmerich C, Dommergues YR : Free-living Rhizobium strain able to grow under N_2 as the sole nitrogen source. Appl Environ Microbiol 45:711-713, 1983.
5. Jarvis BDW, Gillis M, De Ley J : Intra- and intergeneric similarities between the ribosomal ribonucleic acid cistrons of Rhizobium and Bradyrhizobium species and some related bacteria. Int J System Bacteriol 36:129-138, 1986.
6. Simon R, Priefer U, Pühler A : Vector plasmids for in-vivo and in-vitro manipulations of Gram-negative bacteria. In: Pühler A (ed) Molecular Genetics of Bacteria-Plant Interaction. Springer Verlag, Berlin, pp 98-106, 1983.
7. Donald RGK, Nees DW, Raymond CK, Loroch AI, Ludwig, RA : Characterization of three genomic loci encoding Rhizobium sp. strain ORS571 N_2 fixation genes. J Bacteriol 165:72-81, 1986.
8. Schmidt J, John M, Kondorosi E, Kondorosi A, Wieneke U, Schröder G, Schröder J, Schell J : Mapping of the protein-coding regions of Rhizobium meliloti common nodulation genes. EMBO J 3:1705-1711, 1984.
9. John M, Schmidt J, Wieneke U, Kondorosi E, Kondorosi A, Schell J : Expression of nodulation gene nodC of Rhizobium meliloti in Escherichia coli : role of the nodC gene product in nodulation. EMBO J 4:2425-2430, 1985.
10. Truchet G, Debellé F, Vasse J, Terzaghi B, Garnerone A-M, Rosenberg C, Batut J, Maillet F, Dénarié J : Identification of a Rhizobium meliloti pSym2011 region controlling the host specificity of root hair curling and nodulation. J Bacteriol 164, 1200-1210, 1985.

ACKNOWLEDGMENT

This work was supported by a grant from the E.E.C. R&D programme "Science and Technology for Development" contract N° TSD-A-124 and by NATO Research Grant N° 600/83. MH is a research associate of the Belgian National Fund for Scientific Research. K.G. is a recipient of an IWONL fellowship.

NOD-LINKED HOST SPECIFIC GENE FOR SOYBEAN (PEKING) NODULATION IN RHIZOBIUM FREDII USDA193

NEELA RAMAKRISHNAN AND ALAN G. ATHERLY
Department of Genetics, Ames, Iowa 50011, U.S.A.

1. INTRODUCTION

Rhizobium fredii strain USDA193 forms nitrogen fixing nodules on the genetically unimproved Chinese cultivar 'Peking' and Fix⁻ nodules on commercial North American cultivars. The symbiotic genes have been located on a large plasmid (pSym) of about 300 kb in size, present in this strain (1,2). Hybridization and functional complementation studies have revealed that the 'common' nod genes, nodABC and nodD are present on two unlinked DNA fragments, 5.2 kb HindIII and 2.8 kb EcoRI, respectively, of the Sym plasmid (3). We have previously reported that the 2.8 kb EcoRI fragment of R. fredii could extend the host range of heterologous strains such as R. leguminosarum, R. phaseoli, R. trifolii, R. meliloti and also Ti plasmid cured strain of A. tumefaciens, for soybean nodulation (3).

2. RESULTS

Tn5 mutagenesis of the 2.8 kb EcoRI fragment was carried out. Fig. 1. shows the regions of Tn5 insertion in the 2.8 kb EcoRI fragment. Mutational hotspots were very rare. All Tn5 insertions were clustered in two regions.

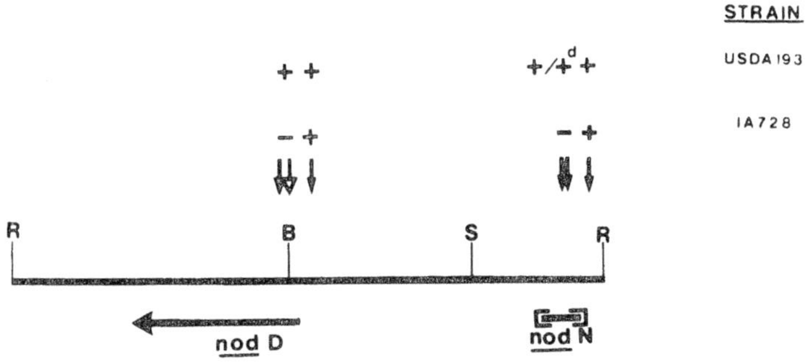

Figure 1. Restriction map of the 2.8 kb EcoRI fragment of pRjaUSDA193 showing Tn5 (closed arrows) and 'omega' (open arrow) insertions. The predicted locations of nodD and nodN genes are also shown. Abbreviations: +, Nod$^+$; +^{d}, Nod-delayed; −, Nod$^-$; R, B, S, Cleavage sites for EcoRI, BamHI and SalI restriction enzymes, respectively.

A Tn5 insertion in the right SalI-EcoRI fragment gave rise to a mutant strain of USDA193 showing delayed nodulation phenotype (see Fig 2). This mutation could be functionally complemented by cosmid clones carrying wild type 2.8 kb EcoRI fragment or an in vitro constructed nodD mutation in the 2.8 kb EcoRI fragment (using 'omega' mutagen) (4). Hybridization studies using nodDABC sequences of R. meliloti as probe showed no DNA homology with the right SalI-EcoRI region of the 2.8 kb EcoRI fragment indicating that this region may code fr a new functional gene(s).

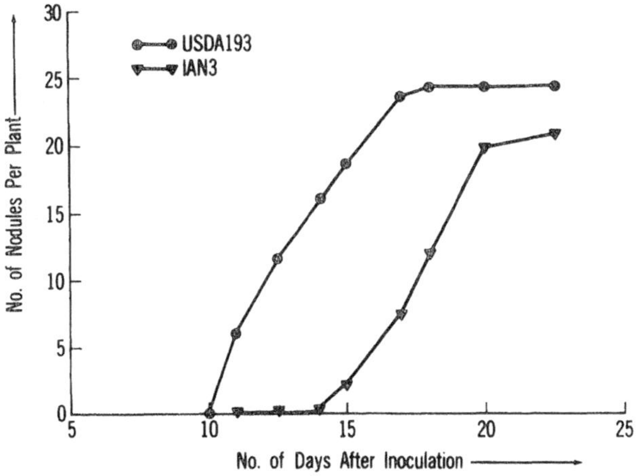

FIGURE 2. A graph comparing the number of nodules formed per plant vs. the number of days after inoculation with the strains USDA193 and IAN3 (mutant strain showing delayed nodulation).

All the Tn5 inserted fragments were recloned into the broad host range cosmid vector pVK101 (5) and conjugally transferred into a pSym deleted derivative of USDA193, IA728. The transconjugants were tested on 'Peking' to study the phenotypic effect of Tn5 insertion on soybean nodulation. The results (included in Fig 1) indicate that both nodD and nodN genes are necessary for soybean nodulation by the 2.8 kb EcoRI fragment.

Hybridization studies using nodN gene sequences revealed that these sequences were conserved in different R. fredii strains but not present in heterologous strains such as R. leguminosarum, R. phaseoli, R. trifolii, R. meliloti or A. tumefaciens.

3. DISCUSSION

As previously reported, the 2.8 kb EcoRI fragment alone could confer soybean nodulation to heterologous rhizobia (delayed nodules with few bacteria present in the peripheral tissue), suggesting that the host specificity gene(s) for soybean (Peking) nodulation may be contained within this fragment. Of the common nod genes, only nodD is present on this fragment. nodD gene sequences are conserved in different rhizobia. Also, recent studies indicate that nodD gene product may act as a regulator of nod gene expression. Hence, it is very unlikely that nodD gene could code for host specificity function. The data presented in this study reveal a second functional region, nodN gene, which could code for the putative host specificity gene for soybean nodulation. Analysis of DNA sequences and protein product(s) would provide more information regarding the number of genes involved, the direction of transcription and their possible roles in the nodulation process.

REFERENCES

1. Masterson RV, Prakash, RK, Atherly AG: Conservation of symbiotic nitrogen fixation gene sequences in Rhizobium japonicum and Bradyrhizobium japonicum. J Bacteriol 163: 21-26, 1985.
2. Prakash RK, Atherly AG: Reiteration of genes involved in symbiotic nitrogen fixation by fast-growing Rhizobium japonicum. J Bacteriol 160:785-787, 1984.
3. Ramakrishnan N, Prakash RK, Shantharam S, DuTeau NM, Atherly AG: Molecular cloning and expression of Rhizobium fredii USDA193 nodualtion genes: Extension of host range for nodulation. J Bacteriol, 1986. In press.
4. Prentki P, Krisch HM: In vitro insertional mutagenesis with a selectable DNA fragment. Gene 29:303-313, 1984.
5. Knauf VC, Nester EW: Wide host range cloning vectors: a cosmid bank of an Agrobacterium Ti plasmid. Plasmid 8:45-54, 1982.

GENOMIC ORGANIZATION OF NODULATION GENES IN *Rhizobium phaseoli*.

Quinto Carmen, Martínez Javier, Cevallos M.A., Dávalos A., and Peralta Y.

Department of Plant Molecular Biology
Centro de Investigación sobre Fijación de Nitrógeno. UNAM
Cuernavaca, Morelos. MEXICO. Apartado Postal 565-A.

1. INTRODUCTION

The symbiotic association between *Rhizobium phaseoli* and *Phaseolus vulgaris* is of particular interest to us, since in Mexico and most of the latinoamerican countries, bean seed protein together with corn are the major protein sources for the native population.
A rational approach to improve symbiotical nitrogen fixation implies the study of the molecular genetics of both symbionts. Our particular interest is to study the molecular events that lead to the symbiotical state of *Rhizobium phaseoli*.
We have previously found that *R.phaseoli* presents a peculiar organization of nitrogen fixation gene sequences characterized by the presence of DNA reiterations (Quinto et al. 1982; Quinto et al. 1985 a). To find out how general the presence of *nif* gene reiterations is, a screening of *P.vulgaris* strains from different geographical origins was performed. Only one of the 40 strains, CIAT899 did not show reiteration of *nif*H genes (Martínez et al, 1985). The important difference between the "nif reiterated" and "non reiterated" strains isolated from bean nodules is the host range of infection, as strain CIAT899 is able to effectively nodulate *Leucaena esculenta* while "*nif* reiterated" strains were specific for beans (Martínez et al, 1985).
Recently in our group, Dr. Palacios and his co-workers have found that the analysis of genomes of different *Rhizobium phaseoli* strains revealed a large number of repeated DNA sequences (submitted for publication; paper in these proceedings). Therefore we have though that the positive complementation of a *R.phaseoli* pSym-cured strains is the best approach to study the symbiosis, in particular the nodulation in the association *R.phaseoli*-*P.vulgaris*. We have selected two strains of *R.phaseoli*: strain CFN42 and strain CIAT899, a "*nif* reiterated" and a "non-reiterated" strains with a narrow and a broad host range of infection, respectively.
In the present paper we present evidence indicating that sequences homologous to the "common nod genes" are repeated at least two times in strain CFN-42, but not in strain CIAT899.

2. METHODS

Total and plasmid DNA were obtained as described before (Quinto et al, 1985). Conditions for Southern transfers and hybridizations have also been described (Quinto et al, 1985).

Plasmids profiles have been performed according to the Eckhardt procedure.

3. RESULTS

3.1. SEQUENCES HOMOLOGOUS TO ALL THE COMMON NOD GENES ARE REPEATED IN R.phaseoli STRAIN CFN-42.

We have constructed a genomic library of strain CE-3 (a CFN-42 streptomycin resistant derivative) in the pSUP205 cosmid vehicle (Quinto et al, 1985 b). Cosmid containing the nif genes were selected by hibridization with a R.phaseoli nifH probe. Representative clones of the different reiterated nifH genes: a (pSM991), b (pSM828) and c (pSM367) were selected and hybridized with the R.meliloti common nod genes (Kondorosi et al, 1984). Internal probes derived from pKSK5 were obtained by several restriction digests to render nod AB, nod C and nod D (Kondorosi et al, 1984). EcoRI digests of pSM991 (nifH a), pSM828 (nifH b) and pSM367 (nifH c) were hybridized with nod AB, nod C and nod D.
All three nick-translated probes hybridized with each cosmid clone. Probes nod AB and C hybridized with the same 7.8 kb EcoRI band in pSM991, 828 and 367. Probe nod D hybridized with a 4.5 kb EcoRI fragment in cosmids pSMA991 and pSM828; in pSMA367 the homologous nod D EcoRI fragment is a band of 4.8 kb. This is in accordance to the result obtained hybridizing these same specific probes with total DNA from strain CE-3 digested with EcoRI and blotted on to nitrocellulose. This result indicates that the "common nod" sequences are reiterated in R.phaseoli strain CE-3. On the other hand, similar experiments were performed with strain CIAT899, which as mentioned above, is a non reiterated R.phaseoli strain for nif genes. We also did not find reiteration of "common nod genes"

3.2. TWO COSMID CLONES CONTAINING THE nif - nod REGION, SHARE THE SAME nod FRAGMENT, BUT HAVE DIFFERENT nif REGIONS.

Restriction analysis of the cosmids pSM991, pSM367 and pSM828 with HindIII and BglII restriction enzymes followed by hybridizations with a R.phaseoli subclone carrying the homologous nod A, B and C have lead us to the conclusion that regions pSM991 and pSM367 are sharing all same nod region although they have different nif fragments.
We have also performed homologous hybridizations among these clones digested with EcoRI and hybridized with each of them at one time: pSM991, pSM367 and pSM828. Results indicate that there are several bands in common among pSM991 (nif a) and pSM367 (nif c), even though between pSM991 (nif a) and pSM828 (nif b) the only bands in common are the "nod" and the "nif" EcoRI fragments.

3.3. POSITIVE COMPLEMENTATION OF A SYM-CURED STRAIN.

We have reported, previously that cosmid pSM991 was able to induce ineffective nodules in a CE-3 Sym-cured strain (Quinto et al, 1985). Efforts are now in progress to construct vehicles with compatible replication origins thus allowing us to continue positive complementation of the symbiotic process.

4. CONCLUSIONS

We conclude that sequences homologous to the "common nod genes" from R.meliloti are reiterated at least two times in R.phaseoli strain CE-3, being at least one of them functional in an R.phaseoli pSym-cured strain although this is not the case for strain CIAT899, which has no reiterations of nif structural genes. Results allow us to conclude that two of the three nif reiterations are sharing their common nod genes despite the fact that they have different nif regions

5. REFERENCES

Eckhardt, T. Plasmid (1978) 1, 584-588.

Kondorosi E. Banfalvi Z, Kondorosi A (1984) Mol. Gen. Genet. 193, 443-452.

Martínez E, Pardo M.A., Palacios R. and Cevallos M.A. (1985) J. Gen. Microbiol 131, 1179-1786.

Quinto C., De la Vega H., Flores M., Fernández L., Ballado T., Soberón G., and Palacios R. (1982). Nature 299, 724-726.

Quinto C., De la Vega H., Flores M., Leemans J., Cevallos M.A., Pardo M.A., Azpiroz R., Girard M.L., Calva E., and Palacios R., (1985 a) Proc. Natl. Acad. Sci. 82, 1170-1174

Quinto C., Cevallos, M.A., Peralta Y., Espín G., and Dávalos A. (1985 b). Proceedings of the 6th International Symposium on Nitrogen Fixation. Evans, H.J., Bottomley, P.J. and Newton, W.E. (eds): Nitrogen Fixation Research Progress. 123.

COMMON AND HOST SPECIFIC NODULATION GENES IN RHIZOBIUM MELILOTI AND THEIR CONSERVATION IN OTHER RHIZOBIA

A. KONDOROSI[1] , E. KONDOROSI[2] , B. HORVATH[1] , M. GOTTFERT[1] , C. BACHEM3 , F. RODRIGUEZ-QUINONES[1] , Z. BANFALVI[1] , P. PUTNOKY[1] , Z. GYORGYPAL[1] , M. JOHN[3] , J. SCHMIDT[3] , J. SCHELL[3]

Institutes of Genetics[1] and Biochemistry[2], Biological Research Center, Hung. Acad. Sci., P.O.B. 521 H-6701 Szeged, Hungary
Max-Planck-Institut fur Zuchtungsforschung[3], D-5000 Koln 30, Federal Republic of Germany

1. INTRODUCTION

Recognition of the appropiate legume host and nodule induction are controlled by two sets of Rhizobium genes, common nodulation (nod) and host-specific nodulation (hsn) genes. These genes have been identified in several Rhizobium species, including R.meliloti, the symbiotic partner of alfalfa (Medicago). Here we present our studies on the organization and regulation of R.meliloti nodulation genes. Moreover, these genes were used to identify and analyse genes of similar function in other rhizobia which may help us to elucidate the basis of host-specificity of Rhizobium-legume interaction and the genetic control of these processes.

2. ORGANIZATION AND REGULATION OF NOD AND HSN GENES IN R.MELILOTI

The common and host specific nodulation genes of R.meliloti are carried by a symbiotic megaplasmid and located in the vicinity of the nitrogen fixation (nif) genes. Using various molecular and genetic techniques we have identified nine genes of two classes in R.meliloti strain 41 (Fig. 1): the common nodulation genes nodA,B,C,D1 and D2 (1, 2, 3) and the host-specific nodulation genes hsnA,B,C and D (4, 5), (formerly designated as nodF,E,G and H, respectively; 5). Further characterization of these genes provided information concerning their location in the bacterial genom and their possible function in the nodulation process (6). The common nodABC genes code for general nodulation functions: root hair curling and nodule induction, while the hsnD gene probably determines the plant-host specificity of these processes. The hsnABC genes are required for efficient host-specific infection via infection threads within the root hairs (4).

The expression of nodulation genes are controlled by plant factor(s) and the product of the regulatory gene nodD (7, 8, 9). Our studies suggest that the coordinate regulation of nod and hsn genes is exerted on the 47 bp long highly conserved promoter sequences (we designated as nod-box) identified in the promoter regions of the nodABC and hsnABC and D transcriptional

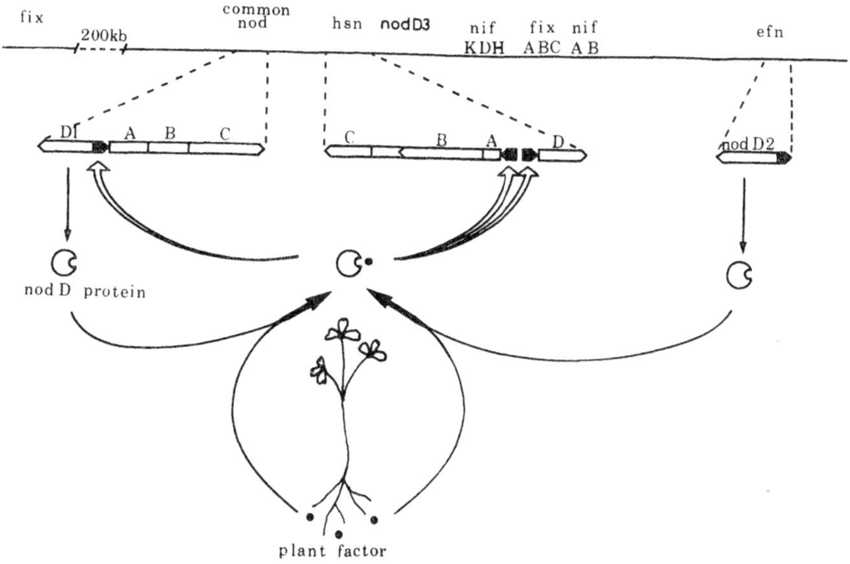

FIGURE 1. Organization and regulation of nod and hsn genes in R.meliloti. The black boxes with arrows indicate the location and direction of nod-boxes.

units (Fig. 1) (5). The nod-box is highly conserved in the promoter regions of nodulation genes of other Rhizobium species and our results showed that a 25 bp long synthetic oligonucleotide containing the internal nod-box sequence can be used to identify genes which are coordinately expressed during nodulation.

The nod-box sequences represent essential parts of the (nod and hsn) promoters and show homology with some other protein-binding sequences (10). Based on these data, a model for the regulation of nodulation genes is presented in Fig. 1. The coordinate activation of the nod and hsn genes may be mediated by the binding of the nodD product (perhaps modified by plant factor) to the nod-box which may result in the binding of RNA polymerase to the nod or hsn promoters and thereby starting transcription (11).

3. OPTIMAL NODULATION OF ALFALFA REQUIRES AT LEAST TWO COPIES OF NODD IN R.MELILOTI

R.meliloti contains more than one copy of the regulatory gene nodD. Using an internal fragment of the nodD1 gene, two additional copies were detected in R.meliloti 41 (nodD2 and nodD3, Fig. 1). Mutations in either nodD1 or nodD2 resulted in a slight delay in nodulation of Medicago while nodD1-nodD2 double mutants were more severely delayed in the nodulation process (2). It is not unlikely that the residual nodulation activity of the double mutant is due to the action of nodD3.

The amino acid sequences of both nodD1 and nodD2 products, deduced from the nucleotide sequences exhibited more than 80% homology. The double mutant could be complemented either by nodD1 or nodD2 cloned into pRK290, indicating their functional equivalence.

4. INTERSPECIES HOMOLOGY OF R.MELILOTI NOD AND HSN GENES

Using internal sections from the structural nodC, nodD, hsnB and hsnD genes, respectively, as hybridization probes against genomic DNA from different Rhizobium strains and species, the interspecies homology of these genes was investigated. As summarized in Table 1 hybridization to both the nodC and nodD probes were found in all species with one exception (R.sp. OR571), indicating that they are widely spread genes, though the nodD gene hybridized only weakly with total DNA from Bradyrhizobium species. Reiteration of the nodD sequences was observed in almost all Rhizobium species, with the exception of R.leguminosarum (Table 2). Interestingly, all the tested Rhizobium strains carrying multiple copies of nodD, contained more than three copies of the nod-box (5), suggesting that the larger number of nod-boxes titrates out the nodD products, unless the nodD gene dosage is increased (see also Section 3).

TABLE 1. Homology of R.meliloti nod and hsn probes to different Rhizobium species, Agrobacterium tumefaciens and Escherichia coli

Species	nodC	nodD	hsnB	hsnD
R.meliloti	++++	++++	++++	++++
R.fredii	+++	++	++	++
R.sp. MPIK3030	++++	+++	+++	+++
R.leguminosarum	+++	++	++	+/-
R.trifolii	+++	++	++/+	- -
R.phaseoli	+++	+	-	-
R.lupini	+	+/-	-	-
B.japonicum	++	+/-	-	-
R.sp. ORS571	- -	-	-	-
A.tumefaciens	- -	- -	- -	- -
E.coli	- -	- -	- -	- -

For R.meliloti, fredii, leguminosarum, trifolii, phaseoli and B.japonicum several (3-10) strains were tested. For hybridization probes internal fragments of the protein coding regions were used. +: intensity of hybridization.

TABLE 2. Reiteration of *nod*D in different *Rhizobium* species

Species	Number of *nod*D copies
R.meliloti	3
R.sp. MPIK3030	2
R.fredii	2
R.leguminosarum	1
R.trifolii	2
R.phaseoli	2

Strains were used as in Table 1.

The *hsn*B gene is present only in some fast-growing *Rhizobium* species and *hsn*D hybridized only to a few strains other than *R.meliloti*, indicating their specific involvement in *R.meliloti*-*Medicago* symbiosis. Our data suggest that *hsn*D hybridization is highly specific for *Rhizobium* strains, that are able to nodulate *Medicago*.

5. THE *NOD* AND *HSN* GENES ARE ORGANIZED INTO SEPARATE CLUSTERS ALSO IN THE *RHIZOBIUM* SPECIES MPIK3030

Previously, we have cloned the common *nod* region from the broad host range *Rhizobium* species MPIK3030 by using the *nod*C gene of *R.meliloti* as hybridization probe (12). More recently, we were able to clone the DNA region determining nodulation specificity for one of the natural plant hosts, *Macroptilium atropurpureum* (siratro) (13). In these experiments a pLAFR1 library of MPIK3030 was mass-conjugated into *R.meliloti* and the transconjugant population was used to inoculate *M.atropurpureum*. Bacteria reisolated from the nodules contained a recombinant plasmid with the *hsn* genes for *Macroptilium*. Interestingly, these plasmids did not confer host specificity for other natural hosts of MPIK3030, suggesting that other *hsn* genes may code for the ability to nodulate these latter hosts. The common *nod* and the *M.atropurpureum*-specific *hsn* genes are closely linked on the large Sym-plasmid (Fig. 2).

6. FUNCTIONING OF DIFFERENT *NODD* GENES MAY REQUIRE DIFFERENT PLANT FACTORS

As shown in Fig. 2, the *hsn* genes, specific for *M.atropurpureum*, are organized into 3 regions in MPIK3030. Tn5 insertions in one region cause Nod$^-$ phenotype (13). This region was shown to hybridize with the *R.meliloti* *nod*D1 probe and from nucleotide sequencing about 75% homology between the *R.meliloti* *nod*D1 and the MPIK3030 *nod*D gene products were found. The *nod*D gene from MPIK3030 was able to restore the Nod$^+$ phenotype of the *R.meliloti* *nod*D double mutant. The *R.meliloti* *nod*D1 or *nod*D2 genes, however, were unable to restore the nodulation ability of the *nod*D mutants of MPIK3030 when assayed

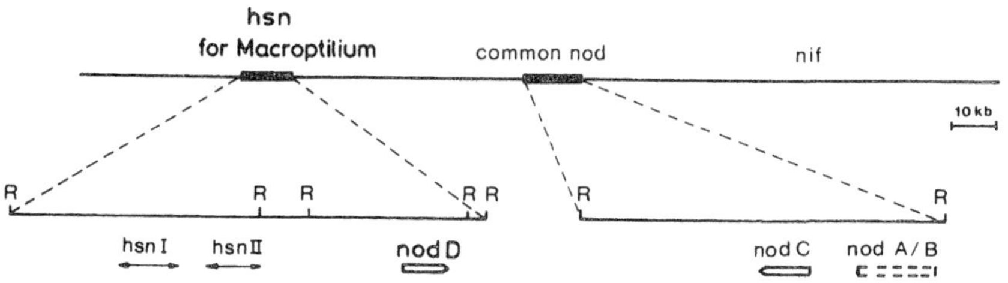

FIGURE 2. Organization of nodulation genes identified in Rhizobium sp. MPIK3030

on Macroptilium. We have shown (3) that the R.meliloti nod genes can express in MPIK3030 when tested for nodulation of Medicago, therefore it is possible that the R.meliloti nodD or its product cannot act in conjunction with the factors excreted by Macroptilium while the nodD gene from MPIK3030 is functional in the presence of plant factors received from Medicago. We suggest that different plant factors may exert their positive regulatory role with different nodD genes when turning on the nod and hsn transcriptional units. This is suggested by our recent finding that the R.meliloti strain L5-30 which is also able to nodulate Macroptilium atropurpureum, unlike most R.meliloti strains, including R.meliloti 41, carries a nodD copy which can suppress the nodD mutations of MPIK3030. Transfer of this nodD region to other R.meliloti results in the host range extension to Macroptilium atropurpureum.

In conclusion: since different legume plants may have different sets of inducing factors, the presence of multiple copies of nodD may help optimal nodulation of different natural hosts.

REFERENCES

1. Torok I, Kondorosi E, Stepkowski, T, Posfai J, Kondorosi A: Nucl. Acids Res., 12, 9509-9524, 1984.
2. Gottfert M, Horvath B, Kondorosi E, Rodriguez-Quinones F, Kondorosi A: J. Mol. Biol., (in press), 1986.
3. Putnoky P, Kondorosi A: J. Bacteriol., (in press), 1986.
4. Horvath B, Kondorosi E, John M, Schmidt J, Torok I, Gyorgypal Z, Barabas I, Wieneke U, Schell J, Kondorosi A: Cell, (in press), 1986.
5. Rostas K, Kondorosi E, Horvath B, Simoncsits A, Kondorosi A: Proc. Natl. Acad. Sci. USA, 83, 1757-1761, 1986.
6. John M, Schmidt J, Wieneke U, Kondorosi E, Kondorosi A, Schell J: EMBO J. 4, 2425-2430, 1985.

7. Mulligan JT, Long SR: Proc. Natl. Acad. Sci. USA, 82, 6609-6613, 1985.
8. Innes RW, Kuempel PL, Plazinski J, Canter-Cremers H, Rolfe BG, Djordjevic MA: Mol. Gen. Genet. 201, 426-432, 1985.
9. Rossen L, Shearman CA, Johnston AWB, Downie JA: EMBO J. 4, 3369-3373, 1985.
10. Ames Ferro-Luzzi G, Nikaido K: EMBO J. 4, 539-547, 1985.
11. Kondorosi E, Kondorosi A: Trends in Biochem. Sci. 11, 296-299, 1986.
12. Bachem C, Kondorosi E, Banfalvi Z, Horvath B, KondorosiA, Schell J: Mol. Gen. Genet. 199, 271-278, 1985.
13. Bachem C, Banfalvi Z, Kondorosi E, Schell J, Kondorosi A: Mol. Gen. Genet. 203, 42-48, 1986.

HOST SPECIFIC NODULATION: EFFECTS OF MULTIPLE nodD GENES OF Rhizobium meliloti

Mary A. Honma and Frederick M. Ausubel
Department of Molecular Biology
Massachusetts General Hospital
Boston, MA 02114, USA

I. INTRODUCTION

In both Rhizobium and Bradyrhizobium species, the common nodulation genes nodA,B,C,D are required for early events of nodule formation. In R.leguminosarum, R. trifolii, and R. meliloti, the expression of the nodABC genes requires root exudate and nodD. (Mulligan, et. al. PNAS 82, 6609-6613. 1985; Rossen, et. al. EMBO J. 4, 3369-3373. 1985; Innes, et. al. MGG 201, 426-432. 1985). Recently, it has been shown that the host specificity genes of R. leguminosarum (nodEF) are expressed if nodD and root exudates are present (Shearman, et. al. EMBO 5, 647. 1986). Mutations in the nodA,B,C genes totally block nodulation in all Rhizobium species tested so far. In R. leguminosarum and R. trifolii, a mutation in nodD also results in a completely Nod- phenotype. Surprisingly, an R. meliloti nodD mutant still forms nodules on alfalfa (Jacobs, et. al. J. Bact. 162, 469. 1985).

The R. meliloti nodD gene was used as a hybridization probe against a Rm1021 genomic blot of EcoRI digested DNA. Two bands hybridized in addition to the fragment containing the original nodD gene. These nodD homologous fragments were 6.8 and 15.5 kb in size and will be referred to as nodD2 (6.8 kb) and nodD3 (15.5 kb). Since a mutation in the nodD gene does not block nodulation, it seemed likely that these homologous regions contained additional functional copies of the nodD gene. Both nodD homologous regions (nodD2, nodD3) map within an 80 kb segment of the nod-nif region.

II. RESULTS

We have sequenced nodD2 and compared it to the published sequence of nodD (Egelhoff, et. al. DNA 4, 241. 1985). Preliminary results show that nodD2 codes for a polypeptide of 310 amino acids in length, while nodD has a predicted size of 308 amino acids. There is ~87% conservation in both nucleotide and predicted amino acid sequence. More differences between nodD and nodD2 are found in the C-terminal half. In addition, nodD2 has a 10 bp deletion in the region upstream from the coding sequences.

A deletion mutant of the nodD homologous region in the 6.7 kb EcoRI fragment (nodD2) was constructed. nodD mutant TJ9B8 was obtained from Sharon Long, and a double mutant (RmD1D2) in nodD and nodD2 was constructed. The mutant and wild type strains were inoculated onto two R. meliloti hosts, Medicago sativa and Melilotus alba, 50 plants for each strain. Each experiment was repeated four times.

Nodules first appeared after 5 days. Time of appearance and number of nodules were scored every day from day 5 and every 2 days after day 16. Single mutants in nodD or nodD2 did not show any clearcut delay in nodulation tests on Medicago. However,the double mutant RmD1D2 formed nodules with about a 5 to 6 day delay compared to wild type. On Melilotus, the RmD2 mutant showed no delay in nodulation. In contrast, Melilotus nodules elicited by mutant TJ9B8 (nodD) appeared about 3 days after wild type nodules. The double mutant RmD1D2 nodulated Melilotus with the same 3 day delay as the single nodD mutant.

These results suggest that both nodD and nodD2 have some role in nodulation of Medicago; however, nodD appears to be more important than nodD2 in Melilotus nodule formation.

III. DISCUSSION

In contrast to other fast-growing Rhizobium species, R. meliloti has 3 copies of a regulatory gene nodD. Other labs have shown that expression of a number of nodulation genes is dependent on nodD and root exudates. One hypothesis is that NodD interacts either directly or indirectly with plant factors to activate expression of the nodulation genes.

We have observed different phenotypes of nodD, nodD2, and nodD1nodD2 mutants on different host plants. Some possible explanations are:

1. Different levels of expression of nodD genes. (combined with different thresholds for "nodulation factors" by different hosts)

2. Ability to interact (directly or indirectly) with the same plant factor, but with different affinity.

3. Interaction of NodDs with different plant factors.

Are all three nodD genes functional? Double mutants in nodD and nodD2 still nodulate, although with some delay. This suggests that nodD3 is functional, since it is the only nodD copy remaining. We are constructing a triple mutant in all three nodD genes.

Of particular interest is determining if there are any functional differences between these three nodD genes. Since we observed different phenotypes on Medicago versus Melilotus, we plant to test the mutants on a third R. meliloti host plant Trigonella.

NODULATION GENES OF RHIZOBIUM LEGUMINOSARUM

J.A. Downie[1], B.P. Surin[1], I.J. Evans[2], L. Rossen [2], J.L. Firmin[2], C.A. Shearman[2] and A.W.B. Johnston[2]

[1]CSIRO Division of Plant Industry, Canberra, ACT 2601, Australia
[2]John Innes Institute, Colney Lane, Norwich NR4 7UH, UK

The host specific nodulation of peas, *Vicia*, *Lens* and *Lathyrus* is encoded by a series of genes present upon the "symbiotic" plasmids of *Rhizobium leguminosarum*. These genes are clustered within a relatively short region of DNA which was cloned upon the overlapping region of two cosmid clones called pIJ1085 and pIJ1089 (Downie *et al*. 1983). The phenotypes of strains carrying mutations within this region fall into two broad classes: (a) those which inhibit root-hair-curling and totally block nodulation - these mutations are within the *nodABC* or *D* genes; and (b) mutations within a series of other genes which reduce nodulation efficiency (Downie *et al*. 1985). This nodulation inhibition can be observed as a reduction of nodule numbers and as a delay in the time of appearance of nodules, but the severity of the effects of the mutations varies markedly depending upon the gene affected, e.g. a mutation within the *nodE* gene strongly reduced nodulation of peas whereas mutations in genes (*nodI* and *nodJ*) downstream of *nodABC* had only a small effect on the nodulation of peas.

In parallel with these two classes of mutant phenotypes, the early steps of the legume - *Rhizobium* interaction can be considered to fall into two braod stages: an initial step in which plant-bacterial contact is not essential and other steps in which intimate contact between plant and bacteria must occur. Thus, it has been shown that plant-bacterial contact is not necessary for the induction of *Rhizobium nod* genes since aseptically grown legume plants produce a series of low molecular weight molecules which induce the nodulation genes of *Rhizobium* (Mulligan & Long 1985; Innes *et al*. 1985; Rossen *et al*. 1985; Shearman *et al*. 1986).

By fractionating pea exudate by HPLC it was found that it contained several flavonoids that could activate transcription of the *nodABCIJ* and *nodFE* operons of *R. leguminosarum* (Firmin *et al*. 1986). Several commercially available flavonoids were found to induce *nodABCIJ* and *nodFE* transcription at very low concentrations, the most potent of those tested being hesperitin (3', 5, 7-trihydroxy-4'methoxyflavanone) which induced the *nod* genes at concentrations as low as 10nM. By comparing the structures of those flavonoids that acted as inducers with those that did not, the following deductions were made concerning the requirements for the molecule to be active. Both flavones and flavanones were active if the molecule had an OH group at the 3' or 4' position of the B ring and if the 7-position contained a hydroxyl group or a glycoside. Flavonoids substituted at the 3-position of the C ring (e.g. flavonols or isoflavonoids) did not act as inducers at low concentrations.

Based on chemical, chromatographic and spectroscopic data it was apparent that one of the major authentic inducers in pea exudate is the flavone glucoside, apigenin 7-O-glucoside. Predominant inducing molecules from alfalfa and clover were identified as the flavone luteolin (Peters *et al.* 1986) and 7,4'dihydroxyflavone (Redmond *et al.* 1986) respectively. It was further shown that several flavonoids such as flavonols and isoflavonoids antagonised this induction of *nod* gene expression (Firmin *et al.* 1986). The most potent antagonist identified was the isoflavonoid genestein which at 5uM inhibited *nodABCIJ* expression by 87%. At higher concentrations (100uM), acetophenone compounds (some of which induce the *vir* genes of *Agrobacterium tumefaciens* (Stachell *et al.* 1985)) also counteracted *nod* gene activation. Thus it is possible that the 'balance' of these *nod* gene activators and antagonists may be imputant in the nodulation process.

nod Genes identified by DNA sequences

In an attempt to define the stages of the host-specific interactions which presumably occur as a result of some direct cell to cell contact, we have sequenced a series of genes in which mutations only partially inhibit nodulation. Using protein homology searches, sequence comparisons and analysis of mutant strains we attempted to ascribe potential functional roles to those *nod* gene products.

Previously it was shown that either of two clones pIJ1085 and pIJ1089 could transfer nodulation of peas to a strain of *R. phaseoli* cured of its symbiotic plasmid. Since pIJ1085 and pIJ1089 overlap by 10kb of DNA we set out to sequence this region, but bearing in mind the observation that some *nod* genes may be absent from pIJ1085 since it induced less efficient nodulation than pIJ1089 (Downie *et al.* 1983).

Initially a 10kb region of DNA was sequenced and eight *nod* genes identified in this region were called *nodABCDEFIJ* (Fig. 1). Within this region two almost identical short sequences were identified preceeding the *nodA* and *nodF* transcriptional units ("repeat sequences", Fig. 1). Since these sequences preceded two inducible *nod* operons (Rossen *et al.* 1985; Shearman *et al.* 1986), it was concluded that this sequence was involved in the expression of inducible *nod* genes (see also Rostas *et al.* 1986). When a short DNA fragment containing one of these regions (preceeding *nodF*) was used as a hybridisation probe to pIJ1089 DNA digested with *EcoR1*, it was observed that hybridisation occurred to a 2.2kb fragment which had been mapped as being adjacent to the *EcoR1* fragment containing the *nodABCDEF* genes. This region (absent from pIJ1085) was also sequenced and two open reading frames were identified which have been called *nodL* and *nodM* (see Figure 1). Between these two open reading frames and preceeding *nodM* a sequence was identified which was very similar to that preceeding the *nodA* and *nodF* genes indicating that *nodM* probably forms (at least part of) an additional inducible *nod* operon.

nodI and nodJ genes

As shown in Figure 1, the *nodI* and *nodJ* genes are immediately downstream of the *nodABC* genes and there is only a short distance between the end of *nodC* and the predicted translational start site of *nodI*. These two genes appear to be in the same operon as the *nodABC* genes since, using plasmid constructions containing the *nodJ* gene fused in frame to the *E. coli*, -galactosidase activity was observed only if (a) the construct contained the region upstream of *nodA* and (b) the *Rhizobium* strain was preincubated with a *nod* gene inducer or legume root exudate.

When the predicted protein sequence of the *nodI* gene was compared with a protein sequence database (Protein Identification Source, NBRF, Georgetown Univ. Med. Center), using a protein homology search programme, the predicted *nodI* amino acid sequence was found to be similar to that of a series of bacterial gene products, *hisP*, *malK*, *oppD* and *pstB* which are involved in the active transport of histidine, maltose, oligopeptides and

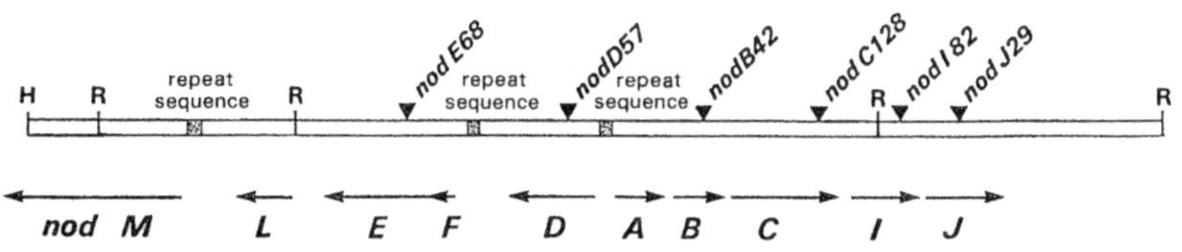

FIGURE 1.

phosphate respectively. These proteins are membrane-associated ATP-binding proteins, all involved in active transport. The homology of the nodI gene product was at least as good as the homologies observed among these other proteins.

The gene product of nodJ, downstream of nodI, was found to be strongly hydrophobic, typical of an integral membrane protein. Therefore it would appear likely that these two genes are involved in some way in the ATP-dependent transport of some molecule(s) which may be involved in nodulation.

Host specific nod genes

In order to determine gene(s) involved in host-specific nodulation, plasmids pIJ1085 and pIJ1089 and derivatives of them containing mutations within nod genes, were transferred to a wild type strain of R. trifolii and the transconjugants tested for nodulation on peas and V. hirsuta. Both pIJ1085 and pIJ1089 conferred upon R. trifolii strain ANU843 the ability to nodulate peas. This observation implies that the nodL and nodM genes may not be host-specific nod genes, at least when judged within the context of this comparison. When a radioactive probe of the nodLM region was hybridised to R. trifolii DNA, strong hybridisation was observed indicating that equivalent genes to nodL and nodM exist in R. trifolii.

Derivatives of pIJ1089 containing Tn5 in the nodC, nodD or nodE genes were all found to confer pea nodulation ability to R. trifolii indicating that all of these genes appear to have functional homologs in R. trifolii. Paradoxically, however, pIJ1085 carrying Tn5 within the nodE gene was unable to confer pea nodulation upon R. trifolii, indicating that (in contrast to the observations made with the same allele in pIJ1089) the nodE genes was essential for pea nodulation. Therefore it appears that the effects of mutations within nodE may be overcome at least partially by genes present on pIJ1089 but absent from pIJ1085.

The amino acid sequences of the nodFELM genes were compared to the protein sequence database. Homology was found between the nodF gene product and acyl-carrier-protein which is involved in fatty acid biosynthesis. This homology was strongest around the active centre of the authentic acyl carrier proteins where the pantetheine cofactor is known to bind. Since Rhizobium strains lacking nodF grow normally, this gene is clearly not involved in normal fatty acid biosynthesis, but it could have a role in some aspect of fatty acid metabolism, e.g. in synthesis of lipopolysaccharide.

The open reading frame corresponding to the nodM gene was found to be strongly homologous to two analogous enzymes from E. coli and B. subtilis, particularly around the amino-terminal regions. These proteins are amido-phosphoribosyltransferases which catalyse an early step in purine metabolism, and catalyse the transfer of an amino group from glutamine to the ribose sugar, phosphoribosyl pyrophosphate and it is known that the amino-terminal region forms part of the catalytic site (Tso et al. 1982; Makaroff et al. 1983). Since deletion of the nodM gene does not affect growth of the R. leguminosarum it clearly plays some alternative role. One possibility could be that it is involved in the synthesis of an amino ribose sugar which could form a component of the cell wall of R. leguminosarum and which could be involved in recognition by the host plant. R. leguminosarum exopolysaccharide does not contain amino-sugar residues (Dudman 1984). Therefore it is reasonable at this stage to propose that the nodM gene may be involved in the formation of a novel lipopolysaccharide, possibly even at the level of forming a new lipidA moiety by functioning in conjunction with the nodFE genes. Such a hypothesis is attractive since the lipopolysaccharides of Gram-negative bacteria are highly antigenic, a property which could be important in recognition, possibly by plant lectin molecules.

References

Downie, J.A., Hombrecher, G., Ma, Q-S., Knight, C.D., Wells, B., and Johnston, A.W.B. (1983). Mol. Gen. Genet. 190, 359-365.

Downie, J.A., Knight, C.D. , Johnston, A.W.B. and Rossen, L. (1985). Mol. Gen. Genet. 198, 255-262.

Dudman, W.F. (1984). Adv. Nitrogen Fixation Research (Ed. C. Veeger, W.E. Newton). pp.397-404, Nijhoff/Junk.

Fisher, R.F., Tu, J.K. and Long, S.R. (1985). Appl. Env. Microbiol. 49, 1432-1435.

Firmin, J.L., Wilson, K.E., Rossen, L. and Johnston, A.W.B. (1986). Nature (in press).

Innes, R.W., Kuempel, P.L., Plazinski, J., Canter-Cremers, M., Rolfe, B.G. and Djordjevic, M.A. (1985). Mol. Gen. Genet. 201, 426-432.

Makaroff, C.A., Zalkin, H., Switzer, R.L. and Vollmer, S.J. (1983). J. Biol. Chem. 258, 10586-10593.

Mulligan, J.T. and Long, S.R. (1985). Proc. Nat. Acad. Sci. 82, 6609-6613.

Peters, K. and Long, S.R. (1986). Science (in press).

Redmond, J.W., Batley, M., Djordjevic, M.A., Innes, R.W., Kuempel, P.L. and Rolfe, B.G. (1986). Nature (in press).

Rossen, L., Shearman, C.A., Johnston, A.W.B. and Downie, J.A. (1985). EMBO J., 4, 3369-3373.

Rostas, K., Kondorosi, E., Hovarth, B., Simoncsits, A. and Knodorosi, A. (1986). Proc. Nat. Acad. Sci. 83, 1757-1761.

Shearman, C.A., Rossen, L., Johnston, A.W.B. and Downie, J.A. (1986). EMBO J. 5, 647-652.

Stachel, E.S. Messens, E., van Montague, M. and Zambryski, P. (1985). Nature, 318, 624-629.

Van Brussel, A.A.B., Zaat, S.A.J., Canter-Cremers, H.C.J., Wijffelman, C.A.A., Pees, E., Tak, T. and Lugtenberg, B.J.J. (1986). J. Bact. 165, 517-522.

INTERACTIONS BETWEEN *RHIZOBIUM MELILOTI* AND *RHIZOBIUM TRIFOLII* NODULATION GENES: WHAT IS THE BASIS FOR DOMINANCE BY *R. MELILOTI*?

ROGER INNES[1], MICHAEL DJORDJEVIC[2], BARRY ROLFE[2], JEAN DENARIE[3], CHARLES ROSENBERG[3] AND PETER KUEMPEL[1]
[1]Department of MCD-Biology, University of Colorado, Boulder, CO 80309, USA
[2]Genetics Department, Research School of Biological Sciences, Australian National University, Canberra City, ACT 2601, Australia
[3]Laboratoire de Biologie Moleculaire CNRS-INRA, BP 27 F-31326, Castanet-Tolosan Cedex, France

Host-specificity of *Rhizobium* nodulation is controlled, at least in part, by host-specific nodulation (HSN) genes. Several lines of evidence suggest that *R. meliloti* HSN genes are dominant to *R. trifolii* HSN genes. Introduction of a plasmid (pGMI149) encoding *R. meliloti* host-specific and common *nod* genes to *R. trifolii* confers alfalfa nodulation ability to *R. trifolii*[1,3]. In contrast, introduction of a similar plasmid (pRt032) encoding *R. trifolii* common and host-specific *nod* genes to *R. meliloti* does *not* extend the host-range of *R. meliloti*[2]. Here we report that introduction of *R. meliloti* HSN genes to *R. trifolii* strongly inhibits *R. trifolii*'s ability to nodulate clover. The work described below addresses two specific questions relating to this apparent dominance of *R. meliloti* HSN genes.

1. Is the dominance of *R. meliloti* HSN genes due to effects of the HSN gene products, or are the changes in host-range caused by inhibition of *R. trifolii* gene transcription?

2. Is pRt032's failure to function in *R.meliloti* caused by interference from the *R. meliloti* HSN genes?

In order to further characterize the interference between *R. trifolii* and *R. meliloti* HSN genes, strains carrying one or both HSN regions were tested for nodulation phenotype on *Trifolium repens* (white clover), *Trifolium subterranean* (subterranean clover), and *Medicago sativa* (alfalfa). Several significant observations were made:

1. Introduction of pGMI515[3] (encodes *R. meliloti* HSN genes) into *R. trifolii* markedly inhibited *R. trifolii*'s ability to nodulate both white and subterranean clover, while conferring a Hac^{+}, $Nod^{+/-}$ phenotype on alfalfa (nodule-like structures with infection threads appeared after 2-3 weeks, no bacteria could be recovered from nodule squashes).

2. Tn*5* insertions in the *nodFE* and *nodH* genes, located on pGMI515, partially, but not completely, abolished the inhibitory effect of pGMI515. (Tn*5* insertions in *nodFE*, when present in pSym of *R. meliloti* 2011, extend strain 2011's root hair curling ability to white clover[1].)

3. Introduction of pRt032 (encodes *R. trifolii* host-specific and common *nod* genes) to *R. meliloti* 2011 did *not* confer clover nodulation ability upon strain 2011.

4. Deletion of *R. meliloti* host-specific and common *nod* genes (strain GMI255[3]) did *not* increase the effectiveness of pRt032 in an *R.meliloti* background.

The results of the phenotype analysis described above indicated that *R. meliloti* HSN genes (encoded on pGMI515) interfered with the functioning of *R. trifolii* HSN genes. It was therefore important to determine whether the *R. trifolii* genes were being properly expressed in the presence of pGMI515. Using pSym located *lac* fusions we determined that: 1. Introduction of pGMI515 into *R. trifolii* ANU843 did not significantly repress expression of *nodA*, a common *nod* gene. 2. Expression of *nodF* (HSN gene) and a *lac* fusion in region IV (also involved in host-range control) was partially repressed (18% and 35% respectively; see Table 1).

It is doubtful that these differences in gene expression can account for the poor nodulation of *R. trifolii* ANU843(pGMI515) on clover.

TABLE 1. β-galactosidase expression from *Rhizobium trifolii nod* gene-*lac* fusions in the presence of *R. meliloti* HSN genes.[a]

Gene or region	Fusion number	Treatment[b]	β-galactosidase activity[c]	
			- pGMI515	+ pGMI515
nodA	218s	dH_2O	21±2	23±1
nodA	218s	exudate	1922±68	1776±161
nodF	1027s	dH_2O	19±1	24±1
nodF	1027s	exudate	459±25	374±10
reg IV	822s	dH_2O	15±2	15±1
reg IV	822s	exudate	260±7	170±4

[a] *lac* fusions were created by insertion of mini-Mu-*lac* transposon Mu dI1734 into the indicated genes and are located in pSym of *R. trifolii* ANU843. *R. meliloti* 2011 HSN genes were introduced as a 26kb fragment using the RP4-based plasmid pGMI515[3].

[b] late log phase cultures (OD_{600} 0.7-0.8), grown in minimal medium, were diluted 10x into either dH_2O or white-clover root exudate. The diluted cultures were incubated at 21°C for 21hr without shaking, at which time β-galactosidase activity was assayed. White-clover root exudate was obtained by growing freshly germinated seedlings in dH_2O (1 seedling/ml) for four days (18hr day, 6hr night light cycle). After this time, plants were removed and the water (containing exudate) autoclaved.

[c] units were determined according to Miller (1972) using the chloroform/SDS lysis procedure. Each value represents the average of 3 separate cultures ± standard deviation.

The nodulation tests also indicated that *R. trifolii* HSN genes failed to function in an *R. meliloti* 2011 background. We therefore wished to determine if the *R. trifolii* genes were being expressed in this strain background. Using pRt032 located *lac* fusions, we determined that: 1. Induced levels of both common and host-specific *nod* genes were nearly as high in *R. meliloti* as in *R. trifolii* (Table 2). 2. Uninduced levels were higher in *R. meliloti*, particularly in the case of the constitutively expressed gene, *nodD*.

These results indicate that the failure of pRt032 to confer clover nodulation ability to *R. meliloti* cannot be attributed to lack of expression of the *R. trifolii* genes.

The data summarized in Table 1 strongly suggest that the interference between *R. trifolii* and *R. meliloti* host-range determinants is caused by the presence of specific gene products, rather than changes in gene transcription. This hypothesis fits a model in which prospective host plants detect both "positive" and "negative" signal molecules produced by *Rhizobium*. pGMI515 may cause *R. trifolii* to produce a "signal" which causes clover plants to evaluate the strain as not being *R. trifolii*. pRt032's failure to confer clover nodulation ability to *R. meliloti* could be attributed to production of a "negative" signal (as perceived by clover plants) by *R. meliloti*. Because removal of the native *R. meliloti* HSN genes did not increase pRt032's effectiveness, this "negative" signal may not be a product of an HSN gene. This signal would be absent from *Agrobacterium*, to which

TABLE 2. β-galactosidase expression from *Rhizobium trifolii nod* gene-*lac* fusions in pSym+ *R. trifolii* and *R. meliloti* backgrounds.[a]

Gene or region	Fusion number	Treatment[b]	β-galactosidase activity[c]	
			R. trifolii	*R. meliloti*
nodA	121	dH_2O	95±16	151±10
nodA	121	exudate	4268±748	2877±202
nodD	545	dH_2O	142±5	571±60
nodD	545	exudate	189±7	687±46
nodF	1027	dH_2O	58±4	157±37
nodF	1027	exudate	1615±238	1379±108
reg IV	810	dH_2O	51±10	176±20
reg IV	810	exudate	3121±160	2395±250

[a]*lac* fusions were created by insertion of mini-Mu-*lac* transposon Mu dI1734 into the indicated genes which were present on the broad host-range plasmid pRt032. The resulting plasmids were introduced into *R. trifolii* ANU843 and *R. meliloti* GMI708 (a rifampicin resistant derivative of *R. meliloti* 2011), both of which have intact Sym plasmids and are Nod+Fix+ on their respective host plants.
[b]see footnote to Table 1.
[c]see footnote to Table 1.

pRt032 *does* confer clover nodulation ability. Thus, host-range may be controlled by a combination of both positive and negative effecter molecules produced by *Rhizobium*, some of which may not be nodulation gene products.

REFERENCES

1. Batut et al: In: Evans HJ, Bottomley PJ, Newton WE (eds):Nitrogen Fixation Research Progress. Martinus Nijhoff, Dordrecht, pp109-115, 1985.
2. Djordjevic et al: Plant Mol Biol 4:147-160, 1985.
3. Truchet et al: J. Bacteriol 164:1200-1210, 1985.

MULTIPLE HOST-SPECIFICITY LOCI IN THE BROAD HOST-RANGE *RHIZOBIUM* NGR234

A. LEWIN*, C. ROSENBERG**, J. STANLEY*, D.N. DOWLING*, J.-F. MANEN*, F. DEBELLE** AND W.J. BROUGHTON*
* Laboratoire de Biologie Moleculaire des Plantes Superieures, Universite de Geneve, 1 chemin de l'Imperatrice, 1292 Chambesy, Geneve, Suisse;
** Laboratoire de Biologie Moleculaire, CNRS-INRA, B.P. 27, 31326 Castanet-Tolosan, Cedex, France

1. ABSTRACT

Speciation within the family Rhizobiaceae is based on host-range yet this is also dependent on the macrosymbiont. We sought an index of the diversity in legume host-range by using a collection of fast-growing rhizobia isolated from 26 different genera of tropical legumes to inoculate *Aeschynomene*, *Arachis*, *Cajanus*, *Desmodium*, *Glycine*, *Lablab*, *Leucaena*, *Lotus*, *Macroptilium*, *Mimosa*, *Psophocarpus*, *Sesbania*, and *Vigna*. *Vigna unguiculata* possessed the broadest spectrum being nodulated by about 70% of all isolates while *Aeschynomene* and *Sesbania* could only nodulate with their homologous strains. *Lablab*, *Leucaena*, *Arachis*, *Macroptilium*, *Psophocarpus*, *Glycine*, etc., displayed a decreasing host range. Similarly, a strain isolated from *Lablab purpureus* (NGR234=MPIK3030) was able to nodulate nine additional legume genera, while isolates from *Gliricidia* and *Leucaena* could nodulate seven extra legumes. We identified DNA encoding broad host-range genes of NGR234 by mobilizing the whole plasmid, (pNGR234a), then cosmid subclones of pNGR234a, followed by subclones of the cosmids into heterologous Nod$^+$ *Rhizobium* strains incapable of nodulating *V. unguiculata*. Three independent sets of host-range loci (Hsn) were found that extended the host range of the *R. loti* and *R. meliloti* recipients to include *Vigna*. One of these Hsn-loci is linked to *nod*D, a second Hsn is linked to *nif*, while the third Hsn is linked to the *nod*C gene. Furthermore, the HsnIII locus complements a mutation in the species-specific *nod*H gene of *R. meliloti* strain 2011 (wildtype NGR234 nodulates *Medicago sativa cv.* Cardinal ineffectively).

2. INTRODUCTION

Essential nodulation genes of rhizobia required for induction of root-hair curling are physically and functionally conserved in different *Rhizobium* species (1,2,3,4). This conservation implies that they are not the only determinant of host specificity--another set of genes must exist that allow specific rhizobia to recognize certain hosts. We sought the components of host-specificity by mobilizing cosmid clones of the broad host-range Sym plasmid pNGR234a (as well as subclones of the cosmids) into heterologous wildtype rhizobia (i.e., not able to nodulate NGR234 host plants). In this way, we found three separate loci that were able to confer on *R. loti* and *R. meliloti*, the ability to nodulate *Vigna unguiculata* amongst other legumes. An index of the diversity of the host-range components of legume plants was sought by inoculating 13 different legume species with a collection of rhizobia isolated from 26 genera of tropical legumes. A spectrum of plant host-range was observed which varied from *V. unguiculata* (nodulated by ca. 70% of the isolates) to *Aeschynomene* and *Sesbania* which could only be nodulated by a single homologous *Rhizobium*. For this reason, *V.*

unguiculata was chosen for primary screening of bacterial host-range genes. Other NGR234 hosts were used to further define the functions of these genes.

3. MATERIALS AND METHODS

Bacterial strains and plasmids used, culture of bacteria, plasmid mobilizations, DNA isolation, recombinant DNA techniques, mapping procedures, microscopy and plant tests have all been described previously (5).

4. RESULTS

4.1. Diversity in plant host range

Forty-nine fast-growing rhizobial strains isolated from 26 different genera of tropical legumes (6) were used to inoculate *Aeschymonene, Arachis, Cajanus, Desmodium, Glycine, Lablab, Leucaena, Lotus, Macroptilium, Psophocarpus, Sesbania, Mimosa,* and *Vigna.* Of these hosts, *V. unguiculata* displayed the broadest host range being nodulated by about 70% of the isolates, followed by *Lablab, Leucaena, Arachis, Macroptilium, Psophocarpus, Glycine,* etc. (Figure 1). At the other end of the host-range spectrum were *Aeschymomene indica,* as well as *Sesbania punctata* and *Sesbania rostrata.* Almost all isolates capable of nodulating *Leucaena* produced effective nodules with their host, while only 5% of the ineffective isolates produced effective nodules on *Arachis* and *Lablab. Macroptilium atropurpureum,* the only small-seeded legume with a fairly broad host-range, was only nodulated by about 45% of the isolates. Thus, *M. atropurpureum* is comparable to *Arachis hypogaea* and *Psophocarpus tetragonolobus* in having a relatively narrow host range, while *V. unguiculata* is the best plant for screening purposes. For this reason, all constructions involving putative host-specificity loci were first tested on it.

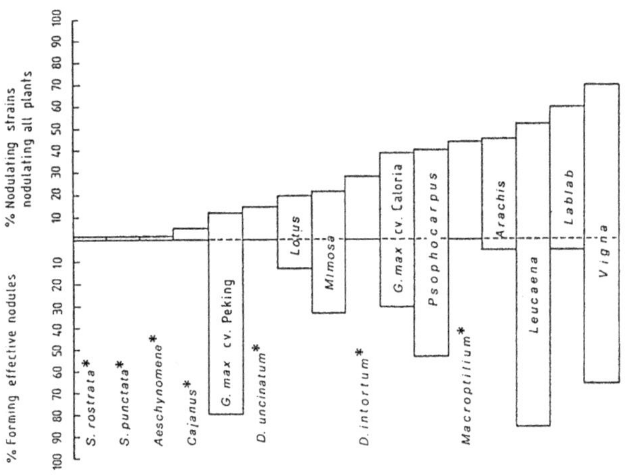

FIGURE 1. Upper half: Diagrammatic representation of the percentage of the 49 strains isolated from 26 different legume genera nodulating the chosen hosts. Lower half: Effectiveness of the various isolates on the chosen legumes. Hosts marked with an asterisk were grown on agar slants and for this reason acetylene-reduction was not measured.

4.2. Diversity in bacterial host range

To define the variability in bacterial host range, we tested the ability of particular isolates to nodulate other legume genera. In Figure 2 the original host-plant is listed along with the number of additional legume genera nodulated by the most infective isolate from the particular host. NGR234 (MPIK3030), the only isolate from *Lablab purpureus* tested, nodulated nine genera of legumes in addition to *Lablab*. Thus, its host range includes *Arachis hypogaea*, *Calopogonium caeruleum*, *Desmondium intortum*, *Glycine max*, *Lablab purpureus*, *Leucaena leucocephala*, *Macroptilium artropurpureum*, *Medicago sativa*, *Mimosa pudica*, *Moghania congesta*, *Psophocarpus tetragonolobus*, *Stylosanthes hamata*, *Tephrosia candida*, and *Vigna unguiculata* as well as the non-legume *Parasponia andersonii* (7,8, this study). As far as we know, this is the broadest host range so far reported for a *Rhizobium*, and as study of broad host range was our primary objective, all further work was performed on NGR234 (MPIK3030).

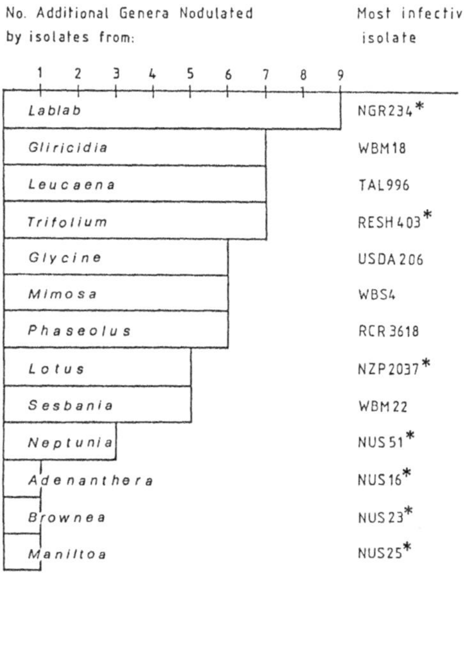

FIGURE 2. Number of additional hosts nodulated listed in terms of host of isolation. Strains marked with an asterisk represent the only isolate tested from that particular genus.

4.3 Identification of host-range genes of NGR234

Symbiotic genes on the Sym plasmid, pNGR234a, can be expressed in a *Bradyrhizobium* species, *R. leguminosarum*, *R. loti*, *R. meliloti* and *Agrobacterium tumefaciens* (5,6). Assuming that cloned *hsn* genes from pNGR234a would also be expressed in these hosts, we chose *R. loti* strain NZP4010 (=NZP2037 cured of the non-symbiotic plasmid pNZP2037a (9)), and *R. meliloti* strains L5.30 and 2011 as recipients for subclones of pNGR234a. *R. loti* and *R. meliloti* strains did not react with the roots of *M. atropurpureum* or *V. unguiculata*. Hence, nodulating transconjugants of these strains must result from acquisition of dominant *hsn*. Mobilization of three non-overlapping sets of cosmids conferred upon the chosen hosts the ability to nodulate *V. unguiculata*. Furthermore, all sets of cosmids conferred upon their rhizobial recipients, the ability to cause root-hair curling of *M. atropurpureum*. The cosmids carrying the HsnII and HsnIII loci and the central portion of the region containing the HsnI do not hybridize with HsnII and HsnIII (5). This shows that there are at least three non-overlapping sets of Hsn-loci. As is shown in Figure 3, the HsnI locus is linked to *nod*D, the HsnII locus to *nif*KDH and the HsnIII locus to *nod*C. A 9.8 kb *Sma*I subclone of pWA90/110 in pMMB22 (10) conferred upon L5.30 and NZP4010 the ability to nodualte *V. unguiculata* as well as to curl the root hairs of *M. artropurpureum*.

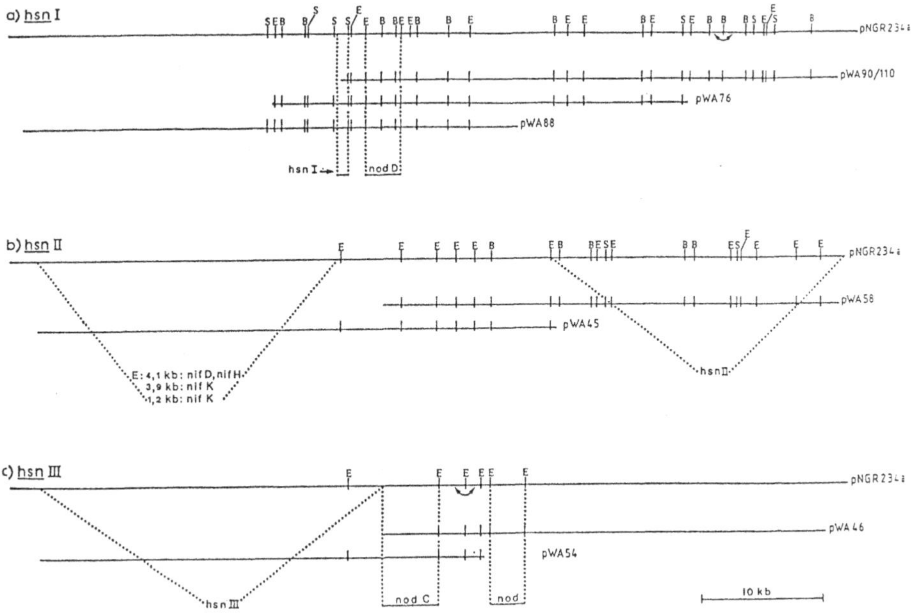

FIGURE 3. Restriction maps of the three Hsn-regions of pNGR234a. Regions marked *nif* are those showing homology with *nif*KDH of *Bradyrhizobium japonicum; nod*-prodes used were from *R. meliloti*. (a) 65 kb-region showing the location of HsnI (localized by subcloning the *Sma*I-fragment of pWA90/110 into pMMB22 (9) and mobilizing it into L5.30 and NZP4010) and *nod*D. (b) 65 kb-region showing the approximate location of HsnII and *nif*KDH. (c) 65 kb-region showing the location of HsnIII, *nod*C and a common *nod*-region that is not homologous to *nod*DABC of *R. meliloti*. B = *Bam*HI, E = *Eco*RI, S = *Sma*I.

4.4 Complementation of a mutation in a *R. meliloti hsn*-gene with HsnIII

As NGR234 can form ineffective nodules on *Medicago sativa*, we asked whether a NGR234 *hsn*-gene would be able to complement mutations in similar genes of *R. meliloti*. Mutants carrying Tn5 mutations in *hsn*-genes of *R. meliloti* strain 2011 (11) were used as recipients in crosses involving cloned Hsn-loci I,II,III of NGR234. The Tn5-Mob system was used to mobilize cosmids from *E. coli*-donors. *R. meliloti* 2011 recipients were Km, Nm-resistant due to Tn5 insertion, and selection was made on the basis of cointegrate-formation with RP4.4, with selection for Tc-resistance in NZP4010. Transconjugants of NZP4010 were used to infect *V. unguiculata*. Reisolates from nodules were then used to transfer the cointegrate to mutants of *R. meliloti* 2011. Alternatively, cosmids were mobilized directly from *E. coli* into 2011 mutants and the complete conjugation-mixture was used to inculate *Medicago sativa*. In both cases, the HsnIII locus of NGR234 complemented (at low efficiencies) a mutation in the *nod*H gene of *R. meliloti* 2011. Microscopic examination of the nodules showed bacteroids. There was no evidence of DNA homology with the *nod*H region of *R. meliloti* 2011, although the HsnIII locus can functionally complement the gene concerned.

5. DISCUSSION

In order to find rhizobia with broad host ranges, we screened 49 fast-growing strains (isolated from the nodules of 26 tropical legume genera) on 13 putative hosts. Our data show that both bacteria and plants exist with variable host-range compatibilities. As our aim was to study broad host range, we choose the *Rhizobium* and the host legume possessing the broadest possible compatibility. NGR234 was able to nodulate the largest number of hosts and *V. unguiculata* to be nodulated by the widest spectrum of rhizobia.

Our strategy to identify the host-specificity gene(s) of NGR234 was to first mobilize pNG234a into wildtype rhizobia that were incapable of nodulating *V. unguiculata*. As nodulating transconjugants arose at high frequencies, we then transferred cosmid subclones of pNGR234a into heterologous rhizobia. Although we have not yet mobilized subclones covering the entire Sym plasmid, three sets of Hsn-loci have been found so far. That there are at least three sets of Hsn genes was shown by restriction mapping, cross-hybridizations and by verifying that the Hsn-clones are non-rearranged clones from pNGR234a. Homology studies have shown that HsnI is linked to *nod*D, HsnII to *nif*KDH and HsnIII to *nod*C. All three sets of *hsn*-genes conferred upon interspecific recipients the ability to nodulate *V. unguiculata*, while transconjugants containing Hsn locus I were also able to nodulate *M. atropurpureum*. Although the restriction enzyme maps diverge, it seems probable that our Hsn locus I is identical with the Hsn regions identified by Bachem et al. (12) and Bassam et al. (13).

6. ACKNOWLEDGEMENTS

We wish to thank N. Catsiyannis, J. Denarie, M. Fontana, F. Haar, H. Hennecke, H. Meyer z.A., F. Maillet, Y. Mark, X. Perret, and U. Samrey for their generous help with many aspects of this work. A.L. is grateful to EMBO for a short-term fellowship. Financial assistance was provided by the European Commission (Contract No. 288), the Universite de Geneve, and the Fonds National Suisse de la Recerchere Scientifique (Contract No. 3.176-0.85).

REFERENCES

1. Djordjevic MA, Schofield RR, Ridge RW, Morrison NA, Bassam BJ, Plazinski J, Watson JM and Rolfe BG. Plant Mol Biol 4, 147-160, 1985.
2. Long SR, Buikema WJ and Ausubel FM. Nature 298, 485-488, 1982.
3. Rossen L, Johnston AWB and Downie JA. Nucl Acids Res 12, 9497-9508, 1985
4. Torok J, Kondorosi E, Stepkowski T, Posfai J and Kondorosi A. Nucl Acids Res 12, 9509-9524, 1985.
5. Broughton WJ, Wong CH, Lewin A, Samrey U, Myint H, Meyer z.A. H, Dowling DN and Simon R. J. Cell Biol. 102, 1173-1182, 1986.
6. Broughton WJ, Heycke N, Meyer z.A. H and Pankhurst CE. Proc. Natl. Acad. Sci. USA 81, 3093-3097, 1984.
7. Trinick MJ. J Appl Bact 49, 39-53, 1980.
8. Pankhurst CE, Broughton WJ, Bachem C, Kondorosi E and Kondorosi A. In: Molecular Genetics of the Plant-Bacteria Interaction (Puhler A, ed), Springer-Verlag, Berlin, pp 69-76, 1983.
9. Pankhurst CE, Broughton WJ and Wieneke U. J Gen Microbiol 129, 2535-2543 1983.
10. Bagdasarian MM, Amann E, Lurz R, Ruckert B and Bagdasarian M. Gene 26, 273-282, 1983.
11. Batut J, Boistard P, Debelle F, Denarie J, Ghai J, Huguet T, Infante D, Martinez E, Rosenberg C, Vasse J and Truchet G. in: Nitrogen Fixation Research Progress (Evans HJ, Bottomley PJ and Newton WE, eds), Nijhoff, The Netherlands, pp 109-115, 1985.
12. Bachem CWB, Banfalvi Z, Kondorosi E, Schell J and Kondorosi A. Mol Gen Genet 203, 42-48, 1986.
13. Bassam BJ, Rolfe BG and Djordjevic MA. Mol Gen Genet 203, 49-57, 1986.

CONSERVED NODULATION GENES ARE OBLIGATORY FOR NONLEGUME NODULATION

Kieran F. Scott[1], Marlene Saad[1], G. Dean Price[2], Peter M. Gresshoff[2], Heather Kane[1], and Kaw Yan Chua[1]
[1] Centre for Recombinant DNA Research, Research School of Biological Sciences. [2]Department of Botany, Australian National University, GPO Box 4, Canberra, ACT 2601, Australia.

The nodulation of the nonlegume plant *Parasponia* by bacterial strains occurs through a mechanism quite distinct from that used by these same strains to nodulate legume species (1,2,3). In contrast to the invasion via root hair curling, infection thread formation and bacterial release observed during legume nodulation (4), bacteria invade *Parasponia* roots through points of bacterium-induced meristematic activity which break the epidermal surface of the root allowing bacteria direct access to cortical cells. These cells become invaded via infection thread formation and nitrogen fixation occurs within these threads (5). the fully developed *Parasponia* nodule resembles a modified lateral root structure with a central vascular system (5). As with legume nodulation, there is also a host-specific component to nonlegume nodulation. Only certain strains of *Rhizobium* and *Bradyrhizobium* are capable of nodulating *Parasponia* and these strains do so with varying efficiency (6). One such organism, *Bradyrhizobium* (sp. *Parasponia*) strain ANU289 is not only capable of biological nitrogen fixation on *Parasponia* but also can effectively nodulate a broad range of tropical legumes. to determine the genetic basis of nonlegume nodulation, we have identified and characterised conserved nodulation genes from ANU289 and demonstrated that the expression of the genes *nodKABC* are obligatory for both legume and nonlegume nodulation. In addition we have identified cosmid clones carrying ANU289 DNA which can confer *Parasponia* nodulation on the narrow host-range strain *Rhizobium trifolii* ANU843.

Conserved Nodulation Genes

We have previously cloned and characterised a nodulation locus from ANU289 by hybridisation with *R. trifolii nod* gene probes (7). This locus has been shown to contain four genes *nodA nodB nodC* and *nodD* which are conserved in all nodulating strains examined to date. In addition, the locus contains a fifth gene, *nodK*, which is linked to *nodABC*. In the intergenic region between the *nodD* and *nodKABC* operons, there is a short sequence (*nod* box) which is conserved in all *Rhizobium* strains so far examined (7). We have now constructed specific mutations in a number of these genes by the *in vitro* insertion of the *nptII* gene, derived from Tn5, into appropriate restriction sites within the cloned *nod* genes. These mutant constructs have been homogenotised back into wildtype ANU289 on the mobilisable suicide vector pSUP202 (8) and their nodulation phenotype assayed on both *Parasponia rigida* plants and the tropical legume *Macroptilium atropurpureum* (siratro). As shown in Table 1 insertions in *nodA* are completely nodulation deficient (nod$^-$) on both *Parasponia* and siratro plants. A deletion of the N-terminal portion of *nodD*, the intergenic region containing the '*nod* box', *nodK* and the N-terminal portion of *nodA* are also completely nod$^-$.

Interestingly, insertions in *nodD*, both at the N-terminal and C-terminal ends are nod$^+$ on both plants. However, ultrastructural analysis of nodules induced by these mutants on siratro indicate that fewer cells contain bacteroids relative to ANU289. No discernible difference in the timing of nodulation, number of nodules or nodule ultrastructure was observed on

Table 1. Nodulation Phenotype of Mutants

Strain	Genotype	Nodulation Phenotype	
		Parasponia rigida	*Macroptilium atropurpureum*
ANU289	Wildtype	nod^+	nod^+
ANU1271	*nptII*::*nodD* (N-term)	nod^+	nod^{+a}
ANU1272	*nptII*::*nodD* (C-term)	nod^+	nod^{+a}
ANU1273	*nptII*::Δ*nodD*,*nod* box, *nodK*, *nodA*	nod^-	nod^-
ANU1274	*nptII*::*nodA*	nod^-	nod^-
ANU1292	*nptII*:: Δ *nodK*-*nod* box	nod^-	nod^-

(a) Ultrastructural analysis indicates few nodule cells contain bacteriods relative to wildtype.

Fig. 1. Cosmids able to confer *Parasponia* nodulation on *R. trifolii*

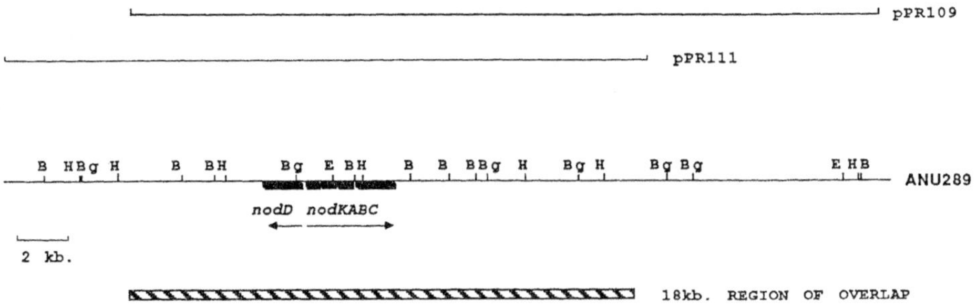

B, *Bam* HI; H, *Hind* III; Bg, *Bgl* II; E, *Eco* RI

innoculation of Parasponia with these mutants. Other Rhizobium strains with leaky nodD mutants have been shown to contain multiple structural homologues of nodD (9). We have not been able to identify additional copies of nodD in ANU289 using nodD probes from several origins, hybridised under several conditions. It seems likely therefore that, either ANU289 contains a functional homologue of nodD which shares very little sequence homology to the copy identified here, or the mutagenesis procedure enables readthrough transcription of the nodD gene from the nptII promotor.

The nodK gene

DNA sequence analysis of the nod locus revealed the presence of an additional ORF preceding nodA which we have designated nodK (7). A mutant generated by deletion of the N-terminal portion of nodK and a portion of the nod box (ANU1292) is nodulation deficient on both Parasponia and siratro. these data support the idea that nodK is located on the same operon as nodABC and has a role in the nodulation process.

Host-specific nodulation genes

In an effort to identify genes involved in the host-specific nodulation of nonlegumes we have constructed a gene bank of ANU289 DNA in the mobilisable cosmid vector pLAFR3 (10). 'En masse' transfer of this gene bank to the narrow-host-range strain R. trifolii ANU843 followed by innoculation of Parasponia rigida plants has led to the identification of two overlapping cosmids pPR109 and pPR111 which can reproducibly confer Parasponia nodulation on this R. trifolii strain. As shown in Fig. 1, these cosmids share 18kb of overlapping sequence which is responsible for the extension of host-range. This 18kb also includes the nodD and nodKABC operons. Therefore, either one or both of these operons themselves allow R. trifolii to nodulate Parasponia or, there is an additional locus linked closely to these genes which confer nonlegume nodulation on R. trifolii. The nodules induced on Parasponia by R. trifolii strains carrying these cosmids are indistinguishable from those induced by ANU289. However, these strains are unable to form nodules on siratro and induce at a very low frequency growths resembling partially differentiated callus tissue on siratro roots. These experiments show that while the pathways of legume and nonlegume nodulation share genetic components in common (viz the expression of the nodKABC operon) the two pathways have components which are unique to either legume nodulation alone or both legume and non-legume nodulation.

References

1. Trinick, M.J. and Gallbraith, J. (1976) Arch. Microbiol. 108: 159-166
2. Lancelle, S.A. and Torrey, J.G. (1984) Protoplasma 123: 26-37
3. Price, G.D. Mohapatra, S.S. and Gresshoff, P.M. (1984) Bot. Gaz. 145: 444-451.
4. Ridge, R.W. and Rolfe, B.G. (1986) J. Plant Physiol. 122: 121-137.
5. Trinick, M.J. (1979) Can. J. Microbiol. 25: 565-578.
6. Trinick, M.J. and Galbraith, J. (1980) 86: 17-38.
7. Scott, K.F. (1986) Nucl. Acids Res. 14: 2905-2919.
8. Simon, R., Priefer, U. and Puhler, A. (1983) Biotechnology 1: 784-791.
9. Honma et al. (1985) In: Nitrogen Fixation Research Progress, Evans, H.J., Bottomley, P.J. and Newton, W.E. eds. Martins Nijhoff p. 120.

CHARACTERIZATION OF SYMBIOTIC GENES AND REGULATION OF THEIR EXPRESSION IN RHIZOBIUM LEGUMINOSARUM PRE.

Jan Hontelez, Rene Klein Lankhorst, Jan-Dirk Jansma, Evert Jacobsen*,
Rommert C. van den Bos and Ab van Kammen
Department of Molecular Biology, Agricultural University, Wageningen, The Netherlands.
*Department of Genetics, University of Groningen, The Netherlands.

In R.leguminasorum PRE we have identified a number of nif and fix genes (1,2) by using heterologous DNA probes from Klebsiella pneumoniae or R.meliloti. Alternatively differential expression in the endosymbiotic state was utilized to characterize these genes. It is well established now that the nif A gene product is involved in the activation of most or all of the Rhizobium nif and fix genes. In this report we describe preliminary experiments undertaken to elucidate the mechanism of the regulation of the R.leguminosarum nif A gene, using a novel Pisum sativum mutant which yields Fix $^-$ nodules upon inoculation with wild type R.leguminosarum (3). We also present a correlated physical and genetic map of two stretches comprising over 55 kb of the R.leguminosarum sym-plasmid.

Characterization of symbiotic genes.

Fragments of the R.leguminosarum PRE sym-plasmid were cloned in cosmid vector pJB8. Hybridization of the resulting sym-plasmid library using ^{32}P.labeled bacteroid mRNA from nitrogen fixing pea nodules or heterologous DNA fragments from K.pneumoniae (nif HD) or R.meliloti (nifA) as probes identified two stretches of sym-plasmid containing clustered sym-genes; one of about 30kb including nifHDK and another of 25kb containing nifA (Figs. 1 and 2). Site-directed Tn5 mutagenesis revealed at least 19 symbiotic genes, several of which were characterized earlier: nifHDK (2), nifA, nifB (3), nodABC and fixF (4). We now have identified a number of other sym-genes: fixABC, fix"23" (5) and nifQ (4) adjacent to nifA as well as nifE (4) downstream of nifHDK.

FixABC AND Fix"23"

The phenotypes of the Tn5 mutants in the region of pRleH21 (Fix$^-$/HDK$^+$; fig. 2) and expression studies in E.coli minicells, yielding polypeptides of 29, 42 and 50kdal., suggest that a fixABC operon is located upstream of nifA. Hybridizations with R.meliloti fixA and fixC probes confirm this conclusion. In the coding regions as well as in the promoter regions homology was found. The orientation of fixABC relative to nifA is similar as in R.meliloti. The expression level of the fixABC genes in 17 days old bacteroids is very high and comparable to that of nifHDK. However, we noted a strong polarity in the expression of this operon. The ^{32}P-mRNA hybridization with fixA was 5-10 times higher than with fixBC.

NifE

pWK26, a plasmid containing K.pneumoniae nifENX (6) as a probe hybridized with a part of pRleH18 (fig. 1), which also bears the C-terminal coding sequences of the nifHDK operon. Tn5 insertions in this region cause a Fix$^-$ phenotype. We thus conclude that pRleH18 carries nifE or nifN (or both) homologous genes downstream of nifHDK. An alternative explanation for the observed hybridization would be homology between K.pneumoniae nifN and R.leguminosarum nifK (Dean and Brigle detected homology between nifEN and nifDK in A. vinelandii; 7). This explanation seems rather unlikely, due to the location of the hybridisation signals and the absence of detectable homology in R.leguminosarum between nifE and nifD (J. Hontelez, unpublished). Furthermore no hybridisation of K.pneumoniae nifENX was detectable with other fragments of the sym-plasmid and in other species nifE homology was also found adjacent to nifK: R.sesbania (8), R.meliloti (9), B.japonicum (10), A.chroococcum (11).
The level of expression of the nifE coding region in 17 days old bacteroids is low (less than 10% of nifHDK activity).

NifQ

Site-directed Tn5 mutagenesis of subclone pRleH33 (an 8.7kb EcoRI fragment cloned in pACYC184. See fig.2) resulted in a phenotype (3311) similar to that of a K.pneumoniae nifQ mutant in that acetylene is reduced at a low level (7-10% of wild type). We assume that this region contains a nifQ-like gene based on this and the following arguments. Mutation 3311 maps close to the C-terminal coding sequences of nifB and in this region a weak hybridisation is detectable with K.pneumoniae nifQ. Expression of pRleH33 in E.coli minicells results in the synthesis of 36 and 41kD polypeptides; the positions of these coding sequences is indicated in Fig.2. Tn5 insertion 3311 in this plasmid does not abolish the expression of these polypeptides. Between nifB and the gene encoding the 41kD polypeptide the space is just sufficient to accommodate a nifQ-like gene supposing it to encode a polypeptide of about 17kD as in K.pneumoniae (Ausubel, F.M. et al., 12). However we have not detected a polypeptide of this size in E.coli minicells expressing pRleH33. The function of the 36 and 41kD proteins is not yet known but they seem not to be involved in symbiotic nitrogen fixation.

Figure 1

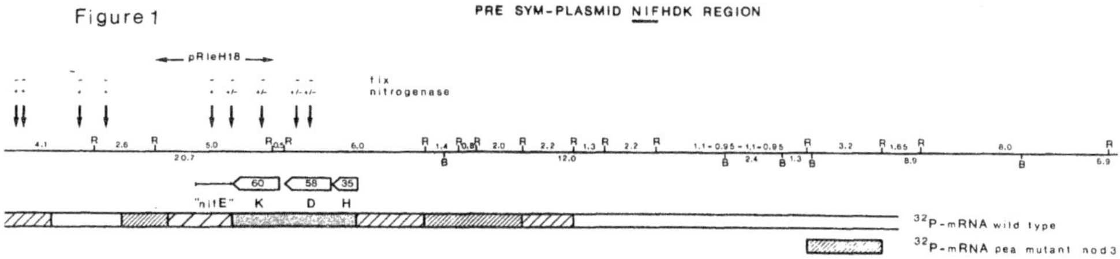

Figure 2

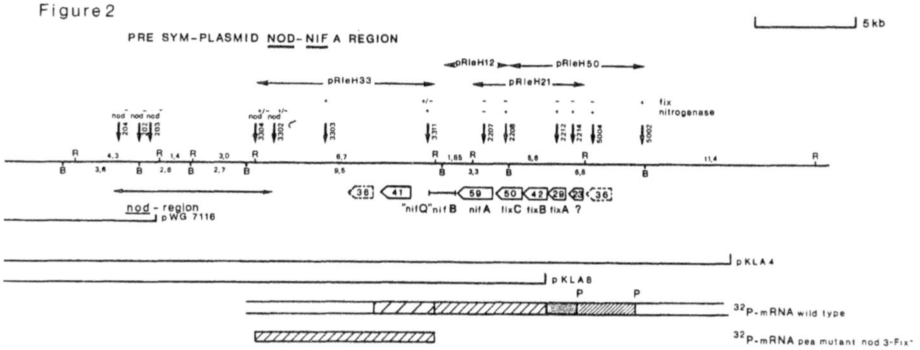

P.sativum Nod_3-Fix^-.

Jacobsen and Feenstra isolated pea mutants, which are defective in nodulation. Previously they have described Pea nod_3, which produces abundant nodulation and in which nodulation is not inhibited by NO_3^- (13) . From this mutant nod_3 Postma et al. (3) now have derived another mutant, Pea nod_3-Fix^-, which produces Fix^- nodules on inoculation with wild type R.leguminosarum. The nodule cells contain fully developed bacteroids and normal amounts of leghemoglobin. Nitrogenase proteins were however not detectable in the bacteroids.

Since it was tempting to assume that the nodules of nod_3-Fix$^-$ lack the capacity to activate the expression of nifA and thus the other nif and fix-genes , we have compared the expression of those genes in respectively wild type pea, nod_3 and nod_3-Fix$^-$ nodules. For this purpose a Southern blot of DNA from the R.leguminosarum sym-plasmid nod- and fix-regions (fig.1 and 2) was hybridised with ^{32}P-labeled bacteroid mRNA from 21 days old nodules of nod_3 and nod_3-Fix$^-$ and from 17 days old wild type nodules. In nod_3-Fix$^-$ no hybridisation with nifA or other fix-genes is detectable. The only detectable transcription from these sym-plasmid regions was located in an 8.7kb EcoRI fragment (pRleH33) downstream of nifB and 3.2kb EcoRI fragment 20kb upstream of nifHDK. The latter fragment is also being expressed in pea nod_3 but not in wild type nodules. The 8.7kb EcoRI fragment is also transcribed in wild type nodules (due to nifB and nifQ), but at a lower level than in nod_3 and nod_3-Fix$^-$. Apparently the expression of the 3.2kb EcoRI fragment is specific for nod_3 and the transcription from the 8.7kb EcoRI fragment is characteristic for the nod_3-Fix$^-$ mutant. Both expressions are apparently not under the control of the nifA product, and therefore do not represent nif genes.

References.

1) Schetgens Th., et al., J. Mol. Appl. Gen. 2 : 406-421 (1984)
2) Schetgens Th., et al., Mol. Gen. Genet. 200 : 368-374 (1985)
3) Postma J.G., et al., These proceedings.
4) Hontelez J., et al., in preparation.
5) Schetgens Th., et al., in preparation.
6) Puehler A. and Klipp W. (1983) in: Nitrogen Fixation (Mueller A. and Newton W. eds.) pp. 111-133
7) Dean D. and Brigle K., Proc. Natl. Acad. Sci. USA, 82,5720-5723 (1985).
8) Norel F., et al., (1985) in Evans H., Bottomley P. and Newton W (eds.), Nitrogen Fixation Research Progress, Nijhoff, Dordrecht, p.140.
9) Reilaender, H., this volume.
10) Hennecke H., et al., (1985) in Evans H., Bottomley P. and Newton W. (eds.), Nitrogen Fixation Research Progress, Nijhoff, Dordrecht, pp. 157-163.
11) Kennedy C., et al., (1985) ibid pp. 469-476.
12) Ausubel F., et al., (1985) ibid pp. 165-171.
13) Jacobsen E. , Plant and Soil 82, 427-438 (1984).

The authors gratefully acknowledge the cooperation of Mrs. F.M. Sullivan of KLM Royal Dutch Airlines office (N.Y., N.Y.) in typing this manuscript.

REGULATION OF THE PROMOTERS IN THE NODULATION REGION OF THE SYMBIOSIS PLASMID pRL1JI OF *RHIZOBIUM LEGUMINOSARUM*

HERMAN P. SPAINK, ROB. J.H. OKKER, CAREL A. WIJFFELMAN, ELLY PEES and BEN J.J. LUGTENBERG

Department of Plant Molecular Biology, University of Leiden, Nonnensteeg 3, 2311 VJ Leiden, The Netherlands

Many of the genes essential for symbiotic nodulation of fast-growing *Rhizobium* species are located on large Symbiosis plasmids. We studied the transcriptional regulation of the nodulation region of the *R.leguminosarum* Sym plasmid pRL1JI.

Using an *Eco*RI fragment of pRL1JI with a nodIII::Tn5 insertion (1) as a probe, we cloned a 6.6 Kb *Eco*RI fragment and the overlapping 11.0 and 5.8 Kb *Bam*HI fragments. These fragments or combinations of fragments, all cloned into *IncP* plasmids, were used for nodulation studies. The clone pMP104 (Fig. 1) was able to nodulate *Vicia hirsuta* and *Pisum sativa*, but not *V.sativa*. However, the clone pMP180 could complement strain RBL5580, which contains the deletion in pRL1JI shown in Fig. 1, for nodulation on *V.sativa*, indicating that loci left from the 6.6Kb *Eco*RI fragment are needed for nodulation on some *R.leguminosarum* hosts. (H. Spaink and E. Pees, unpublished)

To study transcriptional regulation of the *nod* region, subclones of maximally 2.5 Kb were cloned in the promoter probe vector pMP190 and studied for expression after mobilization of the constructs into *Rhizobium*. The promoter probe vector pMP190 of the *IncQ* incompatibility group contains the *E.coli lacZ* gene, the ribosome binding site of the chloramphenicol acetyl transferase gene(2) and a multilinker with 5 unique cloning sites. PMP190 containing cloned *nod* fragments were mobilized from *E.coli* to *Rhizobium* RBL5560 (pRL1JI) or RBL5561 (pRL1JI *nodD*::Tn5) by means of pRK2013 helper plasmid and all constructs were tested in the absence or presence of root exudate of *V.hirsuta*.

Two promoters were found which showed expression in the sence of *nodD* product and root exudate.One of the constitutive promoters appeared to be the promoter of the *nodD* gene itself, consistent with earlier results (3), whereas the other one was located in the *nodABC* region and is tentatively designated as p.*nodC* until a more detailed localization has been carried out. Three promoters were found which showed expression only in the presence of *nodD* product and root exudate (Table 1). Two of them have been described recently (3,4) as p.*nodA* and p.*nodF*, the third inducible promoter, which regulates transcription of unidentified *nod* genes, possibly important for nodulation of *V.sativa*, is designated p.*nodH*, which supposedly corresponds with the *nodM* reading frame reported by Downie et al. (this volume). After induction the level of expression of p.*nodH* was 25% of that of p.*nodA* and p.*nodF*.

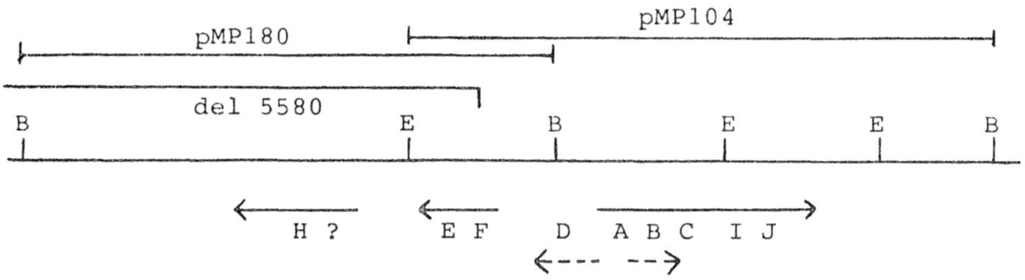

FIGURE 1. Promoters of the nod region of pRL1JI. Clones used for complementation and the location of the deletion RBL5580 are indicated. Plant-inducible promoters and the minimal size of the transcripts read from them are indicated by solid arrows, constitutive promoters are indicated by dotted arrows. The location of p.nodC is indicated tentatively. The location of the indicated nod genes are taken from refs. 3 and 4, whereas the genes of the p.nodH transcript ar not exactly known.

TABLE 1. Expression of inducible nod promoters. Cloned promoters, fused to lacZ were tested in Rhizobium RBL5560 (pRL1JI) and RBL5561 (pRLJI nodD:: Tn5). Exudate (ex.) was derived of V.hirsuta. Naringenin (nar.) has been added in a concentration of 400 nM.

Promoter tested	Units β-galactosidase x 10^{-3}		
	RBL5561 + ex	RBL5560 + ex	RBL5560 + nar.
R.leg. (pRL1JI)			
p.nodA	0.2	14	13
p.nodF	0.3	12	13
p.nodH	0.5	4	4
R.tri (ANU843)			
p.nodA	0.7	5	3
p.nodF	0.5	4	3
R.mel. (SL26)			
p.nodA	1.2	8	7

Fragments which contained the inducible promoters of nodABCIJ, nodFE and nodH were sequenced and showed a highly conserved region, the nod box (Fig. 2), which has been described for other Rhizobium species as well (5,6,7). Of these, the nod box of p.nodH was the least conserved, which may be correlated with the lower level of expression. One of the tested fragments, 114 bp in size, contained complete p.nodA activity and was used for deletion analysis of the promoter activity. Nine Bal31 generated deletions were tested and the results showed that at the left of the nod box the consensus sequence ATCCAY cannot be missed for promoter function and that at the right of the nod box sequences 16 bp downstream of the consensus could not be deleted without loss of promoter activity (Fig. 2). The p.nodA promoter is thus located within the nod box and 16 bp downstream of the nod box.

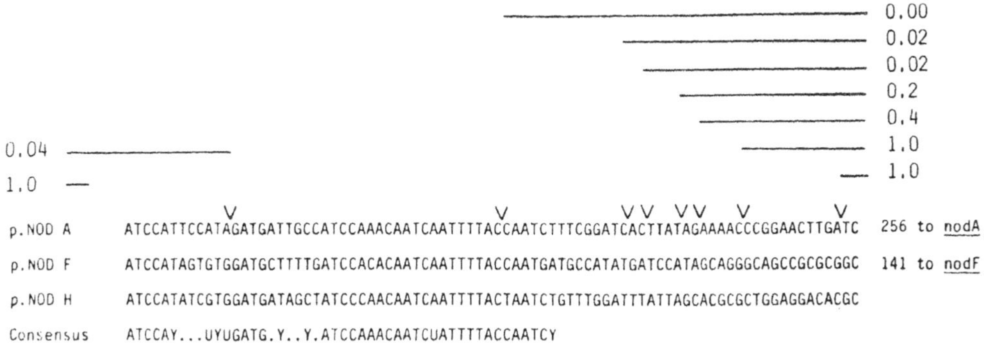

FIGURE 2. Conserved sequences in clones with plant inducible promoters and deletion analysis of p.nodA. The nod boxes of p.nodA, p.nodF and p.nodH are indicated together with a consensus derived of a comparison of the R.leguminosarum, R.trifolii (6), R.meliloti (5) and Bradyrhizobium (7) nod boxes. Bal 31 generated deletions were isolated at the left and at the right site of the nod box of p.nodA and expression levels relative to the expression of the 114 bp fragment are given.

It was found in our laboratory that the flavanoid naringenin can replace V.hirsuta or V.sativa exudate as inducer of the plant-induced nod promoters (Zaat et al., this volume). We tested whether other plant-inducible nod promoters could be induced as well in the isogenic background RBL5560 (pRL1JI). All inducible nod promoters could be induced by naringenin as well as by exudates (Table 1).

REFERENCES

1. Wijffelman, C.A., E. Pees, A.A.N. van Brussel, R.J.H. Okker, and B.J.J. Lugtenberg (1985) Arch. Microbiol. 143: 225-232.
2. Legocki, R.P., A.C. Yun, and A.A. Szalay (1984) Proc. Natn. Acad. Sci. USA. 81:5806-5810.
3. Rossen, L., C.A. Shearman, A.W.B. Johnston, and J.A. Downie (1985) EMBO J. 4: 3369-3373.
4. Shearman, C.A., L. Rossen, A.W.B. Johnston, and J.A. Downie (1986) EMBO J. 5: 647-652.
5. Rostas, K., E. Kondorosi, B. Horvath, A. Simoncsits, and A. Kondorosi (1986) Proc. Natn. Acad. Sci. U.S.A. 83: 1757-1761.
6. Schofield, P.R., and J.M. Watson (1986) Nucl. Acids Res. 14: 2891-2903.
7. Scott, K.F. (1986) Nucl. Acids Res. 14, 2905-2919.

NARINGENIN INDUCES THE nodABC PROMOTOR OF RHIZOBIUM LEGUMINOSARUM AS WELL AS Tsr FACTOR PRODUCTION

S.A.J. ZAAT, A.A.N. VAN BRUSSEL, C.A. WIJFFELMAN, H.P. SPAINK, R.H.J. OKKER, E. PEES and B.J.J. LUGTENBERG

Department of Plant Molecular Biology, University of Leiden, Nonnensteeg 3, 2311 VJ Leiden, The Netherlands

INTRODUCTION

When *Rhizobium leguminosarum* is inoculated on its host plant *Vicia sativa* L. subsp. *nigra* (L.) it causes an altered root morphology, designated as Thick and short roots (Tsr) (7). This phenotype is caused by (a) soluble factor(s). Production of this Tsr factor is dependent on root exudate as well as on the common *nod A*, *B*, *C* and *D* genes (8).

Recently the use of *lac Z* fusions has shown that both common and host specific *nod* genes are induced by root exudate (2,5,6). In the present study we used a *nod* A-*lac Z* fusion to analyse the induction of the *nod ABC* promotor by root exudate and by several plant flavonoids. The most active inducer of the *nod ABC* promotor, naringenin, was found to be able to replace root exudate in inducing the production of Tsr-factor.

MATERIALS AND METHODS

Strain RBL 5560 (pMP 154) harbouring both the Sym plasmid pJB5JI (3) as well as pMP 154, an Inc Q expression vector containing the *nod ABC* promotor of *R.leguminosarum* fused to *lac Z* (Spaink, unpublished), was used to detect *nod* gene inducers. Units of β-galactosidase were determined according to Miller (4).

Root exudate was produced as described earlier (8). For TLC (thin layer chromatographic) analysis, an ethanolic extract of freeze dried exudate was used. After chromatography, fragments of the plates containing the separated compounds were tested for *nod A* - *lac Z* induction.

RESULTS

The R.leguminosarum nodA promoter is efficiently induced by naringenin, eriodictyol, apigenin and luteolin. Preliminary characterization of the *nod A* gene inducer of *V.sativa* root exudate revealed the following characteristics. The activity is heat-stable (at 100°C for 20 min), it has a mol. wt. of less than 1000 d as judged by ultrafiltration, and dissolves in the aqueous phase of a petroleum ether-ethanol-water (10:7:3) mixture. Two dimensional cellulose TLC analysis commonly used for flavonoid pattern analysis, using *tert*.butanol-acetic acid-water (3:1:1) and 15% acetic acid as solvents in the first and second dimension, respectively (1), indicated a possible flavanone or isoflavanone nature of the major inducing activity. A number of commercially available flavanones, isoflavones and related compounds were tested for their nodA promotor inducing ability. Table 1 presents β-galactosidase activity obtained with the four most active compounds after 18 hrs of incubation. All tested isoflavones, among which genistein were inactive. Comparison of the chromatographic behaviour of the natural inducer present in root exudate and the four active flavonoids listed above on TLC cellulose using chloroform-acetic acid-water (10:9:1) as the solvent, revealed that this inducer is not one of these four inducers.

TABLE 1. Induction of the nodA promotor by various structurally closely related flavonoids.[a]

Inducer	Maximal response to inducers (units β-galactosidase x 10^{-3})	Concentration of inducer (nM) required for half-maximal induction
Naringenin	19,6	14
Eriodictyol	18,7	60
Apigenin	19,4	15
Luteolin	18,6	42
Control[b]	0,2	-

[a] The induction time was 18 hrs.
[b] Induction medium (20% B^- medium in Jensen medium (8)) without added inducers was used as control.

Naringenin can replace total root exudate for the induction of Tsr-factor production. The characteristics of induction by naringenin and root exudate are very similar with respect to the kinetics of induction and the genetic requirements for the bacterium. In both cases induction is already detectable after 7 min of incubation, and levels of β-galactosidase increase linearly in time until after 10 hrs a plateau is reached. For induction by either root exudate or naringenin the presence of only the nodA promotor and the nodD gene are sufficient.

R.leguminosarum with or without a Sym-plasmid (pJB5JI) was inoculated in either sterile root exudate or Jensen plant growth medium with or without 700 nM naringenin, at 5 x 10^5 bacteria per ml. After 24 hrs, bacteria were removed by filtration, and the sterile filtrates were tested for the presence of Tsr-factor on sterile V.sativa plants. The Tsr phenotype of the roots was quantified by the main root length. It appeared that naringenin could efficiently replace exudate for Tsr-factor production in the Sym plasmid harbouring strain. The presence of the Sym plasmid was a requirement for Tsr-factor production in both cases.

DISCUSSION

Judged from its chromatographic behaviour in the 2-dimensional TLC test system, the inducing compound in V.stativa root exudate could be of a flavanone or isoflavone nature.

Since commercially available isoflavones appeared to be inactive (e.g. genistein) and since the corresponding flavanones are very active inducers (e.g. naringenin), the inducer in a root exudate is probably of flavanone nature.

Naringenin can replace root exudate for the induction of the nodA gene, as well as for Tsr-factor production. The fact that naringenin can also replace total root exudate for inducing Tsr-factor production by R.leguminosarum means, that Tsr-factor is either an endogeneous bacterial product, or that naringenin, as well as the inducer in exudate can be converted into Tsr-factor. These possibilities are presently under investigation.

ACKNOWLEDGEMENTS

We thank Ms. I. Mulders and Mr. T. Tak for their skilful technical assistance. The investigations were partly supported by the Foundation for Fundamental Biological Research (BION), which is subsidized by the Netherlands Organization for the Advancement of Pure Research (ZWO).

LITERATURE CITED

1. Harborne, J.B. (1971) In J.B. Harborne, D. Boulter, and B.L. Turner (ed.), Chemotaxonomy of the Leguminosae. Academic Press, London and New York, p 31-71.
2. Innes, R.W., P.L. Kuempel, J. Plazinski, H. Canter-Cremers, B.G. Rolfe, and M.A. Djordjevic (1985) Mol. Gen. Genet., 201: 426-432.
3. Johnston, A.W.B., J.L. Beynon, A.V. Buchanon-Wollaston, S.M. Setchell, P.R. Hirch, and J.E. Beringer (1978) Nature (London), 276: 634-636.
4. Miller, J.H. (1972) Experiments in molecular genetics. Cold Spring Harbor Laboratory, New York.
5. Mulligan, J.T., and S.R. Long (1985) Prod. Natn. Acad. Sci. U.S.A., 82: 6609-6613.
6. Rossen, L., C.A. Shearman, A.W.B. Johnston, and J.A. Downie (1985) EMBO J., 4: 3369-3373.
7. Van Brussel, A.A.N., T. Tak, A. Wetselaar, E. Pees, and C.A. Wijffelman (1982) Plant Sci. Lett., 27: 317-325.
8. Van Brussel, A.A.N., S.A.J. Zaat, H.C.J. Canter Cremers, C.A. Wijffelman, E. Pees, T. Tak, and B.J.J. Lugtenberg (1986) J. Bacteriol., 165: 517-522.

An ntrC homologue in B. japonicum

W. Szeto and F. Cannon, BioTechnica International, Inc.,
85 Bolton Street, Cambridge, MA 02140, USA

INTRODUCTION

Bacteria of the species Rhizobium or Bradyrhizobium can elicit nitrogen fixing root nodules during symbiosis with certain legumes. The establishment of the symbiosis is a multi-stepped, interactive process which requires the expression of specific plant and bacterial genes. In. R. meliloti, R. leguminosarum and B. japonicum, the product of a nif-specific regulatory gene nifA is required for activating the transcription of the nitrogenase structural genes nifHDK as well as other nif operons (1,2,3). [In R. meliloti and R. leguminosarum (both fast-growing), nifHDK is transcribed as a single operon (4,5,6). In contrast, in B. japonicum (slow-growing), nifH and nifDK are transcribed as two separate operons (7,8).] The mechanisms regulating nifA expression during symbiosis are unknown.

In the enteric bacterium Klebsiella pneumoniae, nitrogen fixation occurs as a direct response to nitrogen limitation and low oxygen tension. Under these physiological conditions, the product of the general nitrogen assimilatory regulatory gene ntrC (KpntrC) activates the transcription of the nifLA operon. The nifA product in turn activates the transcription of all the other nif genes (for review, see 9). The nifL product is also regulatory. It appears to antagonize the action of the nifA protein in the presence of oxygen (10). In addition to activating the transcription of nifLA, KpntrC also regulates the expression of a variety of nitrogen utilization and assimilation pathways (11). With the exception of nifLA, KpntrC cannot activate the expression of other nif operons.

Recent studies have shown that R. meliloti harbors an ntrC homologue (RmntrC) (12). Although RmntrC is similar to KpntrC in its general nitrogen regulatory properties, RmntrC differs from KpntrC in its action toward nitrogen fixing genes. The RmntrC protein directly activates the transcription of several R. meliloti nif promoters (including nifH, fixA and nifB) in free-living cells grown in nitrogen-limiting media. Because nitrogen starved RmnifA, but not RmntrC mutants, can induce nif gene transcription ex planta, such nif gene expression is absolutely dependent on the RmntrC (but not the RmnifA product). In contrast, nodules produced by RmntrC mutants are fully Fix$^+$ whereas those elicited by RmnifA mutants are completely Fix$^-$. These results demonstrate that both RmntrC and RmnifA proteins can activate the transcription of R. meliloti nif genes; which regulatory gene is operative (RmntrC or nifA) depends on whether the cells are free-living or have entered into the symbiotic state. Furthermore, RmnifA expression during symbiosis is not dependent on the action of RmntrC.

Whether rhizobial species in general have separate, developmentally dependent regulatory pathways for activating nif genes is unknown. The role of centralized nitrogen regulation, if any, in nif gene expression in soil symbionts is also unclear. Indeed, the presence of an ntrC homologue in rhizobial species other than R. meliloti has not yet been demonstrated. The results of studies described below suggest that B. japonicum has an ntrC homologue (BjntrC). Unlike RmntrC, BjntrC does not appear to be required for the expression of other B. japonicum nif genes (including nifA) during ex planta nitrogen fixation.

RESULTS AND DISCUSSION

Activation of R. meliloti nif promoters in nitrogen-starved B. japonicum cells

The R. meliloti (Rm) and B. japonicum (Bj) nif fusion plasmids (12,13) [RmnifH::lacZ (pRmH), RmfixA::lacZ (pRmA), BjnifH::lacZ (pBjH) and BjnifD::lacZ (pBjD)] were used to investigate whether B. japonicum cells produce an ntrC-like activity during nitrogen deprivation. Our studies showed that when B. japonicum cells containing plasmids pRmH or pRmA were grown aerobically in nitrogen-limiting medium, they produced 8 to 10 times more beta-galactosidase (assayed as described in ref. 12) than when growth occurred under the same conditions in nitrogen-excess medium. In contrast, B. japonicum cells harboring plasmids pBjH or pBjD produced low levels of beta-galactosidase whether grown in nitrogen-limiting or nitrogen-excess medium. These results are summarized in Table 1.

TABLE 1. Activity of R. meliloti and B. japonicum nif promoters in nitrogen-starved B. japonicum cultures.

Plasmid	Beta-galactosidase activity $+O_2$, +N	$+O_2$, -N
RmnifH::lacZ	-	++
RmfixZ::lacZ	-	++
BjnifH::lacZ	-	-
BjnifD::lacZ	-	--

$+O_2$, +N = grown aerobically in nitrogen-excess medium
$+O_2$, -N = grown aerobically in nitrogen-limiting medium
nitrogen-excess medium = minimal medium + 0.2% glutamine
nitrogen-limiting medium = minimal medium + 0.0]% glutamine
Minimal medium contains 2% sodium gluconate as carbon source and mineral salts and trace elements as described in ref. 4.

The factor synthesized in nitrogen-starved B. japonicum cells that can activate R. meliloti but not B. japonicum nif promoters cannot be the B. japonicum nifA protein. More likely, it is the product of the B. japonicum ntrC homologue. This is because ntrC is induced by nitrogen limitation in enteric bacteria (11), as well as in R. meliloti (12), and therefore also likely to be the case in B. japonicum. Furthermore, studies have shown that the R. meliloti nifH and fixA promoters can be activated by both the ntrC and nifA products (12) whereas the B. japonicum nifH and nifD promoters can only be activated by the nifA protein (13).

The R. meliloti nif promoter upstream sequences are not required for activation by nitrogen-limitation in B. japonicum

The data summarized in Table 2 show that various deleted forms of RmnifH and RmfixA (15) can be activated to the same levels as the promoters which contained large lengths of promoter upstream sequences in nitrogen-deprived B. japonicum cultures.

TABLE 2. Requirement for R. meliloti nif promoter upstream sequences during activation by nitrogen-limitation in B. japonicum.

Promoter	Bp upstream sequence (from mRNA start)	Beta-galactosidase activity $+O_2$, +N	$+O_2$, -N
RmnifH	795	-	++
	135	-	++
	95	-	++
	35	-	++
	25	-	--
	10	-	--
RmfixA	503	-	++
	168	-	++
	88	-	++
	73	-	++
	8	-	--
	3	-	--

See legend to Table 1 for description of symbols.

Earlier studies have shown that activation of Rm nif promoters by ntrC does not require promoter sequences upstream of -35 (16). Therefore, the results of the studies described above reinforce the notion that the B. japonicum ntrC homologue is the factor responsible for activating the R. meliloti nif promoters in nitrogen-starved B. japonicum cells.

Activation of *nif* promoters in free-living *B. japonicum* cultures by oxygen limitation

Neither *B. japonicum* nor *R. meliloti nif* promoters can be activated in aerobic cultures of *B. japonicum* in nitrogen-excess medium. However, in the same medium, if the cells are shifted to an oxygen-limited environment, the transcription of both the *B. japonicum* and *R. meliloti nif* promoters can be activated (3- to 5-fold above uninduced level). The results of these studies are summarized in Table 3.

TABLE 3. Activation of *R. meliloti* and *B. japonicum nif* promoters in *B. japonicum* cultures by oxygen deprivation.

Plasmid	Beta-galactosidase activity $+O_2$, +N	$+O_2$, -N
Bj*nif*H::*lac*Z	-	+
Bj*nif*D::*lac*Z	-	+
Rm*nif*H::*lac*Z	-	+
Rm*fix*A::*lac*Z	-	+

$+O_2$, +N = aerobic cultures in nitrogen-excess medium
$+O_2$, -N = oxygen-limited cultures in nitrogen-excess medium

Because both Bj and Rm *nif* promoters can be activated by oxygen limitation under conditions where *ntr*C is not induced (growth in nitrogen-excess medium), the induction of Bj*nif*A must be independent of Bj*ntr*C (and therefore independent of centralized nitrogen regulation). Rather, *nif*A production in free-living *B. japonicum* cultures is absolutely dependent on oxygen tension. Further studies are required to determine whether or not such is the case during symbiosis.

Hybridization of Rm*ntr*C DNA to Bj genomic DNA

When ^{32}-P-labeled Rm*ntr*C DNA is hybridized to EcoR I restricted Bj genomic DNA, strong hybridization to a 2.8 kb DNA fragment is detected. This DNA fragment is distinct from those that hybridized to Bj*nif*A DNA (data not shown). These results provide physical evidence that *B. japonicum* possess an *ntr*C homologue that is distinct from the *nif*A gene.

CONCLUSION

The results of our studies suggest that in *B. japonicum*, centralized nitrogen regulation plays no role in *nif* gene expression *ex planta*. Rather, nitrogen fixation in free-living *B. japonicum* cells is activated when the cells are put in an oxygen-limited environment.

REFERENCES

1. Szeto WW, Zimmerman JL, Sundaresan V, and Ausubel, FM. 1984. Cell 36: 1035-1043.
2. Schetgens RMP, Hontelez JGL, van den Bos RC, and van Kammen A. 1985. Mol. Gen. Genet. 200: 368-374.
3. Fisher HM, Alvarez-Morales A, and Hennecke H. 1986. Embo J. 6: 1165-1173.
4. Weber G and Puhler A. 1982. Plant Mol. Biol. 1: 305-320.
5. Corbin D, Barren L, and Ditta G. 1985. Proc. Natl. Acad. Sci. USA 80: 3005-3009.
6. Zimmerman JL, Szeto WW, and Ausubel FM. 1983. J. Bacteriol. 156: 1025-1034.
7. Kaluza K, Fuhrmann M, Hahn M, Regensburger B, and Hennecke H. 1983. J. Bacteriol. 155: 915-918.
8. Adams TH, McLung CR, and Chelm BK. 1984. J. Bacteriol. 159: 857-862.
9. Ausubel FM. 1984. Cell 37: 5-6.
10. Buchanan-Wollaston V, Cannon MC, and Cannon FC. 1981. Mol. Gen. Genet. 184: 102-106.
11. Magasanik B. 1982. Ann. Rev. Genetics 10: 135-168.
12. Szeto WW and Ausubel FM. Submitted for publication.
13. Alvarez-Morales A and Hennecke H. 1985. Mol. Gen. Genet. 199: 306-314.
14. O'Gara F and Shanmugam KT. 1976. Biochim. Biophys. Acta 437: 313-321.
15. Better M, Ditta G, and Helinski DR. 1985. Embo J. 4: 2419-2424.
16. Szeto W, Ronson C, and Ditta G. Unpublished results.

GLUTAMINE SYNTHETASES OF RHIZOBIUM LEGUMINOSARUM

S.COLONNA-ROMANO, R.DEFEZ, M.FILSER, M.GUIDA, M.IACCARINO, A.LAMBERTI AND A.RICCIO International Institute of Genetics and Biophysics, CNR
Via Marconi 10, 80125 Naples
A. FUGGI, Dipartimento di Biologia Vegetale, Univ. of Naples, Italy
W. ARNOLD, U. PRIEFER and A. PÜHLER, Univ. of Bielefeld, W.Germany

Rhizobium bacteria use ammonia for growth in the free-living state, but in the nitrogen fixing bacteroids all ammonia is exported to the plant fraction of the nodule (Bergersen and Turner, 1967). Thus the enzymes for ammonia assimilation need to be regulated differently in the two bacterial states. Another peculiarity of Rhizobiaceae is the presence of two glutamine synthetases (GS; EC 6.3.1.2): GSI, which is similar to the GS of enteric bacteria, is regulated by adenylylation and is relatively heat stable; and GSII, which is heat labile and not known to be modified after translation (Darrow et al, 1981; Somerville and Kahn, 1983; Donald and Ludwig, 1984; Carlson et al, 1985). A coordinated nitrogen control system similar to that of enteric bacteria (Magasanik, 1982) has not been described in Rhizobium spp.. We started a study of the GS activities of R.leguminosarum biovar viceae, strain RCC1001, in order to understand nitrogen assimilation and its regulation in this species (Filser et al, 1986).

We assayed GS transferase (Ferguson and Sims, 1971) activity in crude extracts of strain LPR1105 (a rif-r derivative of RCC1001). After preincubation at 55°C a heat stable and a heat labile fraction were observed. Since the heat labile fraction is completely inactivated after 5 min preincubation at 55°C, while the heat resistant fraction is quite stable, it is possible to calculate their specific activity in crude extracts of R.leguminosarum grown on different nitrogen sources. Table 1 shows that on nitrate the activity is mostly heat labile, on glutamine it is mostly heat resistant, while on YMB (a yeast extract-based medium; Hooykaas et al, 1977) 50% of the activity is resistant and 50% is labile. It appears, therefore, that the heat resistant activity is

Nitrogen source	GS activity heat stable	GS activity heat labile
nitrate	20	840
glutamine	100	<10
y. extract	28	27

TABLE 1: Specific activity (nmoles min^{-1} mg^{-1}) of heat stable and heat labile fractions of total GS transferase activity in crude extracts of R.leguminosarum grown in minimal medium supplemented with different nitrogen sources.

repressed by nitrate or induced by glutamine, while the opposite result is observed with the heat labile activity.

Using a gene bank of R.leguminosarum DNA we (Filser et al. 1986) isolated 28 clones capable to suppress the Gln^- phenotype of the Klebsiella pneumoniae $glnA^-$ strain UNF1827. Crude extracts of some of these clones show a heat stable GS activity and we demonstrate below that one lack of growth on nitrate is due to GSI repression. If so, the K.pneumoniae gene product(s) involved in repression should either require a DNA site on which to act, or a R.leguminosarum gene product, missing in the deletion.

We purified GSI and GS-second from UNF1827(p7D9) and from UNF1827 (p4F7). The same purification procedure was successful for either strain, namely precipitation of nucleic acids with streptomycin, adsorption on Blue Sepharose and elution with AMP. When analyzed by SDS-PAGE the enzymes were >95% and >99% pure and showed only one band 60,000 and 63,000 in molecular weight respectively. These bands give a positive signal after immunoblot analysis using an antiserum against E.coli GS (kindly provided by Dr. E. Stadtman) or an antiserum against purified GSI. The signal obtained with GS-second is weaker with either antisera. These experiments show that GSI and GS-second share common antigenic determinants, that are made of a single subunit and that the molecular weight of GS-second subunit is higher than that reported for GSII (Darrow et al 1981). On non-denaturing PAGE, GSI and GS-second show activity bands with apparent molecular weights of 660,000 & 570,000 respectively, comigrating with bands observed in crude extracts of R.leguminosarum. The two enzymes show a different ratio of biosynthetic to transferase activity. Similar differences were found in crude extracts of R.leguminosarum and were reported (Filser et al, 1986) in crude extract of UNF1827(p7D9) and of UNF1827(p4F7). The pH optimum profile of GSI is different from that of GS-second, both in the biosynthetic and in the transferase assay. Pure GSI is stable when preincubated at 56°C while GS-second is unstable ($t_{1/2}$ 3.8 min). Treatment with phosphodiesterase increases the biosynthetic activity of GSI, suggesting that it may be controlled by reversible adenylylation. Phosphinothricin competitively inhibits GSI and GS-second with a K_i of 0.6 uM and 0.06 uM respectively.

Research work supported by CNR, Italy. Special grant I.P.R.A. Subproject 1. Paper N.936. Also partially supported by PIERREL SpA.

REFERENCES

1. Bergersen, FJ and GL Turner (1967) Biochim. Biophys. Acta 141:507-515.
2. Carlson, TA et al (1985) J. Bacteriol. 162:698-703.
3. Chelm, BK et al (1985) in "Nitrogen Fixation Research Progress" HJ Evans, PJ Bottomley and WE Newton, eds, Martinus Nijhoff, p. 217.
4. Darrow, RA et al (1981) in "Current Perspectives in Nitrogen Fixation" AH Gibson and WE Newton, eds., Australian Academy od Sciences, pp. 182-185.
5. Donald, RG and RA Ludwig (1984) J. Bacteriol. 158: 1144-1151.
6. Ferguson, AR and AP Sims (1971) J. Gen. Microbiol. 69: 423-427.
7. Filser, MMK et al (1986) J. Gen. Microbiol. 132, in press.
8. Hooykaas, PJJ et al (1977) J. Gen. Microbiol. 98: 477-484.
9. Magasanik, B (1982) Ann. Rev. Genetics 16: 135-168.
10. Somerville, JE and ML Kahn (1983) J. Bacteriol. 156: 168-176.
11. Tumer, NE et al (1983) Nature 306: 337-342.

of them, p7D9, codes for GSI. A heat labile GS activity for which we report preliminary data, was found in another clone, p4F7. Since it is not clear that the heat labile GS activity of this clone is analogous to the GSII described by Darrow et al (1981) we provisionally call it GS-second. Clone p4F7 does not hybridize to p7D9 (Filser et al 1986) and DNA sequencing confirms that the inserts are different.

Hybridization of a fragment of these two clones to total DNA of R.leguminosarum shows colinearity of several restriction sites, thus demonstrating that the insert is R.leguminosarum DNA and colinear to it. Hybridization to the symbiotic plasmid pSym DNA shows that the two clones are not part of this plasmid.

The insert of p7D9 is 27 kb long. We subcloned 6.5 kb of this DNA and obtained a plasmid, pMG10, that complements strain UNF1827 and hybridizes to the DNA of the K.pneumoniae glnA gene. We generated deletions of pMG10 by digestion with Bal31 nuclease and found that the smallest Gln^+ plasmid, still able to complement strain UNF1827, contains 2 kb of R.leguminosarum DNA. We introduced pMG10 and its deletion derivatives into the minicells producing strain DS998 and, after SDS-PAGE of ^{35}S-methionine labeled minicells, we observed a band, 58,000 in molecular weight, specific of the insert. This band is present in the smallest Gln^+ deletion, is absent when the Bal31 deletion is generated starting from one side of pMG10 and is reduced in size (33,000) when the Bal31 deletion is generated from the opposite side. We conclude that a protein, 58,000 in molecular weight, is encoded by pMG10 and is necessary for the Gln^+ complementation of strain UNF1827. Furthermore, this experiment allows us to predict the direction of transcription. We call glnA the gene present in pMG10 and GSI the protein encoded.

We sequenced all Rhizobium DNA present in the smallest Gln^+ deletion derivative of pMG10 plus some flanking DNA , for a total of 2454 bp. A sequence with good homology to an ntrA-dependent promoter is found at position 174-191, followed by a putative ribosome binding site at position 235-240 and by an open reading frame of 111 codons (ORF 111). This ORF shows no homology to ntrA, ntrB, ntrC or nifA sequences. At its end there are no obvious sequences suggesting transcription termination and at position 564-599 there is a poor homology to the consensus of the standard E.coli promoter. A new putative ribosome binding site is present at position 650-655 followed by an ORF of 469 codons at position 665-2071. Further downstream we do not see any sequence suggesting transcription termination. The molecular weight of the amino acids encoded by the ORF at position 665-2071 is 60,464, in good agreement with the molecular weights of the band specific of the insert of pMG10 and of the pure protein (see below). The protein encoded by this ORF shows a high degree of homology with Anabaena GS (Tumer et al 1983). A more detailed comparison at the NH_2-terminal shows a high degree of homology with Anabaena GS and Bradyrhizobium japonicum GSI, while there is no homology with B.japonicum GSII (Chelm et al 1985).

Since GSI appears to be repressed when R.leguminosarum is grown on nitrate, we tested the behaviour of UNF1827 carrying different plasmids and found complementation of the Gln^- phenotype by p7D9 or pMG10 when ammonia is the nitrogen source, but not on nitrate. Of the 28 Gln^+ clones originally isolated 8 did not grow on nitrate and 3 of them, randomly chosen, were hybridized with an internal fragment of glnA and a positive signal was found. It appears therefore that GSI is repressed by nitrate in K.pneumoniae. A clone carrying a plasmid with a partial deletion expanding in the ORF 111 grows on nitrate indicating that this DNA is required for the putative nitrate repression. Experiments are in progress to check if

MOLECULAR ANALYSIS OF A FIX CLUSTER FROM RHIZOBIUM MELILOTI

D. KAHN, J. BATUT, P. BOISTARD, M.L. DAVERAN, M. DAVID, O. DOMERGUE, A.M. GARNERONE, J. GHAI, C. HERTIG, D. INFANTE AND M.H. RENALIER
LABORATOIRE DE BIOLOGIE MOLECULAIRE, CNRS-INRA, B.P. 27, F 31326 CASTANET-TOLOSAN CEDEX, FRANCE

1. INTRODUCTION

Until recently the study of nitrogen fixation genes from Rhizobium has focused either on nif genes as they have been defined in Klebsiella pneumoniae (Dixon, 1984), or on fix genes closely linked to nif genes (such as fixABC, Pühler et al., 1984 ; Ausubel et al., 1985). However we may expect some non nif functions to be important for nitrogen fixation in planta :

- differentiation of bacteria into bacteroids,
- bacteroid specific metabolic functions,
- Rhizobium-dependent nodule maturation,
- plant dependent regulation of bacteroid functions.

Indeed many more fix genes are found by general mutagenesis of the Rhizobium genome than would be expected for nif genes (Meade et al., 1982). Moreover, Fix^- Nif^+ mutants have been obtained in Bradyrhizobium japonicum (Regensburger et al., 1986). Thus numerous fix genes are still to be characterized.

One approach to the discovery of new fix genes could be to look for sequences from the Rhizobium genome that are expressed specifically during symbiosis. Our strategy has been to focus on the megaplasmid pSym from R. meliloti, which was already known to contain numerous symbiotic genes (nif, fix and nod) (Rosenberg et al., 1981 ; Ruvkun et al., 1982 ; Corbin et al., 1982 ; Long et al., 1982), and to look for regions that would be specifically transcribed inside nodules. One such region has been found on a 300kb region of pSym in addition to the well-characterized nif-fix cluster (M. David et al., in preparation). Site-directed mutagenesis around this region has identified new fix genes 200kb from nifH (Batut et al., 1985ab).

In this paper, we report several new findings on these fix genes :

1. They are clustered over approximately 12kb ; a 5kb region of this cluster is duplicated and contains functional fix gene(s).
2. The 5kb duplication is specifically transcribed during symbiosis, independently of the nifA product.
3. We describe four new fix genes, fixGHIX, which constitute the transcription unit I of Batut et al. (1985b) ; sequence analysis suggests these genes encode a membrane-bound complex involved in a redox process essential for nitrogen fixation in planta.

2. A FIX GENE CLUSTER WITH A FUNCTIONAL DUPLICATION

Since we had indications that a pSym region 200kb from nifH was involved in symbiotic nitrogen fixation, we undertook a thorough genetic analysis of this region by transposon and deletion mutagenesis of pTH2, a pLAFR1-derived episome containing 29kb of this region of pSym. Three fix complementation groups were determined in a recA background (Batut et al., 1985ab) (Figure 1). fix regions I and II are separated by a 5kb spacer in which transposon or deletion mutagenesis yielded a normal Nod$^+$ Fix$^+$ phenotype. For instance, strain HG30P2, deleted of two contiguous 1.9kb and 1.2kb PstI fragments (empty box above physical map in Figure 1), was Fix$^+$. However expression studies showed that this cryptic spacer region was actively transcribed in the nodule (see below), which suggested it was involved in the symbiotic process. Therefore we hypothesized that a functional reiteration of this symbiotically expressed cryptic region might complement the mutagenized copy. Using the 13.8kb HindIII fragment containing the spacer region as a probe we found hybridization to an 11.2kb HindIII fragment which we could map 40kb on the other side of nifH on pSym. A physical map of the reiterated sequence in the 13.8kb HindIII fragment (hatched box below physical map on Figure 1) showed a good correlation with the extent of the genetically silent spacer region. Random deletions were generated in the 11.2kb HindIII fragment following transposition of Mu dII1734 (Castilho et al., 1984) to pGMI467, a pLAFR1 derivative containing 33.4kb of pSym including the 11.2kb HindIII fragment. The deletions were recombined into pSym by marker exchange. All deletions showed a Fix$^+$ phenotype. However when deletions removing part or all of the reiterated region were transduced into strain HG30P2, they resulted in a Fix$^-$ phenotype. Plasmids containing either reiterated region (pTH2 and pGMI467) complemented the doubly deleted strains, which confirmed that both copies were functionally equivalent. Therefore the 5kb reiteration contains a functional duplication of fix gene(s).

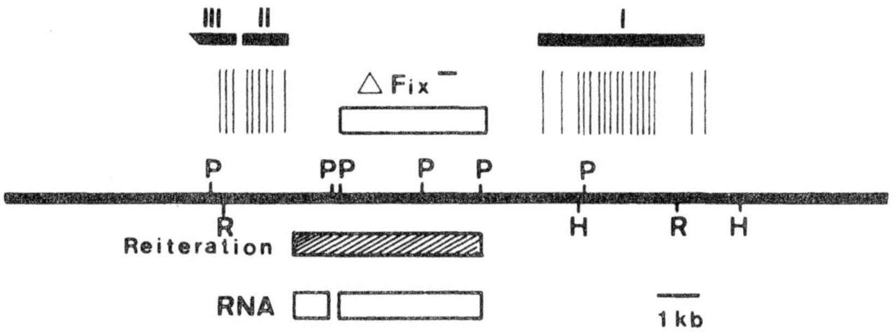

Figure 1. Physical and genetic map of the R. meliloti fix cluster. H = HindIII, P = PstI, R = EcoRI. Vertical bars indicate Fix$^-$ Tn5 insertions. The three fix complementation groups are indicated above the Tn5 insertions.

Few examples of reiterations have been described in procaryotic genomes. However the reiteration of symbiotic genes appears to be quite common among Rhizobium species (Quinto et al., 1985 ; Norel et al., 1985) ; this may complicate considerably the identification of new symbiotic genes. To date we can only speculate about the possible biological significance of such repetitions. They could allow increased synthesis of some gene products. Alternatively different copies may undergo different regulations (see for instance Hulett et al., 1985). However both copies of the fix region appear to be phenotypically expressed during the symbiotic life of R. meliloti.

3. NIFA-INDEPENDENT SYMBIOTIC-SPECIFIC EXPRESSION OF FIX GENES

In a search for new molecular markers of the symbiotic state, we have looked for sequences induced during symbiosis on a 300kb fragment of pSym including the nif-fix cluster. Endosymbiotic RNA from R. meliloti hybridized strongly to the nif-fix cluster and to the 13.8kb HindIII fragment containing fix regions II and III. Precise mapping showed that the active endosymbiotic region (boxed at the bottom of Figure 1) matched the functional duplication in the middle of the fix cluster (hatched box on Figure 1). This region showed very little transcriptional activity in free-living cultures of R. meliloti. To confirm symbiotic induction of the reiterated fix region, we constructed a pSym::lacZ translational fusion in the 1.9kb PstI fragment inside the reiteration using Mu dII1734. The beta-galactosidase activity of this pSym::lacZ52 fusion strain increased from 20U/mg protein in free-living cultures to 310U/mg protein in bacteroids. Therefore the reiterated fix region was induced in the symbiotic state.

It was of interest to know whether this fix region was regulated like the classical nif and fix genes, i.e. whether it depended on nifA-specific activation (Szeto et al., 1984). Therefore we hybridized RNA from bacteroids of the R. meliloti nifA::Tn5 strain 1354 with DNAs from the reiterated fix region and from the nifHDK operon. By contrast with nifHDK, whose transcription was abolished by the nifA mutation, the entire fix reiteration was transcribed at a similar level in the nifA and the wild-type background. This result was confirmed by using a pTH2::lacZ fusion strain in both backgrounds (Table 1).

TABLE 1. Beta-galactosidase activities of lacZ fusions in R. meliloti bacteroids.

Strain	Relevant genotype		Beta-galactosidase (U/mg protein)
GMI5600	lac		11
GMI5600 (pRKP9)	lac,	pRKM1 nifH::lacZ	12000
GMI5601 (pRKP9)	lac nifA1354,	pRKM1 nifH::lacZ	80
GMI5600 (pTH2.52)	lac,	pTH2::lacZ52	1100
GMI5601 (pTH2.52)	lac nifA1354,	pTH2::lacZ52	450

To our knowledge this is the first description of fix genes induced during symbiosis independently of nifA. However, we have found that nifA itself was expressed symbiotically in the absence of a functional NifA product (presumably from the P_{nifA} promoter, M. David and D. Kahn unpublished observation). We are presently investigating the genetic circuitry of these nifA-independent symbiotic regulations.

4. SEQUENCE ANALYSIS OF A CONSERVED FIX OPERON

Among the various operons of this fix cluster, operon I is remarkable because it is highly conserved among fast-growing Rhizobia (in collaboration with P. Hirsch). First, a probe, pDD27, containing 3.2kb of operon I, hybridized strongly with genomic DNA of all fast-growing strains tested. In several cases, it could be shown that the homologous sequences were located on pSym. Second, a deletion mutant in operon I could be complemented by the pSym plasmids pRL1JI from R. leguminosarum and pRtr5a from R. trifolii. Therefore the function of this fix operon has been conserved, which suggests it is important for various fast-growing Rhizobia.

We have performed sequence analysis of this operon to help raise hypotheses on its function. From the 4.3kb of sequence data now available, three ORFs are clearly recognizable, corresponding to fixG, fixH and fixI. Cytological phenotypes of strains carrying Tn5 insertions correlate with the ORFs in which the insertions map (J. Vasse and G. Truchet, personal communication). The initiator codons of fixH and fixI are preceded by purine-rich stretches. However, these codons are not predicted as being initiators by the procedure of Stormo et al. (1982), which screens a 101-nucleotide window along the RNA sequence, and therefore imposes a higher constraint on initiator sequences than the Shine and Dalgarno model (1974). Instead, an interesting ATGA motif is found between fixG and fixH, and between fixH and fixI : the **ATG**(A) initiator codons of fixH and fixI overlap the (A)**TGA** stop codons of fixG and fixH respectively, so that contiguous genes overlap on one nucleotide **A**(TGA), with a -1 frameshift. This is very suggestive of some translational coupling between adjacent genes of the operon. In addition, the same ATGA motif is found at the end of fixI, but it is associated with an initiator codon predicted by the method of Stormo et al. (1982) ; this ATG defines the start of a short ORF (55 amino-acids), which is very likely coding. Since we have no mutant in this short ORF, we can only presume it is a fix gene and we call it fixX.

The amino-acid sequences of part of FixG and of FixH, FixI and FixX have been analysed for potential transmembrane helices by the method of Eisenberg et al. (1984). The analysis predicts that all four proteins contain transmembrane sequences (hatched bars on Figure 2). In addition, the sequences have been compared with protein sequences translated from GENBANK or from NBRF. FixG is found to contain two cysteine clusters typical of iron-sulfur clusters from bacterial ferredoxins (full bars on Figure 2) :

CysxxCysxxCysxxxCys.

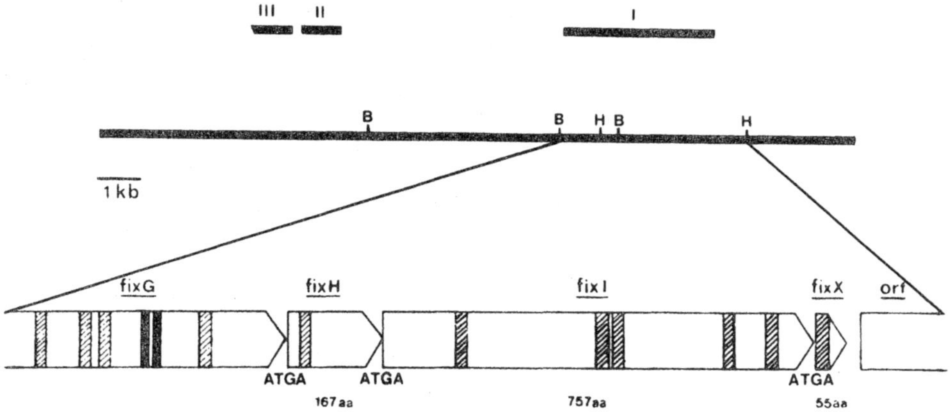

Figure 2. Sequence analysis of operon I. The four ORFs are shown. Predicted transmembrane sequences are indicated by hatched bars. The two solid bars across fixG correspond to potential iron-sulfur centers. B = BamHI, H = HindIII.

Therefore we suggest FixG contains two iron-sulfur centers and is a redox protein. Because of the suggested translational coupling between fixG, fixH, fixI and fixX, we hypothesize that the four proteins FixG, FixH, FixI and FixX participate in a membrane-bound complex involved in a redox process important for symbiotic nitrogen fixation. Three possibilities may be raised concerning the function of such a complex :

1. It could be a high affinity oxidase required for nitrogen fixation in the anoxic conditions that prevail inside nodules (Bergersen and Turner, 1980).

2. Or it could be a dehydrogenase needed to catabolize some organic acid provided by the plant ; however, no homology has been found with succinate dehydrogenase, a typical membrane-located dehydrogenase.

3. Alternatively, it could be a membrane-bound structure required for electron flow to nitrogenase. In this respect, it is worthy to recall that nitrogen fixation by isolated bacteroids is highly sensitive to the transmembrane potential (Laane et al., 1979). As a consequence, there must exist a membrane protein involved in electron transport to nitrogenase in Rhizobium bacteroids.

We will investigate these hypotheses by using plasmid constructs which allow high expression of the fixGHIX operon in free-living cultures.

5. REFERENCES

Ausubel FM, Buikema WJ, Earl CD, Klingensmith JA, Nixon BT and Szeto W (1985) in Evans HJ, Bottomley PJ and Newton WE (eds), Nitrogen Fixation Research Progress, pp 165-171, Martinus Nijhoff, Dordrecht.

Batut J, Terzaghi B, Ghérardi M, Huguet M, Terzaghi E, Garnerone AM, Boistard P and Huguet T (1985a) Mol. Gen. Genet. **199**, 232-239.

Batut J, Boistard P, Debellé F, Dénarié J, Ghai J, Huguet T, Infante D, Martinez E, Rosenberg C, Vasse J and Truchet G (1985b) in Evans HJ, Bottomley PJ and Newton WE (eds.), Nitrogen Fixation Research Progress, pp 109-115, Martinus Nijhoff, Dordrecht.

Bergersen FJ, and Turner GL (1980) J. Gen. Microbiol. **118**, 235-252.

Castilho BA, Olfson P and Casadaban MJ (1984) J. Bacteriol. **158**, 488-495.

Corbin D, Ditta G and Helinski DR (1982) Proc. Natl. Acad. Sci. USA **80**, 3005-3009.

Dixon RA (1984) J. Gen. Microbiol. **130**, 2745-2755.

Eisenberg D, Schwarz E, Komaromy M and Wall R (1984) J. Mol. Biol. **179**, 125-142.

Hulett FM, Wang P, Sussman M and Lee JK (1985) Proc. Natl. Acad. Sci. USA **82**, 1035-1039.

Laane C, Krone W, Konings WN, Haaker H and Veeger C (1979) FEBS Lett. **103**, 328-332.

Long SR, Buikema WJ and Ausubel FM (1982) Nature **298**, 485-488.

Meade HR, Long SR, Ruvkun GB, Brown SE and Ausubel FM (1982) J. Bacteriol **149**, 114-122.

Norel F, Desnoues N and Elmerich C (1985) Mol. Gen. Genet. **199**, 352-356.

Pühler A, Aguilar OM, Hynes M, Müller P, Klipp W, Priefer U, Simon R and Weber G (1984) in Veeger C and Newton WE (eds.), Avances in Nitrogen Fixation Research, pp 609-619, Nijhoff-Junk/Pudoc, The Hague/Wageningen.

Quinto C, De La Vega H, Flores M, Leemans J, Cevallos MA, Pardo MA, Azpiroz R, De Lourdes Girard M, Calva E and Palacios R (1985) Proc. Natl. Acad. Sci. USA **82**, 1170-1174.

Regensburger B, Meyer L, Filser M, Weber J, Studer D, Lamb JW, Fischer HM, Hahn M and Hennecke H (1986) Arch. Microbiol. **144**, 355-366.

Rosenberg C, Boistard P, Dénarié J and Casse-Delbart F (1981) Mol. Gen. Genet. **184**, 326-333.

Ruvkun GB, Sundaresan V and Ausubel FM (1982) Cell **29**, 551-559.

Shine J and Dalgarno L (1974) Proc. Natl. Acad. Sci. USA **71**, 1342-1346.

Stormo GD, Schneider TD, Gold L and Ehrenfeucht A (1982) Nucl. Ac. Res. **10**, 2997-3011

Szeto WW, Zimmerman JL, Sundaresan V and Ausubel FM (1984) Cell **36**, 1035-1043.

6. ACKNOWLEDGMENTS

This work was supported in part by a grant "Fixation biologique de l'azote" from Elf-Aquitaine, EMC, Rhône-Poulenc and CDF-Chimie and by the European Communities Biomolecular Program. Sequence data treatments were performed using computer facilities at CITI2 in Paris on a DPS8 computer with the help of the French Ministère de la Recherche et la Technologie (Programme mobilisateur "Essor des Biotechnologies").

REGULATION OF THE NITROGEN FIXATION (*nif*) GENES IN *RHIZOBIUM MELILOTI*

SHIN-PING WANG, BIN-FU SHEN and SAN CHIUN SHEN
Shanghai Institute of Plant Physiology, Department of Molecular Genetics, Academia Sinica, Shanghai, China

1. INTRODUCTION

Rhizobium meliloti specifically infects alfalfa (*Medicago sativa*) and a few related plants. Nodule formation on the roots of the host plant is an early event that is specified by a set of genes called nod genes (Long, S.R. et al. 1982; Kondorosi, E. et al. 1984). The expression of these nod genes has been currently investigated. The main focus of our research has been the regulation of *nif* genes of *R. meliloti* from the free-living to symbiotic state of this rhizobia.

In *K. pneumoniae*, expression of *nif* genes is integrated with the control of the general nitrogen regulatory system (*ntr* system). No such control circuitry is shown in *Rhizobium* (Ausubel, F.M. et al. 1984). Here we present the findings concerning the property of Rm *nifHDK* operon promoter and its regulation in the free-living and symbiotic state of *R. meliloti*. Particular attention has been paid to the fundamental questions about how the *Rhizobium nifH* promoter turns to be activated via *nifA* in bacteriods and how the expression of *nifA* which constitutes only a basal activity in the free-living rhizobia is induced during symbiosis.

2. RESULTS AND DISCUSSION

2.1 Comparison of The Structural and Regulatory Properties of R. meliloti nifH Promoter (Rm nifH promoter) with that of K. pneumoniae (Kp nifH promoter).

The Kp *nifH* promoter activated by *nifA* and *ntrA* product is categorized as *nifA* activated promoter, while the Rm *nifH* promoter activated by *ntrC/nifA* and *ntrA* product, categorized as *ntrC* activated promoter. The *nif* promoter structure is highly conserved, e.g. the consensus sequence in the -12 region of Kp *nifH* promoter is CCCTGCA and in *R. meliloti* is TTTTGCA. By site-directed mutagenesis we were able to convert the *ntrA* activated Kp *nifH* promoter to the *ntrC* activated one with the alteration of the base pair of the consensus sequence CCCTGCA to TCCTGCA, CCTTGCA or TCTTGCA (Ow, D.E. et al. 1985). We have used the *nifH-lacZ* translational fusions to monitor the expression of these *nifH* promoters in heterologous state. When the plasmid containing Rm *nifH-lacZ* was introduced to *R. meliloti* and *E. coli* respectively, the Rm *nifH* promoter was only moderately active in free-living rhizobia. However, its expression increased about fivefold in *E. coli*. Since *E. coli ntrC* is homologous to Rm *nifA* (Szeto, W.W. et al. 1984), so the result of heterologous expression of Rm *nifH* promoter indicates the higher affinity of Rm *nifA* product than that of Rm *ntrC* product toward the Rm *nifH* promoter. Likewise, the *ntrC* activated Kp *nifH* promoter mutant also showed to be highly activated by *E. coli ntrC* product, but slightly activated by Rm *ntrC* product in free-living rhizobia.

2.2 Expression of Rm nifH Promoter in Free-Living and Symbiotic State of R. meliloti

When the ntrC activated Kp nifH-lac fusion and Rm nifH-lacZ fusion carried plasmids were introduced to wild strain R. meliloti and strains with background of $nifA^-$ or $ntrC^-$ respectively, their activities were followed from the free-living to symbiotic state of rhizobia. The Rm nifH-lacZ was expressed about 20fold higher in the symbiotic state than it was in the free-living state of the wild strain rhizobia. The fusion showed almost the same activity in free-living culture of the wildtype strain and nifA mutant. However, it was not derepressed in the bacteroid of nifA mutant. Contrarily, the Rm nifH-lacZ fusion when introduced to the ntrC mutant was not expressed in the free-living state, yet it was expressed markedly in the bacteriods of ntrC mutant just as it did in the bacteriods of the wildtype rhizobia. The Kp nifH-lacZ was expressed though only a detectable activity in the free-living rhizobia, yet it was also only expressed in the bacteriods of wildtype and ntrC mutant. The data also showed that the basal activity of Rm nifA which activates the Rm nifH promoter in the case of ntrC mutant remained unrepressed by NH_4^+.

2.3 Acceleration of nifH Promoter Expression by the Extracts of Rhizobia Infected Root of Alfalfa

The extracts of rhizobia infected root were prepared after the initial nodulation of the roots. When administrated to the rhizobia cells which harbour the nifH-lacZ fusion grown in nitrogen-limited medium, the extract of rhizobia infected root enhanced the expression of nifH-LacZ markedly in the wildtype strain or the ntrC mutant. The low level of nifH-lacZ expression in ntrC mutant is attributed to the slow growth of rhizobia affected by the defect of ntr system. Since the root extract did not exert any acceleratory effect on the nifH-lacZ expression in the nifA mutant, it indicates apparently that the acting principle residing in the root-extract activates the Rm nifA which in turn activates the nifH promoter of the rhizobia.

REFERENCES

Ausubel, F.M., et al. (1985). Nitrogen Fixation Research Progress--Proceedings of the 16th International Symposium on Nitrogen Fixation, 165-171.

Kondorosi, E., et al. (1984) Mol. Gen. Genet. 193:445-452.

Long, S.R., et al. (1982) Nature. 298:485-488.

Ow, D.W., et al. (1985). Jour. Bact. 161:868-874.

Szeto, W.W., et al. (1984) Cell. 36:1035-1043.

THE UNUSUAL SYMBIOSIS BETWEEN THE NITROGEN FIXING BACTERIUM ORS571 AND ITS HOST SESBANIA ROSTRATA: REGULATION OF NITROGEN FIXATION AND ASSIMILATION GENES IN THE FREE LIVING VERSUS SYMBIOTIC STATE

F. DE BRUIJN, K. PAWLOWSKI, P. RATET, U. HILGERT and J. SCHELL

Max Planck Institut für Züchtungsforschung, 5000 Köln 30, W. Germany

1. INTRODUCTION

The bacterial strain ORS571 assumes a unique position amongst nitrogen fixing species. In addition to its ability to fix nitrogen in aerial stem- as well as root nodules while in symbiosis with its host, the tropical legume *Sesbania rostrata*, it is also capable of fixing nitrogen in the free living state and growth on this fixed nitrogen as primary N-source (1,2). This suggests that in the symbiotic state an uncoupling of bacterial nitrogen fixation and assimilation (for growth) exists, as has been observed with symbiotically nitrogen fixing (brady)rhizobial species, while in the free living nitrogen fixing state these processes may be coupled and coordinately regulated, as has been observed with the diazotroph *Klebsiella pneumoniae* (*Kp*). This makes ORS571 a very interesting strain to study the regulation of nitrogen fixation (*nif*) and other nitrogen assimilation genes in the free living versus symbiotic state.

In *Kp* the processes of nitrogen fixation and assimilation are coordinately regulated by a central nitrogen regulation (*ntr*) and a *nif* specific (*nifLA*) control system. Under conditions of nitrogen deprivation the products of the *ntrC* and *ntrA* genes act in concert to activate the *nifLA* promotor and the promoters of the *glnA*, *aut*, *hut* and *put* operons. The *nifA* product, also acting in concert with the *ntrA* product, activates all other *nif* promoters. The *nifL* product, responding to rising levels of oxygen or fixed nitrogen in the cell, is involved in repression of the *nif* genes (see 4,5). The *ntrC* and *nifA* genes are functionally and structurally related (4,6) but some distinct differences between these two analogous regulatory proteins have evolved. While *nifA* can substitute for *ntrC* in the activation of *ntr* controlled promoters, *ntrC* cannot substitute for *nifA* in fully activating the other *nif* promoters (4,5).

In the (brady)rhizobial species studied so far *nif* and *ntr* regulation appears to have some aspects in common whith *Kp*. The *R. meliloti* (*Rm*) and *B. japonicum* (*Bj*) systems have been examined in most detail. *NifA* like genes have been found in both organisms and *nifA*$^-$ mutations lead to a strict Fix$^-$ phenotype, due to inability to derepress the *nif/fix* genes (7,8). Their effect on derepression of *nif/fix* promoters in the free living state differs: while *Bj nifA*$^-$ mutants fail to derepress the *nif/fix* genes, *Rm nifA*$^-$ mutations appear to have no effect (8,9). In *E. coli* (*Ec*), both the *Rm* and *Bj nifH* promoters can be activated by *Kp nifA*, but only the *Rm nifH* promotor also by *Ec ntrC*. Both *nifA* and *ntrC* mediated activation require the *ntrA* product (4,10).

An ntrC like gene has so far only been described for Rm. $NtrC^-$ mutants of Rm were found to be unable to derepress the nifH promotor in free living state and to be Aut^- and Hut^-, resembling the phenoype of Kp $ntrC^-$ mutants (27). However they were found to be Fix^+ (9, W. Szeto and F. Ausubel, pers. comm.).

Here we summarize our progress in the characterization of the nifA and ntrC genes of ORS571 and in the analysis of ORS571 glt gene(s). We propose a model for the regulation of nitrogen fixation and assimilation genes in the free living versus symbiotic state which is described in detail in Pawlowski et al. (11).

2. RESULTS

2.1. Cloning and characterization of nifA and ntrC like genes.

The Kp nifA gene was used as a hybridization probe to Southern blots of ORS571 genomic DNA. Two EcoRI fragments (15 and 5.2kb) showed a high degree of homology, while five other fragments hybridized to a lesser degree (11). A cosmid clone bank of ORS571 in pLAFR1 (12) was screened with the same probe and two classes of cosmids were identified. Class I cosmids carried the 5.2kb EcoRI fragment and class II the 15kb and a weaker hybridizing 3.0kb EcoRI fragment. The regions of primary nifA homology were narrowed down to ClaI fragments of 3.9 and 4.8kb respectively, and these fragments were subcloned in pACYC184 (13) to yield plasmids pRSA13 and pRSC13 (11; see Fig. 1). Since the nifA and ntrC genes from Kp and the nifA gene from Rm share a high degree of DNA homology (6), we determined the relative homology of the regions cloned in pRSA13 and pRSC13 to nifA and ntrC gene sequences from Kp and Ec. The region cloned in pRSA13 was shown to have a higher degree of homology with nifA, while that cloned in pRSC13 was more homologous with ntrC sequences. Preliminary DNA sequencing data of the nifA homologous region in pRSA13 and comparison with the Kp nifA sequence (6) revealed stretches of 60% homology and suggested the direction of transcription of the ORS571 ´nifA´ locus (Fig. 1), which was found to be linked to the previously identified nifHDK(E) cluster (14). This confirms independent results recently described by Donald et al. (15).

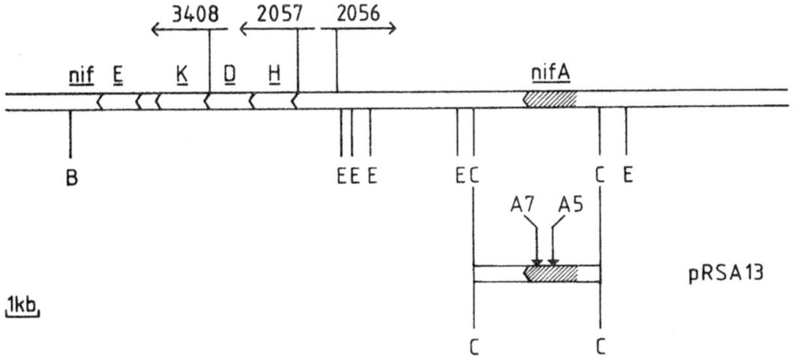

Figure 1. Physical and genetic map of the ORS571 nifHDK(E)-nifA region. B=BamHI, E=EcoRI, C=ClaI. The position of Tn5 insertions (A7, A5) and the position (orientation) of MudIIPR13 fusions are indicated by vertical and horizontal arrows respectively (11).

Plasmids pRSA13 and C13 were mutagenized with transposon Tn5 (16) and selected Tn5 insertions were used for gene replacement experiments (17). Two nifA::Tn5 (A5, A7; Fig. 1) and two ntrC::Tn5 (C6, C7) insertions were characterized by measuring nitrogen fixation (acetylen reduction) levels of free living cultures and detached nodules (Nif, Fix), by measuring GS and GOGAT activities and by examining growth on various aminoacids as primary nitrogen source (Aut, Hut, Put). The results are summarized in Table 1 and described in detail in ref. 11. NifA::Tn5 mutants are strictly Nif^-, Nod^+, Fix^-, Aut^+, Hut^+, and ntrC::Tn5 mutants $Nif^{-/+}$, Nod^+, Fix delayed, $Aut^{-/+}$, $Hut^{-/+}$. Neither class of mutations affected GS or GOGAT activity. These phenotypes, nif specific in the case of the presumptive nifA mutants and more pleiotrophic with regard to nitrogen assimilation genes in the case of ntrC mutants, resemble the phenotype of analogous mutants of Kp and Rm (4,5,9) and support the assignment of nifA and ntrC to these ORS571 loci.

The delayed Fix phenotype of the ntrC::Tn5 mutants was found to be striking. While both nifA and ntrC (as well as other Nif^-, Fix^-) ORS571 mutants induced the formation of unusually large numbers of small, light green, nodules on *S. rostrata* unable to fix nitrogen six weeks after infection, only in the case of the ntrC mutants, 5-10% of these inefficient nodules increased in size, became dark green and started fixing nitrogen by 10-15 weeks after infection. Bacteria were reisolated from such nodules and while in a few case apparent revertants were isolated, as a rule these delayed Fix^+ nodules contained unaltered ORS571 ntrC::Tn5 bacteria, as verified by Southern blotting.

Table 1.	Nif^a	Fix	Aut Hut	GS^a	$GOGAT^a$	nifD-lac β-gal units[b] $-NH_4^+$	$+NH_4^+$ or O_2
ORS571	+ (100%)	+	+	100%	100%	1630 (100%)	<1 (0%)
ORS571 $nifA^-$	- (0%)	-	+	100%	100%	<1 (0%)	<1 (0%)
ORS571 $ntrC^-$	-/+(10%)	delayed	-/+	100%	100%	250 (15%)	<1 (0%)

a. Expressed in % wild type activity.
b. ß-galactosidase units produced by nifD-lac fusion 3404 (Fig. 1).

2.2 Construction and characterization of nif-lac gene fusions.

To construct translational gene fusions of ORS571 genes to lacZ, the miniMu transposon MudIIPR13 (11) was used. The cloned nifHDK(E) region (14) was mutagenized with MudIIPR13 as described (18) and the nifH-lac (#2057), a nifD-lac (3408) as well as a fusion to an unknown gene (ORF) (#2056) were isolated (Fig. 1). Plasmids bearing these fusions were introduced into ORS571, ORS571 nifA::Tn5 and ntrC::Tn5 strains and ß-galactosidase activity was assayed under repressing and derepressing conditions in free living cultures and in the nodules (11). All three fusions are fully repressed by ammonium and/or oxygen (Table 1). In nifA::Tn5 mutants no derepression under any conditions was observed, while in the ntrC::Tn5 mutants derepression levels were reduced 7-fold, which is in agreement with the acetylene reduction levels found in these mutants strains ($Nif^{-/+}$, 10%). Insertion #2056 (Fig. 1) defines a previously unidentified nif/fix locus of ORS571, since the fusion it creates is regulated by NH_4^+ and O_2 via the nifA/ntrC regulatory system. Thus nif regulation by nifA and ntrC in ORS571 occurs at the level of transcriptional control of the nif promoters.

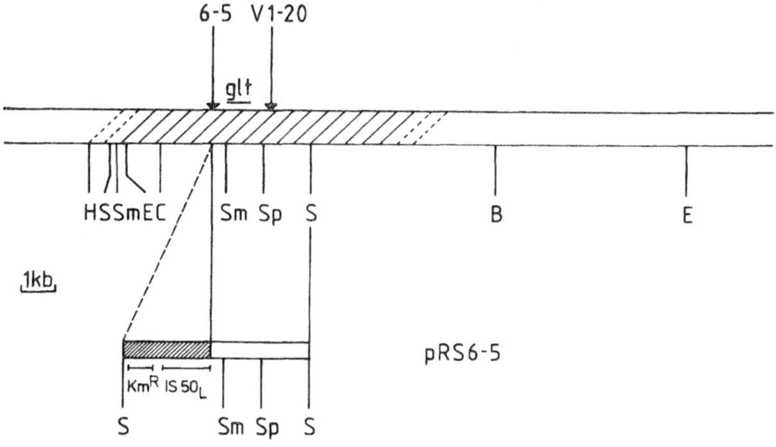

Figure 2. Map of the presumptive ORS571 glt (GOGAT) region. E=EcoRI, B=BamHI, C=ClaI, H=HindIII, S=SalI, Sm=SmaI, Sp=SphI. The positions of Tn5 in the two GOGAT$^-$ insertion mutants are indicated by vertical arrows (6-5, V1-20). The SalI fragment of 6-5 subcloned in pACYC184, including dark cross-hatched Tn5 derived sequences, is as indicated (pRS6-5).

2.3. Isolation of Glutamate synthase (GOGAT$^-$) mutants.

In ORS571 ammonium is assimilated via the GS-GOGAT pathway (19). In order to isolate mutations in this pathway and determine their effect on nitrogen fixation by ORS571, we carried out Tn5 mutagenesis experiments of ORS571, using the suicide vector pGS9 (20). In addition we examined a collection of Tn5 induced mutants of ORS571, generated with the suicide vector pSUP2021 (21) in Gent (22). In addition to Met$^-$, Cys$^-$, Ade$^-$, Leu$^-$ and Phe$^-$ auxotrophs, we identified two independent mutants with an Asm$^-$ phenotype as defined in Kp (23). Both mutants (KA6-5 and GAV1-20) were unable to grow on NH_4^+, but could be supplemented with Gln, Asn, Asp or Glu. The Km^R marker of Tn5 was used to clone part of the mutated KA6-5 locus (plasmid pRS6-5, see Fig. 2) and this plasmid was used to show that mutants KA6-5 and GAV1-20 carry simple Tn5 insertions, approximately 1.5kb apart in the ORS571 genome (Fig. 2). Both mutants were shown to lack (NADPH dependent) GOGAT activity and to have wild type levels of GS activity, suggesting that they carry Tn5 insertions in the structural gene(s) for GOGAT (glt) or in a regulatory gene specifically controlling GOGAT expression. These mutations were also found to lead to a Nif$^+$, Nod$^+$ and delayed Fix$^+$ phenotype, similar to that found for the ORS571 ntrC$^-$ mutants. This contrasts with the phenotype of ORS571 GOGAT$^-$ mutants described by Donald and Ludwig (19; Nif$^-$, Nod$^-$, Fix$^-$), but resembles that found for GOGAT$^-$ mutants of Rm (24) and cowpea strain 32HI (25). Apparently the lack of GOGAT activity in these mutant strains leads to accumulation of glutamine or ammonium to levels which repress nitrogen fixation by feedback inhibition of nitrogenase (26), or repression of the nif operons. That ORS571 GOGAT expression does not appear to be controlled by the ntrC gene described above, was deduced from the fact that ntrC::Tn5 mutants showed wild type levels of GOGAT activity.

3. CONCLUSIONS AND REGULATION MODEL

The ORS571 nifA gene product appears to be essential for the derepression of the nif/fix genes in the free living as well as symbiotic state. This is similar to the nifA functions in Bj (8), but contrasts with that found in Rm, where nifA is only involved in nif/fix gene derepression in the symbiotic state (9). The ORS571 ntrC gene also appears to play a profound role in nif/fix gene derepression in both states and its action is probably via nifA. Moreover, the ntrC gene we identified is involved in the regulation of arginine and histidine utilization genes, resembling in this respect its analogues in Kp (27,5) and Rm (9). In contrast to the Kp ntrC gene, the ORS571 ntrC gene does not appear to (coordinately) regulate the primary ammonium assimilation genes (glnA or glt).

The "leaky" phenotype of the ntrC::Tn5 mutants of ORS571 in the free living state (see Table 1) can be explained by assuming that the ntrC activated promoters have a low level of ntrC independent activity or by postulating that one of the other nifA/ntrC homologous fragments of ORS571 (see section 2.1.) carries an alternative, partially active ntrC gene.

The delayed Fix^+ phenotype of the ORS571 ntrC::Tn5 mutations and $GOGAT^-$ mutants is quite interesting. The effect of ntrC mutations in the symbiotic state may be indirect. Free living nitrogen fixation during the infection process may be required to establish proper nodule development. Therefore, in the absence of ntr-nifA mediated 'priming', poorly fixing/assimilating strains may be delayed in nodule (bacteroid) establishment. In the bacteroids a symbiotic control signal, as has been postulated for Rm (9,29) may replace the ntr system in activation of the nif/fix genes, presumably via nifA. Experiments are in progress to try analyse the ORS571 nifA promoter region and to identify potential (plant derived) signals, by examining the β-galactosidase induction of nifA-lac fusions under specific physiological conditions.

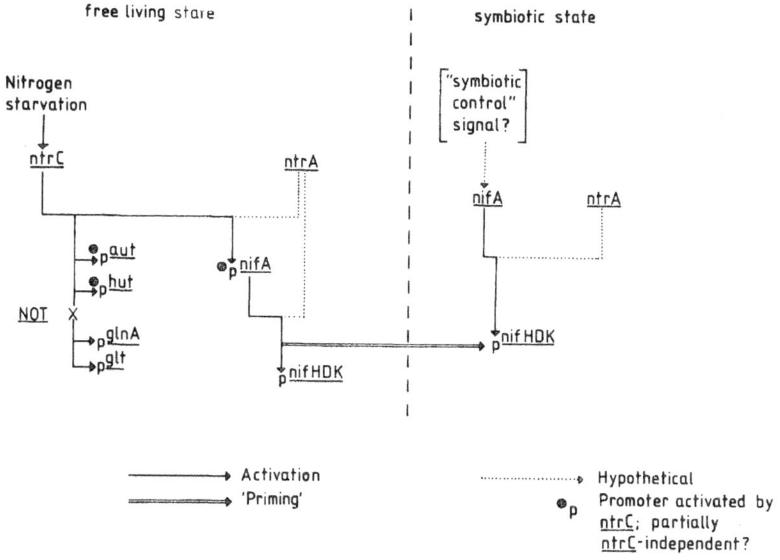

Figure 3. Model for the regulation of ORS571 nitrogen fixation genes.

REFERENCES

1. Dreyfus BL and Dommergues YR (1981) FEMS Microbiol. Let. 10, 313-317.
2. Elmerich C, Dreyfus BL, Reysset G, Aubert JP (1982) EMBO J. 1, 499-503.
3. Jarvis BDW, Gillis M, De Ley J (1986) Int. J. System. Bacteriol. 36, 129-138.
4. Ausubel FM (1984) Cell 37, 5-6.
5. Dixon R (1984) J. Gen. Microbiol. 130, 2745-2755.
6. Buikema WJ, Szeto WW, Lemley PV, Orme-Johnson WH, Ausubel FM (1985) Nucleic Acids Res. 13, 4539-4555.
7. Szeto WW, Zimmerman JL, Sundaresan V, Ausubel FM (1984) Cell 36, 535-543.
8. Fischer HM, Alvarez-Morales A, Hennecke H (1986) EMBO J. 5, 1165-1173.
9. Ausubel FM, Buikema WJ, Earl CD, Klingensmith JA, Nixon BT, Szeto WW (1985) In: Nitrogen Fixation Research Progress, Evans HJ, Bottomley PJ, Newton WE (eds), Martinus Nijhoff (Dordrecht, Boston, Lancaster), 165-171.
10. Alvarez MA, Betancou M, Kaluza K, Hennecke H (1986) Nucl. Acid. Res. 14, 4207-4227.
11. Pawlowski K, Ratet P, Schell J, De Bruijn FJ (1986) submitted to Mol. Gen. Genet.
12. Friedman AM, Long SR, Brown SM, Buikema WJ, Ausubel, FM (1982) Gene 18, 289-296.
13. Chang ACY and Cohen SN (1978) J. Bacteriol. 134, 1141-1156.
14. Norel F, Desnous N, Elmerich C (1985) Mol. Gen. Genet. 199, 352-356.
15. Donald RGK, Nees DW, Raymond CK, Loroch AL, Ludwig RA (1986) J. Bacteriol. 72-81.
16. De Bruijn FJ and Lupski JR (1984) Gene 27, 131-149.
17. Ruvkun, GB and Ausubel, FM (1981) Nature 289: 85-88.
18. Ratet P and Richaud R (1986) Gene 42, 185-192.
19. Donald RGK and Ludwig RA (1984) J. Bacteriol. 158 (3), 1144-1151.
20. Selvaraj G and Iyer VN (1983) J. Bacteriol. 156 (3), 1292-1300.
21. Simon R, Priefer U, Puehler A (1983) In: Molecular Genetics of the Bacteria-Plant Interaction, P{hler A (ed), Springer (Berlin, Heidelberg), 98-106.
22. Van den Eede G, Dreyfus B, Goethals K, Van Montagu M, Holsters M (1986) submitted to Mol. Gen. Genet.
23. Nagatani H, Shimizu M, Valentine RC (1971) Arch. Microbiol. 79, 164-175.
24. Kondorosi A, Svab Z, Kiss GB, Dixon RA (1977) Mol Gen Genet 151, 221-226.
25. Ludwig RA and Signer ER (1977) Nature 267, 245-248.
26. Kush A, Elmerich C, Aubert JP (1985) J. Gen. Microbiol. 131, 1765-1777.
27. De Bruijn FJ and Ausubel, FM (1981) Mol. Gen. Genet. 183, 289-297.
28. Lupski JR, Projan SJ, Ozaki LS, Godson GN (1986) Proc. Natl. Acad. Sci. USA, in press.
29. De Bruijn FJ, Sundaresan V, Szeto WW, Ow DW, Ausubel FM (1984) In: Advances Nitrogen Fixation Research, Veeger C, Newton WE, (eds) Nijhoff/Junk (The Hague), 627-633.

ANALYSIS OF *AZORHIZOBIUM SESBANIAE* ORS571 N_2 FIXATION GENES

ROBERT A. LUDWIG, ROBERT G.K. DONALD, ALBERT I. LOROCH AND DAVID W. NEES

1. ORS571 EXHIBITS FOUR NIF GENE LOCI.

The Vi mutagenesis/cloning technique has been used to clone and physically map the *Azorhizobium sesbaniae* ORS571 N_2 fixation (Nif) genes. Genomic maps have been drawn for four *nif* gene loci. Nif-locus 1 encodes at least three operons: the *nifH$_1$nifDnifK* operon, the *nifE* and additional *nif* gene(s) operon, and the *nifA* operon Fig. 1). Nif-locus 2 carries a second *nifH* (*nifH$_2$*) gene, encoded in one operon, and a cluster of three *nif* genes, homologous to the *fixABC* genes of other rhizobia, in a second operon (Fig. 2). Nif-locus 3 encodes a *nifB* operon (Fig. 3), and Nif-locus 4 carries an operon with a single *nif* gene (Fig. 4). Genetic complementations were conducted with subcloned plasmids carrying various *nif* genes. Although most Nif::Vi mutants were complemented by the appropriate wild-type alleles carried by subcloned plasmids, certain *nifD* and *nifK* mutants were not complemented. Failure to complement may have resulted from heterologous subunit mixing in the assembly of N_2ase component I.

2. NIF mRNA TRANSCRIPTION STUDIES.

The Nif mRNA transcripts have been studied both in culture and *in planta*. Both *nifH$_1$* and *nifH$_2$* genes are actively transcribed. *nifA*::Vi mutants do not show transcription of the *nifH$_1$nifDnifK* operon. From these and Tn*5lac* fusion studies, the *nifA* gene indeed acts as a positive activator of ORS571 *nif* genes.

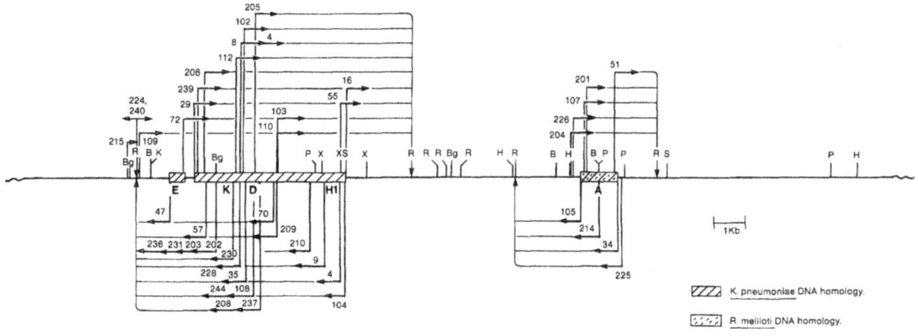

FIGURE 1. ORS571 Nif-locus 1. Sites of Nif::VP2021 insertions, endpoints of DNA inserts of genomically overlapping recombinant phages and plasmid subclones, and regions of heterologous DNA homology are presented. Vi mutants are represented by identifying strain numbers. Abbreviations: B, *Bam*HI; Bg, *Bgl*II; H, *Hin*dIII; K, *Kpn*I; P, *Pst*I; R, *Eco*RI; S, *Sal*I; X, *Xho*I.

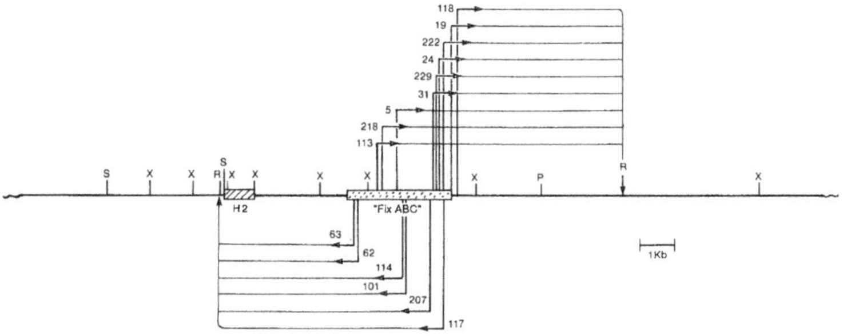

FIGURE 2. ORS571 Nif-locus 2.

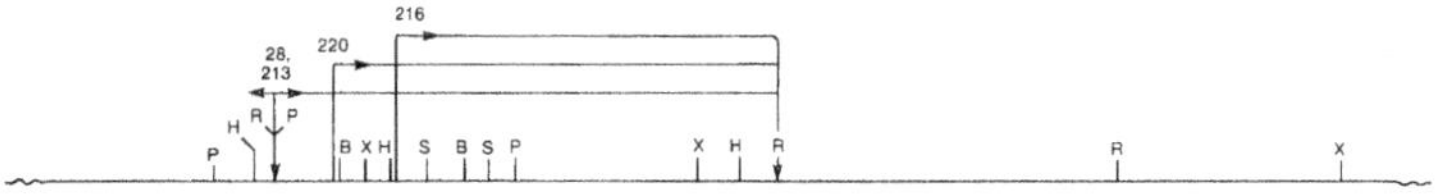

FIGURE 3. ORS571 Nif-locus 3, exhibiting *nifB* homology.

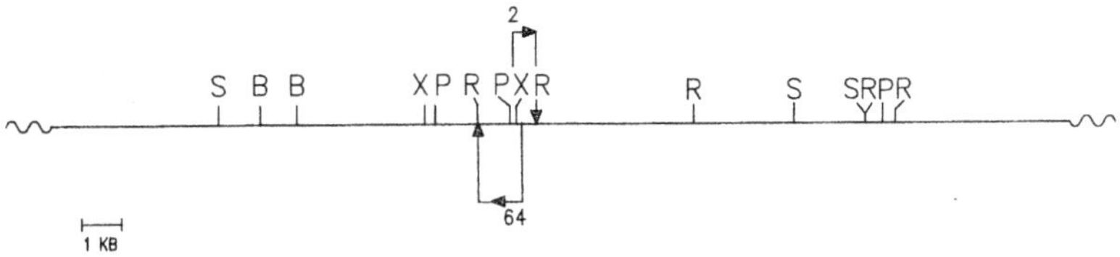

FIGURE 4. ORS571 Nif-locus 4.

3. CONSTRUCTION OF ORS572 STRAINS CARRYING DEFINED, DOUBLE MUTANTS.

Nif::Vi mutants were first resolved to simple Nif::IS*50* mutants by screening for Tc^S derivatives. The two, direct IS*50*R repeats flanking the pVP2021 insertion provide a template for homologous recombination and allow the resolution of each complex, plasmid-genome cointegrate to yield simple IS*50*R insertions. These Nif::IS*50* mutants were verified by genomic blots with IS*50*, *nptII*, and pSUP202 DNA probes. In all cases, resolvants showed hybridization only with the IS*50*-specific probe.

When certain $nifH_1$::Vi mutants, mapping to the promoter region of this operon, were resolved to $nifH_1$::IS*50* mutants, Nif^+, Fix^+ phenotypes were restored. This demonstrated that the $nifH_2$ gene is functional, because it can intragenomically complement $NifH_1^-$ mutants.

Nif::IS*50* mutants were subjected to a second round of site-specific mutagenesis to yield defined, double mutants. As an example, *nifA*::IS*50*, $nifH_1$::Tn*5lac* double mutants were also constructed (Fig. 5). An *E. coli* strain carrying a limited host-range plasmid conferring Tc resistance and also carrying ORS571 $nifH_1DKE$ DNA sequences was mutagenized with Tn*5lac*, resulting in transcriptional *lacZ* gene-fusions. The transposition mutants were physically mapped and these plasmids were then crossed into both ORS571 wild-type and strain 60107R (*nifA*::IS*50*R). Tn*5lac* transposition was found to be very rare in ORS571. All Tc^r, Km^r exconjugants tested arose as a result of a single homologous recombination between the incoming plasmid and the ORS571 chromosome. The single recombinants were grown non-selectively for several generations, Tc^S, Km^r double recombinants (i.e. Tn*5lac* gene

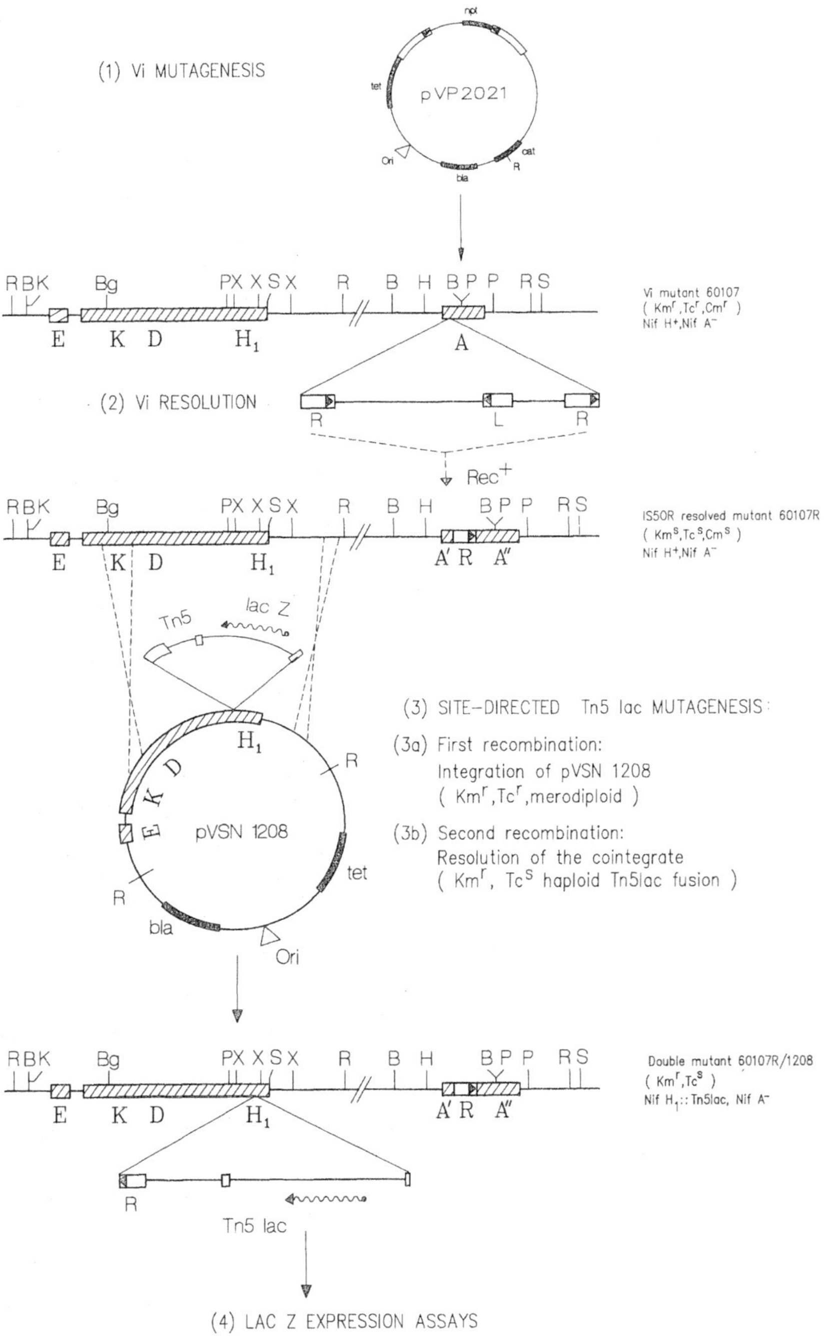

FIGURE 5. Strategy for resolution of Vi mutants to yield IS*50* mutants and for construction of double mutants in ORS571 Nif-locus 1.

replacements) were identified by screening. The double mutants were again verified by genomic DNA hybridization experiments.

The Lac phenotypes (Lac$^+$, Lac^c, and Lac$^-$) of double mutants were assessed under N_2-fixing conditions in culture (Fig. 6). NifA$^+$ strains carrying Tn*5lac* insertions in the *nifH$_1$DK* genes showed that *lacZ* was induced when oriented from right to left as drawn. It was concluded that the *nifH$_1$DK* operon is also oriented from right to left. However in the NifA$^-$ strain 60107, similar Tn*5lac* insertions in this operon failed to induce. These experiments corroborated mRNA transcription studies. It was concluded that the NifA protein is an activator of the *nifH$_1$DK* operon.

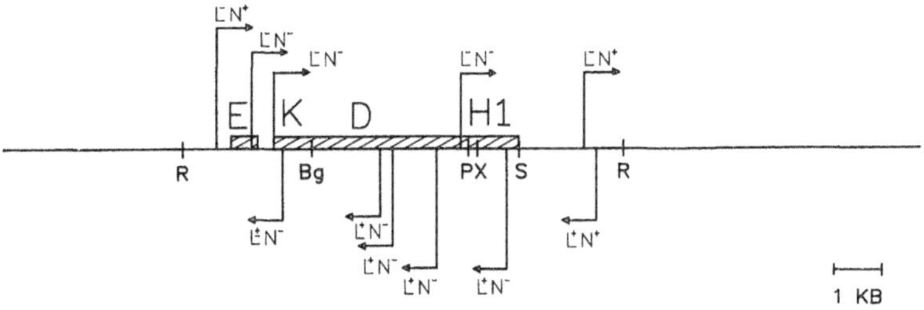

FIGURE 6. ORS571 Nif-locus 1 Tn*5lac* fusions. The orientations of Tn*5lac* elements are indiciated by arrowheads. Abbreviations L$^+$, Lac$^+$; L$^-$, Lac$^+$; N$^+$, Nif$^+$; N$^-$, Nif$^-$.

IDENTIFICATION, CHARACTERISATION AND SEQUENCE ANALYSIS OF THE RHIZOBIUM LEGUMINOSARUM NIFA GENE

S.S.MANIAN, P.GRÖNGER, U.B.PRIEFER AND A.PÜHLER

1. INTRODUCTION

Regulatory genes homologous to Klebsiella pneumoniae nifA have been found in a number of Rhizobium strains. For R.meliloti, it was shown that the nifA (also called fixD) gene is essential for the activation of nif and fix genes and shows not only functional but also sequence homology to K.pneumoniae nifA (1,2,3,4).

In R.leguminosarum, the knowledge about the regulatory gene is still very incomplete. In this study, we have established to complete sequence of the R.leguminosarum nifA gene and its flanking regions. We could show that not only the nifA and nifB genes but also the nifA upstream region of R.leguminosarum are very similar to that of R.meliloti.

2. IDENTIFICATION OF A R.LEGUMINOSARUM REGION HOMOLOGOUS TO R.MELILOTI FIXABC AND NIFA

A cosmid gene bank of pIJ1008 (a recombinant plasmid derived from the symbiotic plasmid pRL6JI and the bacteriocin producing plasmid pVW5JI) was screened by hybridisation to plasmids carrying different R.meliloti symbiotic genes. One cosmid clone, Cos4, hybridised to nodABCD as well as to fixABC and nifA of R.meliloti. The R.leguminosarum fixABC and nifA regions reside on two adjacent EcoRI fragments of sizes 5.45kb and 1.80kb. This region is about 10kb from the common nod region. The organisation of Cos4 is shown in Fig.1. No hybridisation with R.meliloti nifHDK was obtained with this cosmid.

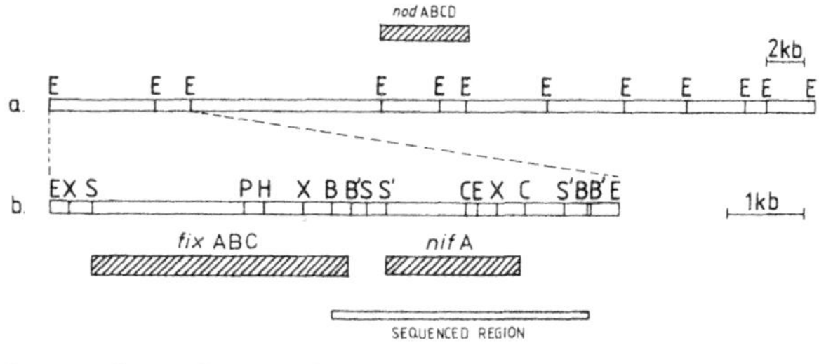

FIGURE 1. Organisation of Cos4. The regions hybridising to R. meliloti fixABC, nifA and nodABCD are indicated. The 3.3kb BamHI fragment shown represents that region which was sequenced. E=EcoRI, X=XhoI, S=SalI, S'=SmaI, B=BamHI, B'=BglII, C=ClaI, H=HindIII, P=PstI

3. ORGANISATION OF THE R.LEGUMINOSARUM NIFA GENE AND ADJACENT REGIONS

To determine the precise location and structure of the nifA gene, the 3.3kb BamHI fragment was cloned and sequenced according to the method of Maxam & Gilbert. Four open reading frames were identified, all of them in the same orientation (Fig.2). One of these shows significant homology to R.meliloti and K.pneumoniae nifA. Thus it appears that this ORF corresponds to the R.leguminosarum nifA gene.

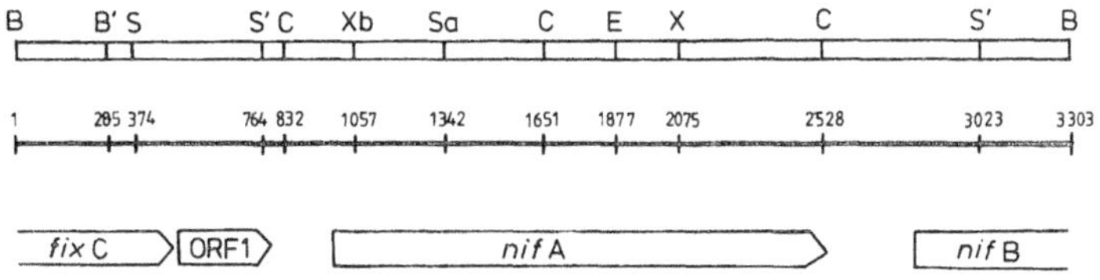

FIGURE 2. Restriction map, location and transcriptional direction of coding regions on the 3.3kb BamHI fragment

Downstream of the nifA gene, we localised the 5'-terminal portion of an open reading frame. Although it matches over wide stretches, this sequence is not completely identical to the N-terminal part of fixZ (5). However, based on its good homology to R.meliloti (H.Reiländer, pers.comm.), this ORF is designated nifB.

Upstream of nifA, there is a very short open reading frame (ORF1) and the C-terminal end of another coding region. Since in R.meliloti, the fixC gene encodes a protein of 43K (6) we do not believe that this short ORF corresponds to the R.leguminosarum fixC. Rather we assume that the coding region preceding ORF1 represents the R.leguminosarum fixC gene. This is also supported by comparison with the 3'-terminal portion of the R.meliloti fixC gene (H.Reiländer, pers.comm.).

4. COMPARISON OF R.LEGUMINOSARUM NIFA TO R.MELILOTI AND K.PNEUMONIAE

The amino acid sequence predicted from the 1157bp long nucleotide sequence of the R.leguminosarum nifA gene was compared to that of R.meliloti and K.pneumoniae.

The N-terminal third shows only weak homology between the three organisms. The first 150 amino acids could be referred to as domain A (4) with a homology of not more than 20% to R.meliloti. This sequence is followed by a block of approximately 20 amino acids with 2 Glu, 4 Gln and 3 Pro residues (domain C). This domain shows homology to R.meliloti at a level of 30% and to K.pneumoniae of 25%. A block of about 240 amino acids shows very strong homology and coincides with domain D. The degree of homology to R.meliloti is 73%, to K.pneumoniae 56%. Another highly conserved region stretches approximately from positions 450-515 (domain E). The homology to R.meliloti is 57%, to K.pneumoniae 44%. It contains a sequence which is strongly homologous to the DNA-binding motif.

5. HOMOLOGY IN THE NIFA UPSTREAM REGION OF R.LEGUMINOSARUM AND R.MELILOTI

Unexpectedly, we found a very small open reading frame between the putative fixC and the nifA gene of R.leguminosarum. This reading frame, called ORF1, seems to be also present in R.meliloti. First hints for the presence of a conserved sequence between fixC and nifA were obtained by heteroduplexing the fixABC/nifA regions of R.meliloti and R.leguminosarum (Fig.3). The region of homology clearly extends beyond the fixC coding regions.

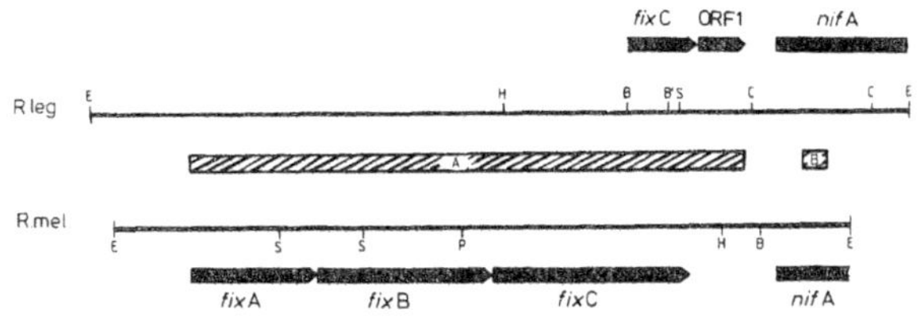

FIGURE 3. Homology between the fixABC/nifA regions of R.leguminosarum and R.meliloti as obtained by heteroduplex studies.

Sequence analysis confirmed the presence of an ORF in R. meliloti in this region (see H.Reiländer et al., this volume) which turned out to be very homologous to that identified in R.leguminosarum.

The existence and a possible role of a polypeptide translated from these ORF's remains unknown and must await further analysis.

REFERENCES

1. Szeto, W.W., J.L.Zimmermann, V.Sundaresan and F.M.Ausubel. Cell 36: 1035-1043, 1984
2. Buikema, W.J., W.W.Szeto, P.V.Lemley, W.H.Orme-Johnson and F.M.Ausubel. Nucleic Acids Research 13: 4539-4555, 1985
3. Weber, G., H.Reiländer and A.Pühler. EMBO J. 4: 2751-2756, 1985
4. Drummond, M., P.Whitty and J.Wootton. EMBO J. 5: 441-447, (1986)
5. Rossen, L., Q.S.Ma, E.A.Mudd, A.W.B.Johnston and J.A.Downie. Nucleic Acids Research 12: 7123-7134, 1984
6. Pühler, A., M.O.Aguilar, M.Hynes, P.Müller, W.Klipp, U.Priefer, R.Simon and G.Weber. In:"Advances in Nitrogen Fixation Research" (eds. C.Veeger and W.E.Newton), Martinus/Junk Pudoc, The Hague, 1984

Analysis of hup DNA and Hup host range of Rhizobium leguminosarum BIO.

H.V. Tichy,[I] C. Schild, H.M. Ripke, L.M. Nelson,[2] H. Fees, W. Lotz

Institut fur Mikrobiologie und Biochemie, Universitat Erlangen-Nurnberg, D-8520 Erlangen, F.R.G.

Rhizobium leguminosarum BIO has been isolated from root nodules of Pisum sativum; it shows a Nod^+ Fix^+ Hup^+ phenotype in symbiosis with pea plants (Tichy, Lotz 1985). Plasmid pRlB505 has been isolated from a cosmid gene bank (vector pMMB34, BamHI fragment 2) of strain BIO. The cloned DNA (BamHI fragments I, 3, 4) carries genes for hydrogen uptake as deduced from the homology of the cloned DNA with the Rhizobium japonicum-specific hup DNA of plasmid pHUI. The hybridization probe used was a subclone of pHUI (Cantrell et al. 1983) containing the 6 kb HindIII fragment. The orientation of the map of pHUI relative to the map of pRlB505 was based upon hybridization of EcoRI subclones of pHUI with the pRlB505 DNA (digested with BamHI and HindIII respectively). All EcoRI fragments of the pHUI insert, except the smallest one, hybridized to pRlB505; the two large EcoRI fragments strongly, the three other EcoRI fragments weakly (Fig. I).

Restriction enzyme cleavage sites have been mapped on pRlB505 for BamHI, HindIII, SstI, SalI and BglII. A more detailed map has been obtained for the HindIII-2 fragment of pRlB505 for ApaI, BglII, ClaI, EcoRI, EcoRV, HpaI, KpnI, MluI, SalI, SmaI and StuI. Under stringent conditions, only a region of 4,3 kb of this fragment (total size of IO,2 kb) hybridized with the HindIII-I fragment of pHUI.

Tn5-insertions have been isolated after site-directed mutagenesis on the subclone pHVTII5 of pRlB505 (vector pACYCI84). Co-integrates of the Tn5-tagged plasmid with pRK290 were mobilized by pRK2OI3 into Rhizobium leguminosarum BIO. One of the insertion mutants assayed, BIO-II5I, showed a Nod^+ Fix^+ Hup^- phenotype with P.sativum. The phenotype was assayed using methylene blue reduction (Tichy, Lotz 1985) and $^3H_2$ uptake (Nelson, Child 198I). The location of the Tn5-insertion in this mutant (in the EcoRI fragment 4) corresponds to a region of low homology of pRlB505 with pHUI.

Four week old Pisum sativum plants nodulated by mutant BIO-II5I evolved threefold more H_2 than those nodulated by strain BIO, but C_2H_2 reduction was similar in both cases (I,2 µmol gfw^{-I} h^{-I}). Nodulation assays with wild type strain R.leguminosarum BIO and different varieties of Vicia faba have resulted in four different symbiosis-specific phenotypes: I. Nod^+ Fix^+ Hup^+, II. Nod^+ Fix^+ Hup^-, III. Nod^+ Fix^- Hup^-, IV. Nod^-. These results demonstrate the significant influence of the V.faba genotype on the expression of symbiosis-specific functions in Rhizobium leguminosarum.

References

Cantrell MA et al (I983) Proc. Natl. Acad. Sci. U.S.A. 80, I8I.
Nelson LM and Child JJ (I98I) Can. J. Microbiol. 27, I028.
Tichy HV and Lotz W (I985) FEMS Microbiol. Lett. 27, I07.

Acknowledgements

This work was supported by grants from the Deutsche Forschungsgemeinschaft and the Bundesministerium fur Forschung und Technologie to W.L. and a grant from the Canadian Dept. of External Affairs to L.M.N.

[I] Institut fur Biologie II, Universitat Freiburg, Freiburg, F.R.G.

[2] Plant Biotechnology Institute, National Research Council, Saskatoon S7N OW9, Sask., Canada.

(see next page for Fig. I)

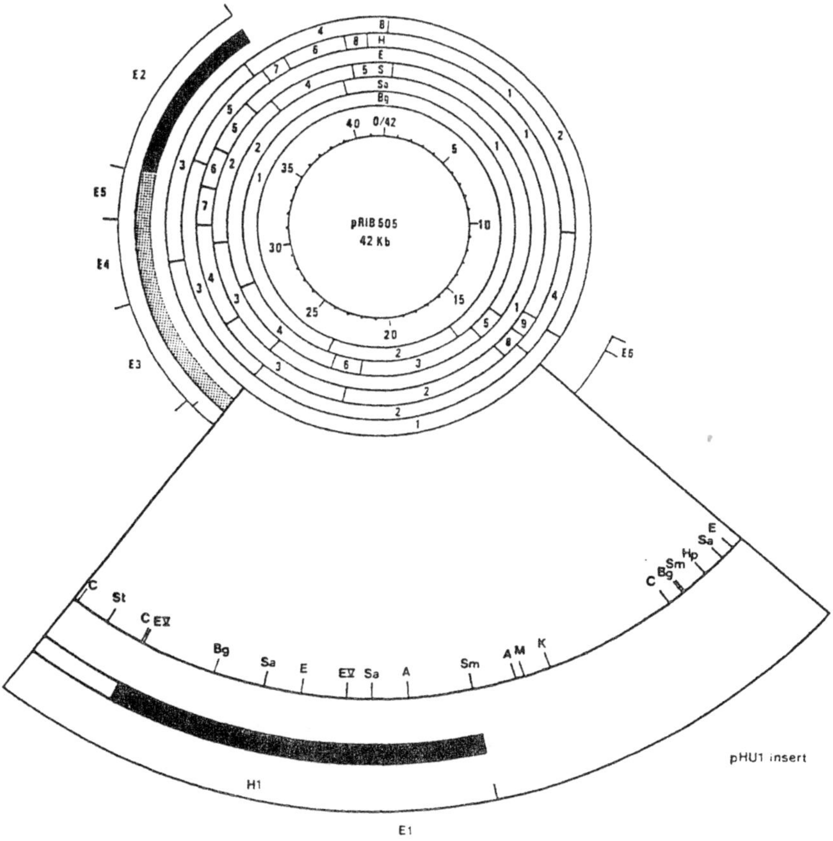

Fig. I

Restriction enzyme cleavage map of pR1B505.
The degree of homology between pHUI and pR1B505 is indicated by the bar outside the map of pR1B505 (strong homology: ▬ weak homology: ▒ no homology under stringent conditions: ▭). A detailed restriction map of the HindIII-2 fragment of pR1B505 is shown.

BIOLUMINESCENCE IN ROOT NODULES OF SOYBEAN CONTROLLED BY NITROGENASE PROMOTERS

Roman P. Legocki, Misuk Legocki, Thomas O. Baldwin[1], and Aladar A. Szalay
Boyce Thompson Institute for Plant Research, Cornell University, Ithaca, New York 14853
[1] Department of Biochemistry and Biophysics, Texas A&M University, College Station, Texas 77843-2128

SUMMARY

We have introduced nitrogenase nifD and nifH promoter-luxAB fusions into the Bradyrhizobium japonicum chromosome and showed symbiotically regulated bioluminescence in soybean root nodules. B. japonicum transconjugants containing a single copy per genome of the nif promoter-controlled luciferase structural genes luxAB from Vibrio harveyi did not produce light in free-living cultures, but expressed bioluminescence in root nodules that was strong enough to be detected by a naked eye.

We postulate the use of luxAB genes for monitoring gene expression. In contrast to the lacZ gene of E. coli, the bacterial luciferase system functions in unrestricted host background and unrestricted medium composition, and its activity can be quantified in vivo using standard laboratory equipment.

RESULTS

We have previously shown in stem nodules of Aeschynomene scabra that a chromosomally integrated nifH promoter-lacZ gene fusion in the stem Rhizobium bacteroids is activated in trans with the native nif region [1]. We now report that the activity of nifD and nifH nitrogenase promoters of Bradyrhizobium japonicum can be measured in a single root nodule of soybean using liquid scintillation counter, based on bioluminescence.

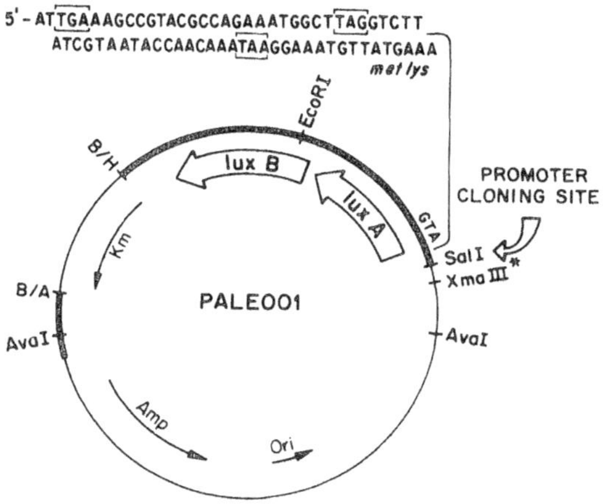

FIGURE 1. Promoter-search vector pPALE001. Heavy line shows V. harveyi DNA. The 70 bp nucleotide sequence upstream of the ATG contains translational stop codons (brackets) in all 3 translational reading frames. (*) at XmaIII of pPALE001 shows inactivation of this site by insertion of a trpA transcription terminator. B/H and B/A indicate ligation sites of a HindIII-AvaI fragment containing the NPTII gene (Km) into a BglII site of pPALE001.

The nifD promoter region of B. japonicum was excised from pAY6 [2] as a 0.8 kb ClaI fragment containing approximately 0.3 kb of the N-terminal region and 0.5 kb of the 5′ upstream sequences. The nifH promoter region

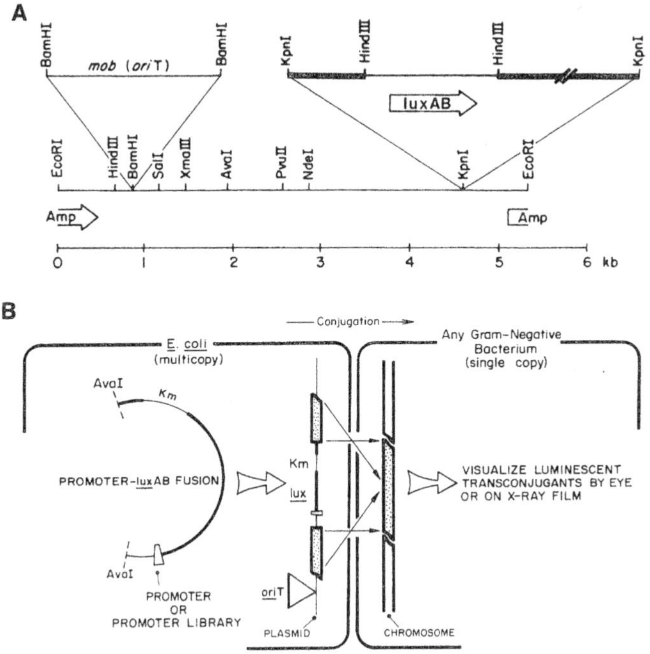

FIGURE 2. (A) Restriction map of the mobilizable plasmid pMR19. A 13.5 kb fragment of B. japonicum chromosome (thick line) contains a single cloning site, HindIII, suitable for cloning luxAB and other genes into the genome of B. japonicum. (B) A scheme outlining the general method of E. coli-B. japonicum conjugation. A double cross-over event between the selected region of chromosomal homology and its native copy leads to a stable integration of the nifD promoter-luxAB gene fusion into the B. japonicum genome. The presence of NPTII gene (Km) downstream of luxB provides an independent selectable marker.

was located on a 0.75 kb SmaI-XhoI fragment of pAY8 [2] containing approximately 0.3 kb of the N-terminus and 0.45 kb of the 5′ upstream sequences. The nifD and nifH promoter-containing fragments were purified by electro-elution, and each blunt-end ligated into the SalI site of a promoter-search vector carrying the luciferase genes luxAB, pPALE001. As indicated in Fig. 1, plasmid pPALE001 carries a 70 bp sequence of V. harveyi containing translation stop codons in all three reading frames prior to the ATG codon of luxA. This sequence will block any translational fusion upstream of luxA without causing transcriptional termination.

The nifD and nifH promoter-luxAB fusions were excised from pPALE001 with AvaI (see Fig. 1) and placed in the middle of a 13.5 kb fragment of B. japonicum chromosome, located on a mobilizable plasmid pMR19 (Fig. 2A). Following conjugation from E. coli SM10 (Fig. 2B), Bradyrhizobium transconjugants were selected for kanamycin resistance, and colonies containing a single copy of the promoter fusion per genome were identified as "double cross-overs" by DNA hybridization, as demonstrated previously in stem Rhizobium BTAil [1]. No bioluminescence was detected in free-living transconjugants containing the nif promoter-luciferase gene fusions, as judged by extended exposures to X-ray film.

For plant tests, soybeans were divided into three inoculation groups: wild-type, nifD promoter-luxAB fusion, and nifH promoter-luxAB fusion. Beginning at day 18 after inoculation, single nodules were removed from roots, weighed, and measured for acetylene reduction in a

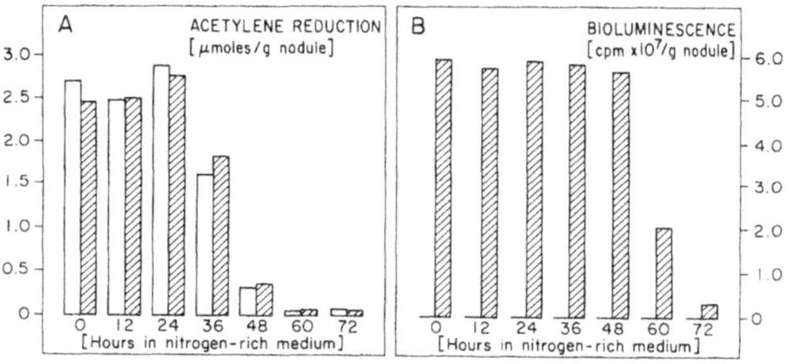

FIGURE 3. Measurements of acetylene reduction (A) and bioluminescence (B) in soybean nodules formed by the wild-type strain of B. japonicum (open bars) and by the transconjugant strain containing the nifD promoter-luxAB gene fusion (hatched bars). Soybeans were transferred to nitrogen-rich medium at day 18 after inoculation and both assays were performed every 12 hours for 4 days.

0.5 cc volume of 10% (v/v) acetylene [3]. Each nodule was homogenized in 200 μl of the luciferase assay buffer using a small mortar and pestle, and total extracts were examined for bioluminescence in a liquid scintillation counter. Figure 3A shows that nodules formed by the wild-type B. japonicum and those formed by transconjugants containing the nifD promoter-luxAB fusion fix nitrogen at approximately equal rates. About 36 hr after addition of nitrogen-rich medium [4], which is known to inhibit expression of nif genes in rhizobia, a rapid decline in nitrogen fixation was observed. Measurements of bioluminescence performed on the same tissues (Fig. 3B) showed high levels of luciferase activity in nodules containing the nifD promoter-luxAB fusion and a similar decline of bioluminescence at 60 hr. There was no bioluminescence detected at any stage in the wild-type nodules.

We did not fully understand why the decline in bioluminescence did not coincide with the decline in nitrogen fixation, but rather the two events occurred 24 hr apart. In contrast to the nitrogenase complex, the bacterial luciferase from V. harveyi has been described as a relatively stable protein, i.e. its structure and enzymatic activity were not affected by repeated freezing and thawing [5], and thus it was possible that this protein had a long half-life in B. japonicum bacteroids. To determine levels of the luciferase protein in nodules treated with nitrate, bacteroids were isolated at 12 hr intervals, and their total protein extracts analyzed electrophoretically. Figure 4A shows that the amount of luciferase in bioluminescent nodules is very small and that its presence cannot be detected by conventional staining. A nitrocellulose replica reacted with antibodies against purified luciferase (Fig. 4B) indicates virtually the same level of the enzyme between 0 and 48 hr following the addition of nitrate, and a steady decrease in the amount of luciferase thereafter. It thus appears that the turnover rate of the luciferase polypeptides in nodules may indeed be very low, and that the stability of this protein may cause the observed delay in the decline of bioluminescence.

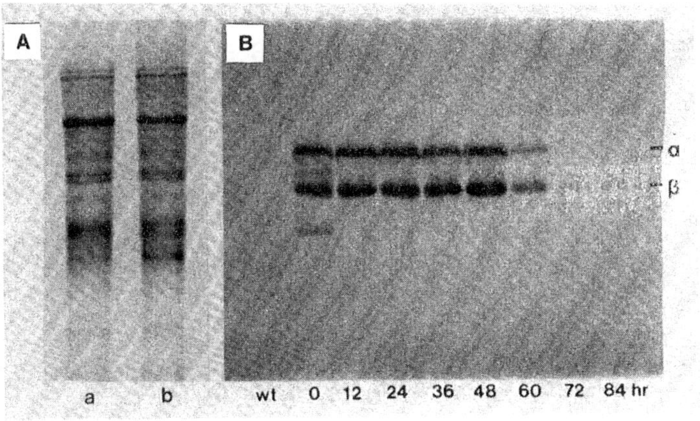

Fig. 4. (A) SDS-polyacrylamide gel electrophoresis of total bacteroid proteins from mature wild-type (a) and bioluminescence (b) nodules (Coomassie blue stain). After the addition of nitrate, bacteroids were isolated every 12 hours (panel B, lanes 0-84 hr) and their total protein extracts electrophoresed, transferred to nitrocellulose, and reacted with an antiserum against purified luciferase. Antigen-IgG complexes were visualized using a goat anti-rabbit serum conjugated with peroxidase (Boehringer Mannheim). wt: bacteroids from wild-type nodules, α and β: subunits of luciferase.

To determine if in the early stages of nodule development the expression of nitrogenase is in fact accompanied by the expression of bioluminescence, total extracts of two types of nodules were compared using the luxdot assay: nodules formed by B. japonicum containing a single chromosomal copy of a P1 [6] promoter-luxAB fusion, and nodules equipped with the nifD promoter-luxAB fusion. It is clear that bioluminescence controlled by the constitutively expressed P1 promoter is detectable even in very young nodules (7 days after inoculation, Fig. 5A), whereas expression of bioluminescence in nodules containing the nifD promoter-luxAB fusion occurs only at 3-4 days later (Fig. 5B), and it coincides with the nitrogenase activity (Fig. 5C). The initiation of

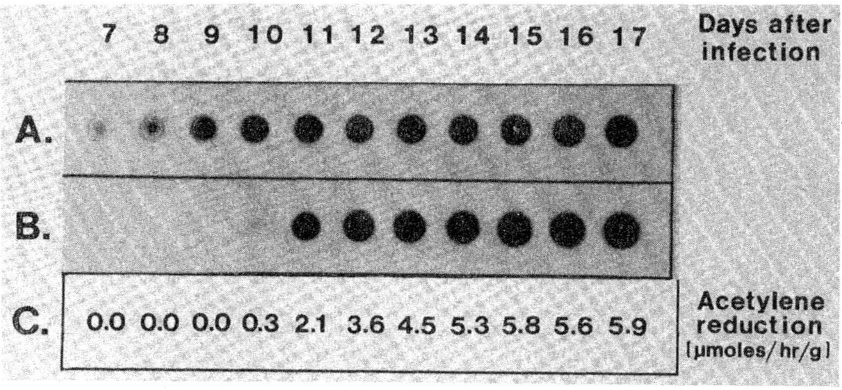

Fig. 5. Expression of bioluminescence in soybean nodules controlled by a constitutively expressed P1 promoter (A) and the symbiotically regulated nifD promoter (B), as monitored by the luxdot assay [8] on X-ray film. Five µl droplets of total nodule extracts (0.1g nodule per 0.5 ml of the assay buffer) were exposed to X-ray film in the presence of n-decanal vapors. The increase in luxdot intensity in panel A is due to an increase of the bacteroid number from about $1x10^7$/mg tissue on day 7 to approximately $1.5x10^{10}$/mg on day 14, as determined in a hemocytometer. Panel C shows nitrogen fixation activity in the bioluminescent nodules measured prior to their homogenization.

nitrogen fixation in soybean nodules at day 10 of the symbiosis is consistent with previous reports [7], and the appearance of bioluminescence at that time shows that the nifD promoter-luxAB fusion is activated coordinately with the native nif locus of B. japonicum. Similar results to those shown for the nifD promoter were obtained for nodules containing the nifH promoter-luxAB fusion.

It is interesting to note that while bioluminescence in total extracts from wild-type nodules showed on the average 30-40 cpm per nodules, the same size nodule containing the nifD promoter-luxAB fusion showed typically a peak value of as much as $6x10^7$ cpm. Bioluminescence in these nodules in the presence of n-decanal vapors was visible to the naked eye, but only if the tissue was cut open. This may be due to two factors related to nodule structure. The light generated by bacteroids, located in the central cortex, could be blocked by several uninfected cell layers of the peripheral cortex and/or the luciferase reaction, known to require oxygen, is enhanced upon exposure of bacteroids to atmospheric oxygen.

It is also noteworthy that extracts from a single nodule containing one copy of luxAB per bacteroid yielded as much as 2.0×10^5 to 9.0×10^5 cpm in a liquid scintillation counter without addition of exogenous aldehyde. These results indicated the presence in soybean nodules of aldehyde(s) that can serve as substrate for the bioluminescence reaction. The levels of bioluminescence are two orders of magnitude lower than those with n-decanal added exogenously, but nevertherless, the observation is worthy of further study.

Bioluminescence as a measure of gene expression can be quantified by a variety of simple methods, including the luxdot assay [8].

REFERENCES

1. Legocki R.P., Yun A.C., Szalay A.A. (1984) Proc. Natl. Acad. Sci. USA 81:5806-5810.
2. Yun A.C., Noti J.D., Szalay A.A. (1986) J. Bacteriol., in press.
3. Hardy R.W.F., Holsten R.D., Jackson E.K., Burns R.C. (1968) Plant Physiol. 43:1185-1207.
4. Summerfield R.J., Dart P.J., Huxley P.A., Eaglesham A.R.J., Minchin F.R., Day J.M. (1977) Exp. Agric. 13:129-142.
5. Hastings J.W., Baldwin T.O., Nicoli M.Z. (1978) Methods Enzymol. 57:135-152.
6. Stueber D., Bujard H. (1981) Proc. Natl. Acad. Sci. USA 78:167-171.
7. Bergersen F.J., Goodchild D.J. (1973) Australian J. Bio. Sci. 26:729-741.
8. Legocki R.P., Legocki M., Baldwin T.O., Szalay A.A. (1986) Proc. Natl. Acad. Sci. USA, in press.

IN VIVO CLONING OF GENES FROM BRADYRHIZOBIUM JAPONICUM*

KETAN S. SHAH AND L. DAVID KUYKENDALL

1. INTRODUCTION

The Tn5 introduction vector pGS9, developed by Selvaraj and Iyer (11), was recently used for obtaining both auxotrophs and symbiotically defective mutants of fast and slow growing soybean rhizobia (9,10). Site-directed Tn5 mutagenesis technique was succesfully applied to *Bradyrhizobium japonicum* by Hahn and Hennecke (5). *In vivo* constructed R-prime plasmids carrying *R. meliloti* chromosomal DNA regions have been reported (6,7). Gene mobilizing R-plasmid pJB3JI, a kanamycin sensitive derivative of pR68.45 (2), was used by Banfalvi et al. (1) to isolate a Tn5-tagged 90 kb region of the megaplasmid DNA carrying symbiotic genes of *R. meliloti*. R-prime plasmid formations between pRL180 (RP1 derivative) and nodulation plasmid of *R. fredii* has recently been reported (4). An R-factor transfer system was demonstrated in slow-growing *B. japonicum* I-110 (8). Recently, we have described mobilization of Tn5 insertions from *R. fredii* by pJB3JI (12). This paper describes a successful strategy for *in vivo* construction of gene libraries of *B. japonicum* strain I-110.

2. MATERIALS AND METHODS

2.1. Strains. *B. japonicum* strain I-110 and genetically marked sublines were from this laboratory. *E. coli* WA803(pGS9) was obtained from Dr. V. Iyer. *E. coli* HB101(pJB3JI) was obtained from Dr. J. Beringer.

2.2. Media and culture condition. *B. japonicum* strains were grown in AlE medium at 30°C as described previously (8). *E. coli* strains were grown in LB medium at 37°C. Antibiotics (Sigma Chemical Co.) were prepared in either aqueous solution for streptomycin sulfate and kanamycin sulfate, 0.1 N sodium hydroxide for rifampin and nalidixic acid, or 50% ethanol for tetracycline hydrochloride. Fresh solutions were filter sterilized.

2.3. Mating condition. Bacterial matings for plasmid transfer and Tn5 introduction were performed as previously described (12) except these matings, involving *Bradyrhizobium*, were on AlE medium.

2.4. Plasmid isolation and detection. Plasmid profile on agarose gels were obtained following essentially the method described by Crosa and Falkow (3).

3. RESULTS AND DISCUSSION

We used plasmid pGS9 for transposon Tn5 introduction into *B. japonicum*. Counterselection of both donor and wild type recipient cells with appropriate concentration of antiobiotic(s) allowed us to obtain Tn5 introduction frequencies as high as 1.5×10^{-6} and as low as 1.6×10^{-7} (Table 1). These transfer frequencies were about 100 fold higher than background frequencies for antibiotic resistance. Bonafide Tn5 exconjugants were clearly distinguished from background Str^R mutants of the recipient by replica plating to medium containing kanamycin (100 ug/ml). Pools of 1,000 Tn5 transconjugants of *B. japonicum* I-110 and 500 of *B. japonicum* I-110 FN were used as recipients of the broad-host-range plasmid pJB3JI. Frequencies of about 2×10^{-7} were obtained for pJB3JI transfer from

TABLE 1. Frequencies of transposon Tn5 introduction into B. japonicum

Recipient strain[a]	Antibiotic resistance[b]	Tn5 transfer frequency[c]
B. japonicum I-110	Str (200)	1.5×10^{-6}
B. japonicum I-110 FN[d]	Str (100)	1.6×10^{-7}
Bradyrhizobium strain 61N[e]	Str (100) + Kan (100)	1.0×10^{-6}

a. Matings between about 10^9 cells of E. coli WA 803 (pGS9) and about 10^9 cells of recipients were done on membrane filters placed on AlE agar at 30°C for 16-20 h. Matings were interrupted by suspending the mixtures in HM salts +0.10% Tween 20.
b. Numbers in parenthesis represent ug/ml of antibiotic in AlE medium.
c. Exconjugants were tested for coinheritance of Kan to verify their status
d. Resistant to 5 fluorouracil (10 ug/ml) and nalidixic acid (500 ug/ml).
e. Resistant to nalidixic acid (500ug/ml).

E. coli donor strain to Tn5-labelled pools of B. japonicum. Background tetracycline resistant mutants arose at a frequency of about 100 fold less than the frequency of pJB3JI transfer in B. japonicum (Table 2). High-levels of intrinsic carbenicillin and tetracycline resistance in Bradyrhizobium strain 61N posed the problem of positive selection for pJB3JI. This strain has antibiotic resistance characteristics of Elkan's DNA homology group II of Bradyrhizobium (unpublished).

TABLE 2. Frequencies of pJB3JI transfer into Tn5-labelled B. japonicum

Recipient strain[a]	Selection for	Antibiotics (ug/ml)	Background	pJB3JI transfer frequency
B. japonicum I-110	pJB3JI transfer	Tet (75) Str (50)	$\leq 10^{-9}$	2.2×10^{-7}
B. japonicum I-110 FN	pJB3JI transfer	Tet (100) Str (50)	$\leq 10^{-9}$	2.0×10^{-7}

a. HB101 (pJB3JI) was the donor strain.

Intergeneric mobilization of genomic Tn5 insertions employed using B. japonicum I-110::Tn5 (pJB3JI) as donor and E. coli HB101 as recipient. Selection for both Kan^r and Tet^r E. coli HB101 at 37°C on LB agar provided Tn5-carrying pJB3JI exconjugants at frequencies betweem 10^{-7} and 10^{-8} (Table 3). Control plates were clear. Simultaneous coinheritance of Kan^R and Tet^R, clearly indicated that the colonies obtained from the mating mixtures were true recombinants. Transfer of Tn5-bearing pJB3JI was about four to five orders of magnitude lower than the transfer frequency of pJB3JI alone.

Gel electrophoresis of plasmid DNA from E. coli exconjugants indicated recombinant plasmid DNAs of different sizes (Fig. 1). E. coli transconjugants designated as no. 27, 79, 81 and 82 gave recombinant plasmid with DNA insert of at least 30 kb (the size of Tn5 is 5.7 kb).

*Scientific Article A-4481, contribution no. 7474 of the Maryland Agric. Exp. Stn., Dept. of Agronomy, College Park, MD 20742.

TABLE 3. Mobilization of Tn5 from B. japonicum (pJB3JI) to E. coli HB101.

Donor strain	Selection for	Antibiotics (ug/ml)	Transfer frequency
B. japonicum I-110 FN TDP[a]	pJB3JI transfer	Tet (10)	10^{-3}
	both pJB3JI & Tn5 transfer	Tet (10) + Kan (10)	10^{-8}
B. japonicum I-110 TDP[b]	pJB3JI transfer	Tet (10)	10^{-3}
	both pJB3JI & Tn5 transfer	Tet (10)+ Kan (10)	10^{-7}

a. TDP (Transposon Donor Pool) consisting of 1435 clones.
b. TDP comprised of 370 clones.

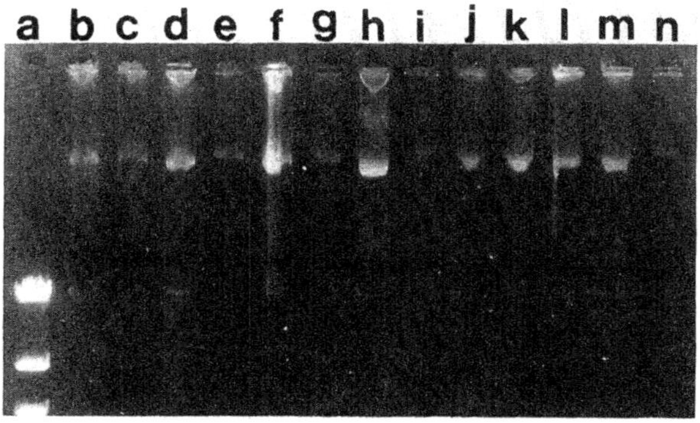

FIGURE 1. Agarose gel electrophoresis of plasmid DNAs of E. coli exconjuants, number shown in parenthesis. Lane a. λHind III digest, Lane b.(76), Lane c.(77), Lane d.(78), Lane e.(79), Lane f.(80), Lane h. HB101 (pJB3JI), Lane i.(82), Lane j.(83), Lane k.(84), Lane l.(23), Lane m.(26) and Lane n.(27). Only the three largest fragments of λ Hind III DNA are shown.

In this study, we used pJB3JI to mobilize random genomic Tn5 insertions from B. japonicum to E. coli to provide for in vivo genetic cloning of symbiotic determinants. Use of in vitro constructed genomic libraries of I-110 DNA to complement Fix^- mutants, has not been successful due to instability of the cloning vehicle. In vivo generated R prime libraries, however, may be very useful due to stability and ability to insert large sizes of DNA. We have observed pR68.45 being stable in B. japonicum I-110 isolated from soybean root nodules (unpublished). Using in vivo cloning strategy, larger recombinant plasmids could be obtained and would be useful for mapping and analysis of the symbiotic genes from B. japonicum.

ACKNOWLEDGEMENTS
We thank Michael Behler for technical assistance. This work was supported by the Presidential INDO-US Science and Technology Initiative. Ketan Shah is employed under a cooperative agreement with University of Maryland, College Park, MD.

REFERENCES

1. Banfalvi Z, Randhawa GS, Kondorosi E, Kiss A, Kondorosi A: Mol. Gen. Genet. 189:129-135, 1983.
2. Brewin NJ, Beringer JE, Johnston AWB: J. Gen. Microbiol. 120:413-420, 1980.
3. Crosa JH, Falkow S: Plasmids. In: Philipp Gershardt (Ed. in Chief) Manual of Methods for General Bacteriology, pp 269-270. ASM. Washington D.C., 1981.
4. Engwall FS, Atherly AG: Plant Mol. Biol. 6:41-51, 1986.
5. Hahn M, Hennecke H: Mol. Gen. Genet. 193:46-52, 1984.
6. Johnston AWB, Setchell SM, Beringer JE: J. Gen. Microbiol. 104:209-218, 1978.
7. Kiss GB, Dobo K, Dusha I, Breznovits A, Orosz L, Vincze E, Kondorosi A: J. Bacteriol. 141:121-128, 1980.
8. Kuykendall LD: Appl. Environ. Microbiol. 37:862-866, 1979.
9. Rostas K, Sista PR, Stanley J, Verma DPS: Mol. Gen. Genet. 197:230-235, 1984.
10. Sadowsky MJ, Rostas K, Sista PR, Bussey H, Verma DPS: Arch. Microbiol. 144:334-339, 1986.
11. Selvaraj G, Iyer VN: J. Bacteriol. 158:580-589, 1983.
12. Thomas PM, Kuykendall LD, Angle JS: Appl. Environ. Microbiol. 52: 206-208, 1986.

GENES FOR THE CATABOLISM AND SYNTHESIS OF A NODULE-SPECIFIC, OPINE-LIKE COMPOUND ARE CLOSELY LINKED AND ON THE SYM PLASMID OF RHIZOBIUM MELILOTI.

Peter J. Murphy (1), Nina Heycke (1), Zsofia Banfalvi (2), Adam Kondorosi (2), Jacques Tempé (3) and Jeff Schell (1)

(1) Max-Planck-Institut für Züchtungsforschung, 5000 Köln 30, W. GERMANY. (2) Institute of Genetics, Biological Research Center, Hungarian Academy of Science, H-6701 Szeged, P.O. Box 521, HUNGARY. (3) Institute de Microbiologie, Bat. 409, Université de Paris-Sud, F-91405 Orsay, FRANCE.

INTRODUCTION

The biological rationale for the *Agrobacterium*-crowngall interaction is the production of opines (1,2). The redirection of plant metabolites into a form which the inducing bacteria, but few others, can utilize ensures a competitive advantage to the bacteria.

Since *Rhizobium* and *Agrobacterium* are taxonomically related (3) and as the plant-parasitic and plant-symbiotic states have a number of characteristics in common (see, 4,5) we investigated whether some *Rhizobium* species produced opine-like compounds.

Previously we have reported that nodules on alfalfa (*Medicago sativa*) induced by *Rhizobium meliloti* strain L5-30 produce a nodule-specific, opine-like compound (6,7). This compound is defined as opine-like as, of the strains tested, only those that can induce its synthesis can also catabolize it as a sole carbon source. The compound has been identified as L-3-*O*-methyl-*scyllo*-inosamine (3-*O*-MSI, M. E. Tate, unpublished data).

As a contribution to answering the question as to whether opines play a more general role in plant-bacterial interactions we have isolated the genes for the induction of synthesis (*mos* genes) and the genes for the catabolism (*moc* genes) of 3-*O*-MSI.

RESULTS AND DISCUSSION

Isolation of the moc genes.

To isolate the *moc* genes we prepared a total DNA clone bank of L5-30 in the broad host-range cosmid vector pLAFR1. These plasmids were mated into a Moc^- deletion strain of L5-30 and catabolizing clones selected by the ability to confer growth on 3-*O*-MSI as a sole carbon source. Initially a clone with a 33 kb insert was isolated which after sub-cloning was reduced to a 15.1 kb fragment required for 3-*O*-MSI catabolism.

The presence of at least two functional regions on the 15.1 kb catabolic fragment has been shown by Tn*5* and deletion mutagenesis. This was demonstrated as 3-*O*-MSI catabolism could be inhibited by insertion of Tn*5* into a number of sites on a 5.4 kb *Eco*R1 fragment or by removal of a terminal 2.4 kb *Kpn*1-*Eco*R1 fragment. Tn*5* mutants isolated between these two regions did not inhibit 3-*O*-MSI catabolism (Fig. 1).

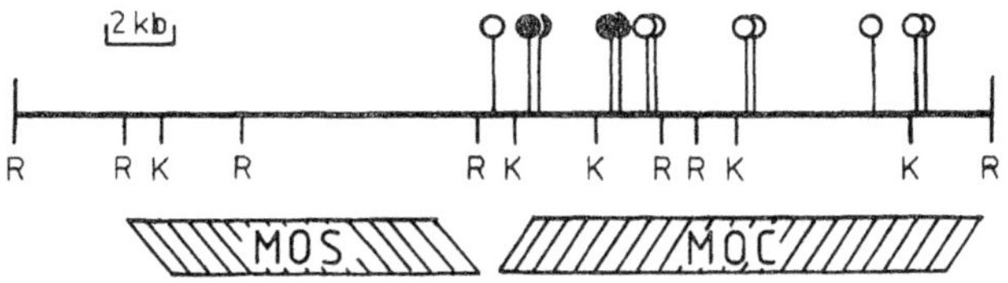

Fig. 1 moc-mos cosmid (pPM1071). The vector is pLAFR1. R, EcoR1; K, Kpn1; ●, Moc$^-$ phenotype; ○, Moc$^+$ phenotype; mos, genes for the induction of synthesis of 3-O-MSI; moc, genes for the catabolism of 3-O-MSI.

Isolation of the mos genes.

To isolate the mos genes we assumed that functionally related genes would be closely linked. Accordingly, we marked the initially isolated catabolism clone with Tn5 (in a region which is not essential for catabolic functions) and homogenotized this fragment back to the L5-30 genome. R-prime plasmids were prepared in this region. A number of R-primes isolated in this way conferred both 3-O-MSI synthesis and catabolism functions. To further isolate these genes we made a "mini" clone bank of one of these R-primes, again using the vector pLAFR1. These clones were mated into AK631 (a wild type strain of R. meliloti) which is Mos$^-$ and tested for the production of 3-O-MSI in nodules. Using this procedure a clone which has both moc and mos functions was isolated. By further sub-cloning and testing for moc and mos phenotypes we have shown these functions to be coded by different genes which are closely linked (Fig. 1).

Localization of the moc-mos genes.

To localize the moc-mos genes we utilized the Tn5-Mob vector system (8) to individually mobilize the L5-30 plasmids into a Nod$^-$, Moc$^-$, Mos$^-$ background. A transconjugant so obtained simultaneously acquired nod, moc and mos functions indicating the moc-mos genes are on the sym plasmid of L5-30. This was further supported by hybridization of a moc probe to plasmids extracted from L5-30.

Are opine-like compounds present in other Rhizobium strains and species?

Although few Rhizobium can catabolize 3-O-MSI this ability is not the exclusive property of L5-30. Besides L5-30, one out of 20 other R. meliloti strains tested could also catabolize 3-O-MSI. When DNA from this strain was probed with the 15.1 kb moc probe bands common to those in L5-30 hybridized. This strain also produced scyllo-inosamine which is a breakdown product of 3-O-MSI.

Furthermore, an R. leguminosarum strain tested could also both catabolize and induce the synthesis of 3-O-MSI. This strain also showed homology to the moc probe although hybridization was weaker and to different bands than those in R. meliloti.

Interestingly the presence of an unrelated nodule-specific,

opine-like compound has recently been reported in nodules induced by R. loti (9).

It is quite likely that more such compounds would be found with a thorough search in other Rhizobium. Clearly though, if opine-like compounds in Rhizobium function by giving a particular bacteria a selective advantage then different compounds would be expected to be found in different bacteria.

CONCLUSIONS

We have shown that the genes for both the catabolism of 3-O-MSI (moc genes) by the bacteria and the induction of its synthesis (mos genes) in the nodule are carried by the bacteria. Furthermore, these genes are closely linked and reside on the sym plasmid of R. meliloti L5-30. The moc-mos genes have been cloned into a broad host-range vector which can be mobilized into other R. meliloti strains where they are expressed.

The initial paradox of why a bacteria has both catabolism and synthesis functions for a compound can be explained if, as for Agrobacterium, an input from the plant exists. This may be a subtle way for the bacteria to obtain plant or symbiotic products to enhance its own survival.

The close linkage of these genes suggests that they have co-evolved as a functional unit and their location on the sym plasmid is in line with the suggestion that they are important in the symbiotic state.

At present we do not know how the mos genes function. They could be regulatory genes controlling plant enzymes or code for enzymes which are only expressed in the symbiotic state.

Having isolated the moc-mos genes we now have a powerful tool to analyse both the mechanism by which 3-O-MSI is produced in the nodule and the role of this compound in the symbiotic state.

REFERENCES

(1) Schell, J., Van Montagu, M., De Beuckeleer, M., De Block, M., Depicker, A., De Wilde, M., Engler, G., Genetello, C., Hernalsteens, J. P., Holsters, M., Seurinck, J., Silva, B., Van Vliet, F. and Villarroel, R. (1979) Proc. R. Soc. Lond. B. 204, 251-266.

(2) Guyon, P., Chilton, M-D., Petit, A. and Tempe, J. (1980) Proc. Natl. Acad. Sci. USA 77, 2693-2697.

(3) De Ley, J., Bernaerts, M., Rassel, A. and Guilmont, A. (1966) J. Gen. Microbiol. 43, 7-17.

(4) Vance, C. P. (1983) Annu. Rev. Microbiol. 37, 399-424.

(5) Verma, D. P. S. and Long, S. R. (1983) Int. Rev. Cytol. Suppl. 14, 211-245.

(6) Tempé, J., Petit, A. and Bannerot, H. (1982) C. R. Acad. Sci. Paris (Ser. 111) 295, 413-416.

(7) Tempé J. and Petit, A. (1983) in Molecular Genetics of the Bacteria-Plant Interaction. ed. A. Pühler. (Springer-Verlag, Berlin) pp. 14-32.

(8) Simon, R. (1984) Molec. Gen. Genet. 196, 413-420.

(9) Shaw, G. J., Wilson, R. D., Lane, G. A., Kennedy, L. D., Scott, D. B. and Gainsford, G. J. (1986) J. Chem. Soc. Chem. Commun. 180-181.

MOLECULAR BIOLOGY OF GENES INVOLVED IN CARBON METABOLISM IN *RHIZOBIUM MELILOTI* AND *BRADYRHIZOBIUM JAPONICUM*

F. O'GARA, B. BOESTEN, M. O'REGAN, B. KIELY, B. HIGGISSON, C. CONDON, K. BIRKENHEAD and S. MANIAN, Microbiology Department, University College, Cork, Ireland.

The important role played by TCA cycle intermediates in symbiotic nitrogen fixation has been highlighted by both biochemical and genetic data (1-6). Therefore in the context of maximizing or improving symbiotic nitrogen fixing associations elucidation of the factors controlling the metabolism of carbon in bacteroids is of fundamental importance. During symbiosis the pattern of gene expression in both the bacterial and plant partners undergoes a differential switch to facilitate the provision of the cellular and metabolic activities necessary for symbiosis. cAMP is well recognised as an important effector molecule in controlling gene expression in *E. coli* and other gram negative bacteria (7, 8). However the involvement of cAMP in gene expression in *Rhizobium* is unclear and consequently it is of interest to evaluate the role of the nucleotide in the life cycle of *Rhizobium*.

1. Adenyl cyclase locus of *Rhizobium meliloti*

Although cAMP can be detected in *Rhizobium* strains (9) its involvement in cellular functions during symbiosis is unclear. The *cya* gene encoding adenyl cyclase cloned from *Rhizobium meliloti* (10) was localized to a 0.8 Kb *Pst*1-*Eco*RI by subcloning experiments (Fig. 1).

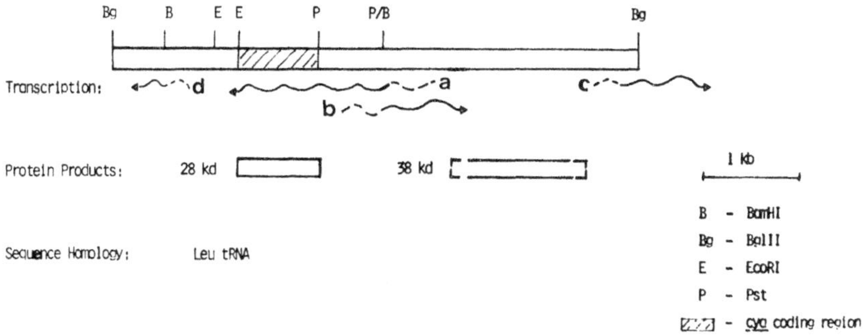

FIGURE 1. Organisation of *cya* locus

A *Rhizobium meliloti cya* gene product of 28 kDa, which is significantly smaller than the corresponding protein from enteric bacteria, was identified using the 'maxicell' technique. The smaller size of the *cya* gene from *R. meliloti* together with the lack of significant DNA sequence homology in hybridization experiments between cloned *cya* genes from *E. coli* and *R. meliloti* may indicate that the organisation of these genes is quite different in these organisms. Using the "maxicell" system a 38 kDa protein encoded by the BglII-BamHI fragment upstream from the cya region was identified. The function of this protein is unknown and results from DNA hybridization experiments indicate that the sequence is not conserved in other Rhizobium species. A cya-lac fusion isolated using MudI demonstrates that the promoter region for the expression of the R. meliloti cya gene in E. coli is localized a considerable distance upstream from the structural gene (Fig. 1). Promoter activity associated with the cloned cya DNA fragment was further analysed using the transcriptional promoter probe vector pGD500 (11) (Fig. 1 transcription orientations indicated by A, B, C, D). Promoter A is responsible for Cya expression in E. coli only. B is active in E. coli and Rhizobium under nutrient shift down conditions. C is active in E. coli and Rhizobium and D is expressed in E. coli only. DNA sequence analysis has shown that a leucine t-RNA gene is located in the 0.25 Kb EcoRI fragment adjacent to cya (O'Gara, Danchin et al. in prep.).

2. Cya mutants in R. meliloti

The cloned cya gene was exploited to construct mutations in this region using site directed mutagenesis. cAMP synthesis activity associated with the cloned cya gene in E. coli was mutated using transposon Tn5 or by creating internal deletions in the coding region and inserting a kanamycin resistance gene as a selectable marker. These constructs were homogenotized back into the wild type R. meliloti genome and the physical location of the Tn5 or the genomic deletion was verified by DNA hybridization analysis. The construction of deletions in this cya region of the genome resulted only in a decrease in cAMP levels (19.6 pmol cAMP/mg protein Vs 34 pmol for wt). These results indicate that R. meliloti contains an additional/alternative system for cAMP synthesis and further work is in progress to identify this system.

3. Role of dct genes in R. meliloti-Alfalfa symbiosis

R. meliloti mutants defective in C_4-dicarboxylate transport isolated by Tn5 and NTG mutagenesis were Nod$^+$ Fix$^-$ and were exploited to clone dct genes (6). The Nod$^+$ Fix$^-$ phenotype of these mutants confirms that dicarboxylic acids are essential to support nitrogen fixation in alfalfa bacteroids. A cloned DNA fragment on plasmid pRK290 encoding succinate transport function(s) was used to create genetically engineered R. meliloti and B. japonicum strains with gene dosage effects in the dct region. Rates of ^{14}C labelled succinate uptake were compared and it was observed that the uptake rate measured over a seven minute period was significantly higher in the strains containing additional copies of the dct region (78 vs 42 nmol/mg protein.min for R. meliloti and 4 vs 2 nmol/mg protein.min for B. japonicum). The R. meliloti strains demonstrating increased succinate transport rates in free living conditions showed increased nitrogen fixation activity in nodules of four week old alfalfa plants. (525 nmol C_2H_4/hr per plant compared to 230 nmol C_2H_4/hr per plant for plants inoculated with the wild type strain). This experimental approach provides a novel system to further investigate whether photosynthate availability in nodules or its subsequent transport and metabolism by bacteroids is limiting N_2-fixation.

4. C_4-dicarboxylates and regulation of sym genes

To support and maintain nitrogen fixation it is clear that the bacteroid has to coordinate the expression of genes necessary for the synthesis of nitrogenase proteins and the ancillary processes (e.g. ATP, reductant). It has been suggested that the presence of a "nif A-regulated" promoter sequence 5' to the dct structural gene of R. leguminosarum may be a mechanism by which dct and nif genes are co-ordinately activated in symbiosis (12). In our programme we are investigating whether carbon compounds provided by the plant to the bacteroid (e.g. C_4-dicarboxylates) could play a role in regulating the expression of sym genes. In this context we investigated the expression of translational fusions of R. meliloti sym promoters P1 and P2 (13) in bacteroids formed by dct mutants (Table 1).

	Phenotype	Promoter	B-Gal (Miller Units)
R. meliloti CM2	Dct^+	-	4
R. meliloti CM2 (pMB210)	Dct^+	P1	4452
R. meliloti CM2 (pMB211)	Dct^+	P2	3320
R. meliloti CM12 (pMB210)	Dct^-	P1	21
R. meliloti CM12 (pMB211)	Dct^-	P2	50

B-galactosidase activity was measured in 5 week old nodules. The percent of Tc^r cells per nodule averaged 60% for the wild type strain CM2. Assuming that the stability of plasmids pMB210 and pMB211 is similar in the dct mutants, then these results suggest that the expression of genes associated with sym promoters P1 and P2 is drastically reduced in nodules formed by a DCT^- mutant strain. Further experiments are in progress to understand the nature of this regulatory relationship between carbon metabolism and Nif gene expression in bacteroids.

References

(1) Bergersen FJ, Turner GL: Biochim Biophys Acta 141: 507-515, 1967.
(2) Glenn AR, Brewin NJ: J Gen Microbiol 126: 237-242, 1981.
(3) Finan TM, Wood JM, Jordan DC: J Bacteriol 154: 1403-1413, 1983.
(4) Arwas R, McKay JA, Rowney FRP, Dilworth MJ, Glenn AR: J Gen Microbiol 131: 2059-2066, 1985.
(5) Ronson CW, Lyttleton P, Robertson JG: Proc Natl Acad Sci USA 78: 4284-4288, 1981.
(6) Bolton E, Higgisson B, Harrington A, O'Gara F: Arch Microbiol 144: 142-146, 1986.
(7) Botsford JL: Microbiol Rev 45: 620-642, 1981.
(8) Ullman A, Danchin A: Adv cyclic nucleotide res 1-53, 1983.
(9) McGetrick A, Goulding C, Manian S, O'Gara F: J Bacteriol 163, 1282-1284, 1985.
(10) Kiely B, O'Gara F: Mol Gen Genet 192: 230-234, 1983.
(11) Ditta G, Schmidhauser T, Yakobson E, Lu P, Liang X, Finlay D, Guiney R, Helinski D: Plasmid 13: 149-153, 1985.
(12) Ronson CW, Astwood PM: In. Evans HJ, Bottomley PJ and Newton WE (eds), Nitrogen fixation research progress 201-207, 1985.
(13) Better M, Ditta G, Helinski DR: EMBO Journal 4: 2419-2424, 1985.

AZORHIZOBIUM SESBANIAE ORS571 CONDUCTS SYNERGISTIC N_2 FIXATION AND NICOTINIC ACID OXIDATION

ROBERT A. LUDWIG, TIMOTHY T. POPIN AND CHRISTOPHER L. KITTS

1. INTRODUCTION TO AZORHIZOBIUM SESBANIAE ORS571

Azorhizobium sesbaniae strain ORS571 is unique among members of the *Rhizobiaceae* in conducting N_2 fixation during growth *ex planta* as well as during symbiosis *in planta*. ORS571 forms nodules on both root and stem lateral-root primordia of *Sesbania rostrata*. As such, ORS571 appears to be a chimeric N_2 fixing bacterium with properties of both orthodox rhizobia and diazotrophic bacteria. We show here that ORS571 needs to catabolize nicotinic acid in order to fix N_2.

2. ORS571 REQUIRES NICOTINATE CATABOLISM TO FIX N_2

ORS571 can grow on nicotinate as sole nitrogen source and 6-OH-nicotinate as sole carbon and nitrogen source. We have isolated Vi mutants in the nicotinate catabolism (Nic) pathway and characterized mutant phenotypes. Vi mutagenesis resulted in two Nic$^-$ phenotypes (Table 1).

TABLE 1. Growth and N_2 Fixation of ORS571 and *nic*::Vi mutants.

Strain	*Nic/Succ*	*6-OH-Nic*	*Nif*	*Nod*	*Fix*	*MNZ*
ORS571	+	+	+	+	+	-
61001	-	-	-	-	-	-
61002	-	-	-	ND	ND	-
61003	-	-	-	-	-	-
61004	-	+	+	+	+	+
61005	-	+	+	+	+	+
61006	-	-	-	+	+	-
61007	-	+	+	+	+	+
61008	-	+	+	+	+	+
61009	-	-	+	ND	ND	+
61021	-	ND	+	-	ND	-
61025	-	-	-	+	-	-
61027	-	+	ND	+	-	-
61034	-	+	+	ND	ND	+
61040	-	+	ND	+	-	ND

Abbreviations: MNZ, metronidazole; ND, not determined

2.1 CLASS I MUTANTS

One group of Nic$^-$ mutants yielded only residual rates of acetylene reduction under N_2 fixing conditions. These mutants were unable to grow on 6-OH-nicotinate as sole source of both carbon and nitrogen. In addition, these strains nodulated *S. rostrata* plants poorly, if at all, and the few small nodules observed were Fix$^-$ (did not reduce detectable amounts of acetylene).

Class I Nic$^-$ mutants are defective for late reactions. Because these strains were Nif$^-$ and Fix$^-$, oxidative catabolism of 6-OH-nicotinate is required for N_2 fixation both in culture and *in planta*.

2.2 CLASS II MUTANTS

The second group of Nic$^-$ mutants grew on 6-OH-nicotinate as sole source of both carbon and nitrogen. Mutant cultures exhibited 1/2 to 2/3 of the acetylene reduction rates observed for wild-type. These mutants produced Fix$^+$ nodules that reduced acetylene at wild-type levels. Class II mutants were also resistant to metronidazole, a metabolic inhibitor activated to a toxic state by an intact electron transport system. Both wild-type and Class I mutants were sensitive to metronidazole.

ORS571 wild-type could not grow on nicotinate as sole carbon and nitrogen source, nor did it grow on nicotinate and fumarate (Table 1). However, ORS571 grew normally on nicotinate as nitrogen source with succinate added as carbon source. If ORS571 nicotinate catabolism progresses as diagrammed (Fig. 1), the inability to grow on nicotinate alone may be ascribed to the lack of an appropriate external reductant for nicotinate hydroxylase. Succinate, but not fumarate, fulfills the requirement for such a reductant.

Class II mutants may be nicotinate hydroxylase mutants or they may contain lesions in some electron transport function(s) that relays electrons from succinate to nicotinate hydroxylase. Because Class II mutants are metronidazole resistant, they are probably defective in electron transport.

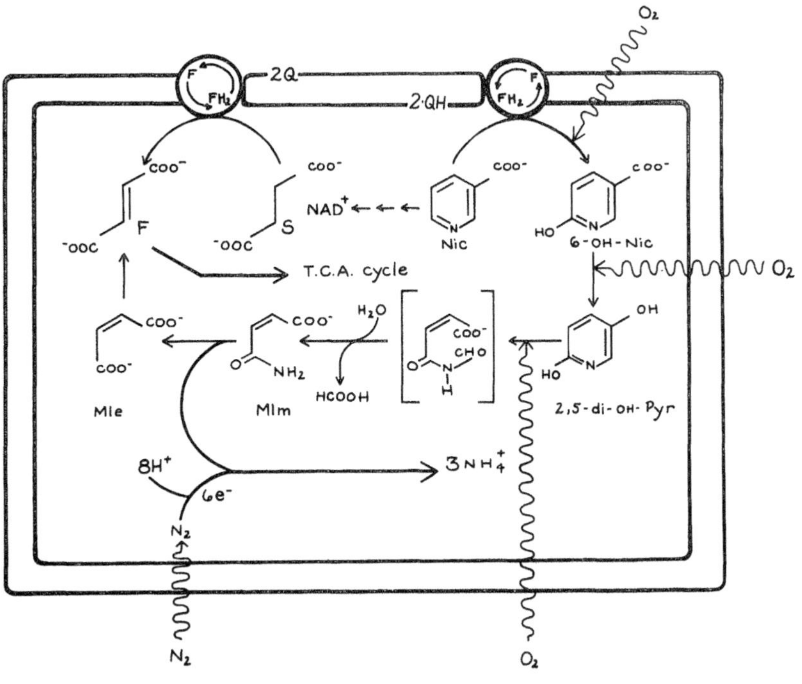

FIGURE 1. Metabolic flow chart for *A. sesbaniae* ORS571 during synergistic N_2 fixation and nicotinate oxidation. Nicotinate is substrate for both catabolic reactions and NAD$^+$ biosynthesis. Both nicotinate and N_2 yields ammonium as nitrogenous growth substituent. Fumarate, formate, and bicarbonate are end products of Nic catabolism. Abbreviations: F, flavin moiety; Q, membrane-bound quinone electron carrier; 2,5-di-OH-pyr, 2,5-dihydroxypyridine; Mle, maleate; Mlm, maleamate; S, succinate; TCA, tricarboxylate cycle.

3. THE NIC⁻ MUTANTS DEMONSTRATE THAT ORS571 IS AN NAD^+ AUXOTROPH, AND THAT NICOTINATE CATABOLISM IS REPRESSED BY GLUTAMINE

Because ORS571 requires nicotinic acid for N_2 fixation, its growth requirement for nicotinate implies NAD^+ limitation, which might be attributable to two metabolic properties. Either, (1) ORS571 is NAD^+ auxotrophic, or (2) it constitutively oxidizes nicotinate and deprives NAD^+ biosynthetic enzymes of substrate. To distinguish between these two possibilities, ORS571 wild-type was cultured in defined media with either glutamine or glutamate as sole nitrogen source and with varying levels of nicotinate added.

Table II. Nicotinate limitation in ORS571 batch cultures growing on glutamine vs. glutamate.

[Nic]	Final A_{600}nm on N-source: Glutamine	Glutamate
0	.030	.017
10 nM	.050	.025
50 nM	.174	.031
100 nM	.203	.044
500 nM	.628	.047
1 µM	.953	.066
10 µM	1.002	.227
100 µM	1.199	.847

Two conclusions were be drawn from these experiments: (1) By regulating the rates of nicotinate catabolism, ORS571 batch cultures grow to varying final densities before encountering NAD^+ limitation, and (2) when employed as N-sources in batch cultures, glutamine, but not glutamate, represses ORS571 nicotinate catabolism. Therefore batch cultures provided with glutamine as N-source produce greater cell yield than those provided with glutamate.

Because class II Nic⁻ mutants also retained the nicotinate growth requirement despite being unable to oxidize nicotinate, ORS571 wild-type must be NAD^+ auxotrophic.

4. CONCLUSIONS

We propose that a nicotinate catabolism pathway, similar to that of *Pseudomonas fluorescens*, exists in ORS571. During nicotinate catabolism, O_2 scavenging permits the maintenance of a sufficiently low intracellular O_2 concentration so as to allow both nitrogenase activity and aerobic respiration.

Because nicotinate and its derivatives stimulate growth of legume roots at micromolar concentrations, the *in vitro* results may mimic symbiotic conditions *in planta*. These results address the paradox of a symbiosis initiated by a bacterium able to supply its own reduced nitrogen needs quite independent of the host plant. The strict NAD^+ auxotrophy of ORS571 may allow the host plant to direct the course of the symbiosis by regulating the supply of nicotinate.

The provision and utilization of nicotinate seems to be an integral part of this symbiosis between ORS571 and *Sesbania rostrata*. The NAD^+ auxotrophy of ORS571 precludes its independent growth. Nicotinate catabolism, N_2 fixation, and ammonium assimilation are all repressed by glutamine, which must be involved in the coordinate control of these processes.

AT LEAST THREE LOCI ENCODE THE LEAF-CURL PHENOTYPE IN RHIZOBIUM STRAIN IC3342

N.M. UPADHYAYA, K.F. SCOTT, W.T. TUCKER, AND P.J. DART
Centre for Recombinant DNA Research and Department of Molecular Biology, Research School of Biological Sciences
The Australian National University, Canberra
A.C.T. 2601, Australia

Pigeonpea, an important grain legume crop in the semi-arid tropics, is nodulated by cowpea group rhizobia. Nodulation by the fast-growing *Rhizobium* strains, IC3342 and IC3324, causes pronounced curling and deformation of leaves (Kumar Rao *et al*., 1984). The symptoms appear first on the third developing trifoliate leaf bud which bends outwards, followed by hypernasty, reduced leaf expansion, suppression of apical dominance, sprouting of axial buds, and generally stunted growth. These strains also effectively nodulate siratro (*Macroptilium atropurpureum*) and *Desmodium* causing leaf-deformation, and ineffectively nodulate *Lathyrus sativus*, *Phaseolus vulgaris* and *Arachis hypogaea* without symptoms. Grafting and sap feeding studies indicate the production of curl-inducing principle in nodules or roots and translocation through the sap to the leaves (Upadhyaya *et al*., 1985). We report on genetic analyses of strain IC3342 aimed at identifying the genes involved in curl induction and their relationship to nodulation and nitrogen-fixation.

The strain IC3342 has nodulation (*nod*) and N_2-fixation (*nif*) genes on one of the three mega-plasmids present, as established by hybridizing with a *nif*HDK gene probe from the fast-growing cowpea strain ANU240 and a *nod* gene probe from *R. trifolii* strain ANU843. A heat-cured, non-nodulating derivative fails to induce the symptoms, implying that nodulation is essential for the curl induction.

Tn5 mutagenesis performed on a double marked (Streptomycin and Rifampicin resistance) derivative of the strain IC3342 by mating *E. coli* strain SM10 containing suicide vector pSUP1011, gave Km^r colonies at a frequency of 2.5×10^{-5} per recipient when the donor population was 7×10^6 cells/ml and that of recipient 4×10^5 cells/ml. Screening of 400 probable transconjugants using siratro as test plant in test-tubes yielded 19 mutants with $fix^- curl^-$ or $fix^+ curl^-$ phenotype. Only five mutants had the stable $fix^- curl^-$ (2) and $fix^+ curl^-$ (3) phenotype in subsequent screening. Plasmid visualization and hybridization with Tn5, *nod* and *nif* gene probes indicated that one of the $fix^- curl^-$ and all the three $fix^+ curl^-$ mutants have Tn5 insertions on the Sym plasmid. All the five mutants appear to have the Tn5 insertions on *Eco*R1 fragments of different size ranging from 7.5 to 12 kb indicating that a minimum of three loci are involved in leaf-curl induction.

All five Tn5-containing fragments have been successfully cloned onto a multi-copy vector pUC19 and plasmid clones (pMNU1, pMNU2, pMNU3, pMNU4, pMNU5) have been amplified in *E. coli* strain RR1. A cosmid library has been constructed from ~20 kb fragments of wild type DNA (obtained trhough partial digestion with Sau3A and subsequent isolation by sucrose gradient centrifugation) using the vector pLAFR3. About 10000 cosmids, with 90% of them having insert DNA, have been pooled to construct the library. Screening of this cosmid library for fragments corresponding to individual Tn5 insert fragments is in progress. The cosmids containing insert DNA corresponding to one or more Tn5 insert fragments will be introduced into a non-leaf-curling strain to see whether leaf-curling can be induced.

Sub-cloning and sequencing of fragments from all the five plasmid clones is being carried out by the Sanger (dideoxy) method using M13mp18 vector. Sequence data analysis of about 350 base pairs around Tn5 insertions in pMNU1 and pMNU2 ($fix^- curl^-$ phenotype) indicate possible long, open reading frames towards either sides of Tn5 insertion.

The mutational analysis presented here shows that the leaf curl symptoms observed on pigeonpea result from bacterial gene expression. A genebank data base search showed that the genes identified here have no significant homology with any previously characterized gene. Additional sequence analysis and complementation experiments should enable us to determine the genetic basis of the contribution of *Rhizobium* to these plant symptoms.

REFERENCES

1. Kumar Rao *et al.* (1984) *Soil Biol. Biochem.* 16, 89-91.
2. Upadhyaya *et al.* (1985) *In* Nitrogen fixation research progress, *ed* H.J. Evans *et al.*, Martinus Nijhoff Publishers *pp* 145.

Section V

MOLECULAR GENETICS OF THE OTHER DIAZOTROPHIC ORGANISMS

USE OF HETEROLOGOUS HYBRIDIZATION IN PHYLOGENETIC STUDIES OF SYMBIOTIC *ANABAENA* STRAINS

C. FRANCHE, B.E.G. GUNNING, B.G. ROLFE* AND J. PLAZINSKI
Departments of Developmental and Molecular Biology*, R.S.B.S., A.N.U., Canberra City, A.C.T. 2601, Australia

1. INTRODUCTION

Azolla is a genus of heterosporous aquatic fern which lives in symbiotic association with a nitrogen-fixing cyanobacterium, *Anabaena azollae*. The symbiont is located in a cavity in the dorsal leaves and develops in synchrony with the pteridophyte (5,7). Due to the high rate of nitrogen fixation in the *Azolla-Anabaena azollae* symbiosis, *Azolla* is of particular interest as a green manure in rice paddy soils.

Six extant species of *Azolla* are presently recognized and divided into two sections, *Euazolla* and *Rhizosperma*. The symbiotic cyanobacteria associated with *Azolla* have been classified as a single species, *Anabaena azollae* (7). However, the difficulties encountered in growing the endosymbionts in culture in the absence of the hosts, has made it impossible to ascertain the relationships between *Anabaena azollae*, even for *Anabaena* within the same species of *Azolla*.

Using DNA probes from the free-living *Anabaena* sp., PCC7120, *Rhizobium trifolii* and *Agrobacterium tumefaciens* strains, containing respectively *nif* (3,9), *nod* (4) and *vir* (1) gene clones, we have compared the DNA hybridization patterns between the *Anabaena azollae* strains extracted from 14 *Azolla* species including seven *Euazolla* and seven *Rhizosperma*.

2. METHODS

Cyanobacterial strains and *Azolla* species used for our studies were reported by Franche and Cohen-Bazire (2).

Isolation of the symbionts from the fern was carried out as described by Peters and Mayne (8).

DNA extraction procedures were those of Maniatis et al. (6). Hybridizations with the DNA probes were carried out at 65°C, 55°C or 50°C for 16 hours with and without 10% dextran sulphate (6).

3. RESULTS

3.1 DNA probes with the free-living cyanobacterium PCC7120

Four probes containing *nif*H,D,K,S and one probe from the 11kb excised fragment from *Anabaena* sp.PCC7120 strain were used. Most of the restriction sites in the *nif*K and *nif*D genes were conserved among endosymbiotic members of both *Azolla* sections. Some restriction sites appeared identical in symbiotic *Anabaena azollae* and symbiotic *Nostoc* strains (*Nostoc cycas* PCC7422 and *Nostoc cycas* NS51) isolated from roots of *Cycas*. This suggests a close relationship between these symbiotic cyanobacteria. The DNA restriction fragments which hybridized to the *nif*H and *nif*S probes were conserved amongst the endosymbionts of *Rhizosperma* section, but they were different from those observed in A. azollae isolates extracted from four Euazolla species. No hybridization was found between DNA from the A. azollae symbionts and the probe carrying a part of the 11kb region separating nifK from nifDH in vegetative cells of Anabaena sp. PCC7120. A negative result was also obtained with DNA from Nostoc cycas whereas the same probe (pAn207.3 (3)) hybridized to the DNA extracted from seven free-living Anabaena-Nostoc strains.

Three of the four indigenous plasmids carried by Anabaena PCC7120 (1)) were used as hybridization probes with the EcoRI digests of the DNA isolated from A. azollae microphylla and A. azollae pinnata imbricata IND strains. Preliminary results indicate a strong homology between the 3.3Mdal plasmid of strain PCC7120 and two EcoRI fragments (3.3kb and 1.3kb) derived from both cyanobacterial strains.

3.2 Non-cyanobacterial DNA probes

When Rhizobium trifolii (strain ANU843) nodulation genes were used as specific probes only part of Regional DNA showed positive hybridization to DNAs isolated from: strain PCC7120 (8kb and 4.9kb HindIII generated fragments), A. azollae microphylla and A. azollae imbricata IND (in both strains 4.9kb HindIII-cleaved fragment showed positive hybridization). Three cyanobacterial isolates (A. azollae caroliniana, A. filiculoides Shepparton and A. filiculoides East Germany) showed positive hybridization to one Agrobacterium pTiC58 vir region probe (Kpn9) but not to three others.

DNA fragments hybridizing to the Kpn9 probe were a 19kb fragment (from BglII-cleaved DNA of A. azollae caroliniana), 18kb and 5.5kb fragments from the endosymbionts of the two A. filiculoides strains.

In conclusion, our data suggests that using DNA hybridization method one can differentiate between A. azollae isolates.

REFERENCES

1. Close TJ et al: Plasmid 12, 111-118, 1984.
2. Franche C and Cohen-Bazire G: Plant Science 39, 125-131, 1985.
3. Golden JW et al: Nature 314, 419-423, 1985.
4. Innes RW et al: Mol Gen Genet 201, 426-432, 1985.
5. Lumpkin TA and Plucknett DL: Azolla as a green manure. West View Press, Boulder, Colorado, 1983.
6. Maniatis T et al: Molecular Cloning: A Laboratory Manual. Cold Spring Harbour Laboratory, Cold Spring Harbour, New York, 1982.
7. Peters GA and Calvert ME: Algae Symbiosis, A Continuum of Interaction Strategies. Cambridge University Press, 1983.
8. Peters GA and Mayne BC: Plant Physiol 53, 813-819, 1974.
9. Rice D et al: J Biol Chem 257, 13157-13163, 1982.
10. Simon RD: J Bacteriol 136, 414-418, 1978.

CHROMOBACTERIUM LIVIDUM NCTC 10590 IS A NITROGEN-FIXING *AGROBACTERIUM RADIOBACTER*

M. H. SOLIMAN* AND G. R. K. SASTRY, Department of Genetics, University of Leeds, Leeds LS2 9JT, U.K.

(*On leave from Department of Genetics, Faculty of Agriculture, Cairo University, Cairo, A.R.E.)

INTRODUCTION

Despite all the upheaval in the use of Ti plasmids and plant genetic engineering, not much consideration has been given to the development of *Agrobacterium* as a nitrogen-fixing bacterium with a wide host range. It is even more surprising since the genus *Rhizobium* and *Agrobacterium* have, for a long time, been known to be taxonomically related (1). Further testimony to such a relationship is provided by the finding that at least some Sym plasmids cause nodule-like structures when transferred into Ti-cured *A. tumefaciens* (2); also, chromosomal genes of *R. meliloti* which are known to affect nodulation properties of the bacterium complement at least one chromosomal nontumorigenic mutant (*ctu1*) of *A. tumefaciens*, as evidenced by the restoration of tumorigenic activity (3).

Lack of interest in developing *Agrobacterium* into a beneficial nitrogen fixing organism is mostly due to what we consider groundless fears that it is a serious plant pathogen. Since the discovery that *A. tumefaciens* C58 and B6 have some kind of *nif* system (4) we have been searching for an 'ideal' *Agrobacterium* which can be used for this purpose. It is claimed here that the bacterial strain NCTC 10590, originally isolated from sterilised, germinating seed of *Psychotria nairobiensis* (5) has a great potential for development into a nitrogen-fixing bacterium with a broad host range.

Although NCTC 10590 has been classified as *Chromobacterium lividum* by the original authors (5) it has been placed in the genus *Agrobacterium* by Holmes and Roberts (6) after a taxonomical reassessment; the main reason for reclassification is that NCTC 10590 is 3-ketolactose positive, a diagnostic feature of *Agrobacterium* (7). The strain is not tumorigenic (and we will therefore refer to it as *A. radiobacter*) but becomes so when provided with the Ti plasmid (Soliman, unpublished).

Nitrogen fixation. ^{15}N incorporation studies showed NCTC 10590 can fix nitrogen (5). We found that it grows both aerobically and anaerobically on nitrogen-free medium and can reduce acetylene. It produces carotene only when grown aerobically, suggesting that the pigment functions as a protection from the effect of oxygen.

Cloning of *nifHDK*. In Southern blots a *nifHD* fragment from *R. meliloti* (in pRmR2) hybridizes with total NCTC 10590 DNA digests. As far as we can judge from the blots, the strain contains only one copy of *nifHD*. Several *E. coli* DH1 colonies containing recombinant plasmids (vector pBR322) which hybridized with the *R. meliloti* probe have been retrieved from an NCTC 10590 gene bank. One of these, pMS1, which contained a 6.4kb insert, was used for further investigation. Details of cloning strategies adopted and the restriction map of pMS1 will be published elsewhere.

The assumption that pMS1 carries NCTC 10590 *nif* genes *H*, *D* and *K* was further confirmed by introducing the plasmid into different *K. pneumoniae* strains carrying mutations for the same genes; complementation was strong as seen by the growth of the transformants (Fig. 1) on nitrogen-free medium.

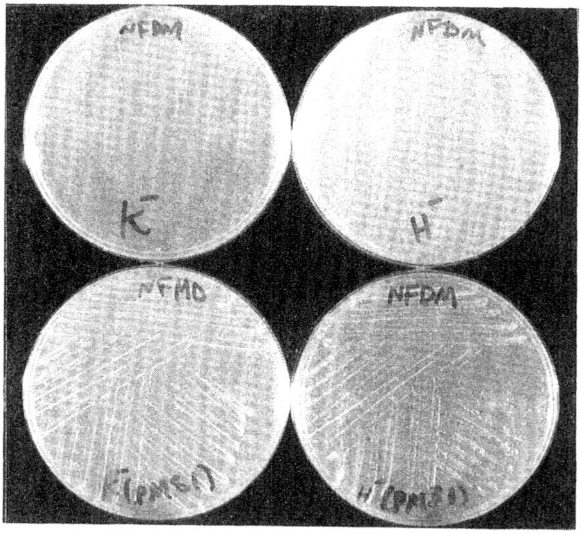

FIGURE 1. *K. pneumoniae nifH*⁻ and *nifK*⁻ with (bottom) and without (top) pMS1 grown aerobically (30°C) on NFDM medium.

Like the nitrogen fixation system in other *Agrobacterium* strains (4) *nif* genes of NCTC also appear to be less sensitive to the aerobic condition. But the intriguing question is how NCTC 10590 *nifHDK* (and maybe one or two other *nif* genes) located on a 6.4kb insert in pMS1 can collaborate with the rest of the *nif* genes of *K. pneumoniae*, which are sensitive to repression by oxygen and produce complementation under the aerobic condition?

ACKNOWLEDGEMENTS

MHS is the recipient of a fellowship from the Government of A.R.E.

REFERENCES

1. Bergey's Manual of systematic bacteriology (1984) *I*: 234-256 (Krieg NR and Holt JG, eds.). Williams and Wilkins.
2. Hooykaas PJJ, Den Dulk-Ras H, Regensburg-Tuink AJG, van Brussel AAN and Schilperoort RA (1985). Plasmid 14: 47-52.
3. Miller IS, Fox D, Saeed N, Borland PA, Miles CA and Sastry GRK (1986). Mol. Gen. Genet. (in press).
4. Kanvinde L, Soliman MH, Wardhan H, Nowell L, Fox D and Sastry GRKS (1986). In these Proceedings.
5. Bettelheim KA, Gordon JF and Taylor J (1968) J. Gen. Microbiol. 54: 177-184.
6. Holmes B and Roberts P (1981) J. Appl. Bact. 50: 443-467.
7. Bernaerts MJ and De Ley J (1963) Nature 197: 406-407.

STUDIES ON THE DIAZOTROPHIC NATURE OF *AGROBACTERIUM*

LALITA KANVINDE, M. H. SOLIMAN*, H. WARDHAN, LISE NOWELL, D. FOX AND G.R.K. SASTRY, Department of Genetics, University of Leeds, Leeds LS2 9JT, U.K.

(*on leave from Department of Genetics, Faculty of Agriculture, Cairo University, Cairo, A.R.E.)

INTRODUCTION

Although *Agrobacterium* and *Rhizobium* are known to be closely related (1) the former is always considered to be a plant pathogen. As a result, no investigation has ever been carried out to see if any *Agrobacterium* strains have an inherent ability to fix nitrogen either in the free-living condition or in crown galls produced by the bacterium. It is true that several successful attempts have been made to transfer *Rhizobium* Sym plasmids into Ti-cured *A. tumefaciens* strains (2) but none of the published reports mention the activity of nitrogenase in either the transconjugants or the recipients.

Growth on nitrogen-free medium. The first indication that *A. tumefaciens* might be able to fix nitrogen was an accidental finding that the bacterium is capable of growing on nitrogen-free medium (NFDM). Since then, systematic experiments have been carried out on standard strains B6, C58 and a Ti-cured C58, NT1, all of which were able to grow on NFDM. Under aerobic conditions they take only two extra days to reach the same sized colonies as cultures grown on minimal medium with ammonium sulphate as nitrogen source (Fig. 1).

FIGURE 1. B6 grown aerobically (left) and anaerobically (right) on NFDM plates made with Noble agar.

That this observation is not an experimental artifact caused by contamination of fixed nitrogen was established using high grades of agar (Noble and Agarose) and incubating the plates in anaerobic jars filled with nitrogen. Detailed growth profiles in NFDM liquid medium have not yet been obtained but all three strains grew well, B6 being the best.

Acetylene reduction. Detailed investigations on the ability to reduce acetylene were carried out on B6, C58 and NT1. In most experiments serine and ammonium sulphate were used as nitrogen sources in $-NH_4^+$ and $+NH_4^+$ NFDM media respectively. All three strains tested reduced acetylene efficiently when grown in anaerobic conditions and in the absence of ammonium in the medium, two signs of nitrogen fixation. However, some qualifications need to be added to this statement. Often, *Agrobacterium* cultures produce ethylene only after 10-16h following acetylene injection. Under the conditions in which we culture, the bacterium produces considerable quantities of exopolysaccharides, often leading to clumping. Attempts were made to improve the situation by changing the carbon and nitrogen courses, with variable results; for instance, often ethylene appeared within an hour after acetylene was injected when proline and arabinose were used as nitrogen and carbon sources respectively (Table 1).

TABLE 1. Acetylene reduction carried out by 3ml cultures grown in NFDM + arabinoase + succinate with proline as nitrogen source under anaerobic and aerobic conditions (nmoles of ethylene/vial/O.D.); data obtained from $(NH_4)_2SO_4$ cultures is not shown as none of them produced ethylene.

	Hours after acetylene injection:						
	0	1	2	3	4	5	24
Bacterium	Ethylene produced:						
A. tumefaciens							
(1) NT-1							
(a) anaerobic	0	65	-	215	420	-	-
(b) aerobic	0	6	-	10	-	25	-
(2) C58							
(a) anaerobic	0	4050	6625	-	7660	-	-
(b) aerobic	0	0	0	35	35	-	-
(3) B6							
(a) anaerobic	0	130	205	-	-	-	16,900
(b) aerobic	0	0	-	10	20	-	25
K. pneumoniae							
UN1179 (*nif*$^+$)							
(a) anaerobic	0	840	1440	-	3030	-	-
(b) aerobic	0	-	15	-	20	-	-

The second enigma was that in the experiment from which the data in Table 1 was obtained showed considerable repression in aerobic cultures; such an effect was not always observed.

An interesting feature of the *A. tumefaciens* system is that once the culture starts to reduce acetylene the process continues for several days. It should be pointed out also that none of the *Agrobacterium* strains examined ever produced ethylene without prior acetylene injection.

Cloning of *nif* genes. Ruvkin and Ausubel (3) reported that neither C58 nor B6 have DNA sequences which can hybridize with a *K. pneumoniae nifHDK* probe. We repeated these tests using a *nifHDK* probe from *Rhizobium* ORS571 with the same negative results.

However, total DNA digests from *Agrobacterium* strain NCTC 10590 (which also grows on nitrogen-free medium and reduces acetylene) gave a clear signal of hybridization with *R. meliloti nifHD*. The *nifHDK* from NCTC 10590 has now been cloned and partially characterised (4). *A. tumefaciens* 18-1 also showed hybridization with *nifHDK* probes from *Rhizobium* ORS571 and NCTC 10590 (Fig. 2).

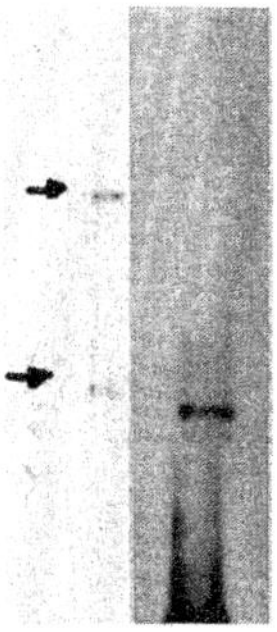

FIGURE 2. Southern blots of 18-1 total DNA digest probed with *nifHDK* from *Rhizobium* ORS571 and NCTC 10590.

(Tests carried out in Leeds showed that 18-1, obtained from Professor W. Heumann, is 3-ketolactose positive and tumorigenic on tomato seedlings; it grows on nitrogen-free medium and reduces acetylene.)

Complementation analyses. As was pointed out (4) NCTC 10590 *nifHDK* complements the respective genes in *K. pneumoniae*. To see whether NT1 (which did not show any hybridization signal on Southern blots with standard probes) contained a different type of genes which facilitate growth on nitrogen-free medium, a cosmid bank was screened; this resulted in the retrieval of six recombinant cosmids (which will be referred to as pLS cosmids, average insert size 25kb) facilitated the growth of *E. coli* DH1 on NFDM at 30°C but not at 37°C. Preliminary evidence suggested that pLS cosmids also enabled DH1 to reduce acetylene. When transferred into *K. pneumoniae nifH*$^{-}$ and *nifK*$^{-}$, these cosmids showed strong complementation, as indicated by the growth on NFDM and tests for acetylene reduction.

DISCUSSION

Broadly speaking, *Agrobacterium* strains seem to fall into two groups. To the first group belong C58 and B6 which, although they carry all the hallmarks of diazotrophy, do not seem to have DNA sequences similar to those of standard *nifHDK* probes. The fact that NT1 was also able to carry similar reactions shows that the Ti plasmid is not involved in this process. Further studies on the 'nif' genes from this group of strains will have to await the elucidation of the molecular biology of pLS cosmids. Since a clear expression was not obtained in *E. coli* it should be possible to use a minicell procedure to investigate the nature of the proteins produced by pLS cosmids.

NCTC 10590 and 18-1, which belong to the second group of strains, have *nif* sequences which hybridize with standard probes. Once the preliminary screening for the presence of the category II type *nif* system in different strains is completed, we propose to construct a cosmid bank of 18-1 for detailed investigation of its genes concerned with nitrogen fixation.

In spite of the preliminary nature of the work reported here, one or two features stand out clearly. First, as has been mentioned, all strains studied grow well on NFDM even when grown aerobically; in this respect they resemble *Rhizobium* ORS571. A second related feature of the present system is that when *Agrobacterium nif* genes were transferred to *E. coli* (pLS cosmids) and *K. pneumoniae* (both pLS and pMS1 of Soliman and Sastry) they enabled both bacteria to grow aerobically on nitrogen-free medium.

Not many experiments have been carried out to investigate the biological significance of the nitrogen-fixing ability of *Agrobacterium*. However, in one test young tomato tumours produced by C58 and B6 caused significant acetylene reduction. But whether crown galls actually assist the plants in fixing nitrogen needs to be carefully investigated.

ACKNOWLEDGEMENTS: A fellowship from the Goverment of A.R.E. to MHS and a Leeds University postgraduate studentship to HW are acknowledged. GRKS is grateful to the Royal Society for a travel grant to attend the Symposium. Research in our laboratory is supported by a grant from the SERC and from Leeds University.

REFERENCES

(1) See Bergey's Manual of systematic bacteriology (1984) I: 234-256 (Krieg NR and Holt JG, eds.) Williams and Wilkins.

(2) For example, Hooykaas PJJ, Den Dulk-Ras H, Regensburg-Tuink AJG, van Brussel AAN and Schilperoort RA (1985). Plasmid 14: 47-52.

(3) Ruvkin GB and Ausubel FM (1980). Proc. Natl. Acad. Sci. USA 77:191-195.

(4) Soliman MH and Sastry GRK, in these Proceedings.

DEVELOPMENTS IN THE GENETIC ANALYSIS OF *AZOSPIRILLUM*

MARK VANSTOCKEM, KRIS MICHIELS, MAGGI MARIS, JOS VANDERLEYDEN AND AUGUST P. VAN GOOL.
F.A. JANSSENS MEMORIAL LABORATORY FOR GENETICS, K.U. LEUVEN, KARDINAAL MERCIERLAAN 92, B-3030 HEVERLEE, BELGIUM.

1. INTRODUCTION

Bacteria of the genus *Azospirillum* (1) are free-living diazotrophs that can readily be isolated from the rhizosphere and roots of grasses (2). No differentiated structures are formed following colonization of the root zone. Bacteria are thightly attached to the root surface and pictures of root hair deformation were reported (for review, see 3). Numerous field experiments, carried out since 1975, have demonstrated that *Azospirillum* inoculations can promote crop yield under certain environmental and soil conditions. Biochemical and physiological studies on *Azospirillum* in laboratory conditions suggest that processes like biological nitrogen fixation and plant hormones production might contribute to plant growth responses upon inoculation (3, 4). Until now little is known about the genetics of these bacteria or the molecular biology of their association with plants (5). Most progress has been made in the biochemistry and molecular genetics of nitrogen fixation (6, 7, 8). Recently, efforts to develop tools for the genetic analysis of *Azospirillum* were reinforced, and research programs to identify bacterial genes in the *Azospirillum*-plant interaction process were initiated.

2. GENETIC TOOLS FOR *AZOSPIRILLUM*

2.1. At random transposon mutagenesis

Plasmids containing Tn5 and composed either of a p15A type replicon and N-type bacterial mating system (9) or a ColE1 replicon and a Mob-region of IncP-type plasmids (10) can be used for transposon mutagenesis in *A. brasilense* Sp7 and *A. lipoferum* SpBr17 (11, 12, Elmerich, personal communication). Km^r *Azospirillum* exconjugants can be isolated at maximum frequencies of 10^{-7} per recipient cell. We have observed that the streptomycin resistance gene of Tn5 is expressed in *Azospirillum* (Vanstockem *et al.*, in preparation) as was demonstrated in *Rhizobium* (13). Tn5 appears to be integrated at random in the genome of *Azospirillum*. Auxotrophic mutants and mutants with a Nif^{-} $^{or\ att}$ phenotype have been isolated.

2.2. Curing and mobilization of indigenous plasmids

All *Azospirillum* strains contain plasmids, varying from one to six molecular species (14). All the indigenous plasmids of *Azospirillum* are cryptic sofar. We were able to label the two megaplasmids (> 300 Mdalton) of *A. brasilense* Sp7 with Tn5-Mob, using the system of Simon (15). Attempts to cure or to mobilize the labelled plasmids from these strains were unsuccessful sofar.

However loss (or cointegration) of one particular plasmid (115 Mdalton) of strain Sp7, was occasionally observed in nitrosoguanidine induced Sm^r clones of Sp7, in a spontaneous Rif^r clone of Sp7, and in a number of Km^r resistant clones, containing Tn5-Mob in the chromosome.

2.3. Cloning vectors and transformation

Derivatives of IncP-type and IncQ-type plasmids, introduced by conjugation stably replicate in *Azospirillum*. The best selectable antibiotic resistance markers are kanamycin and tetracycline. Broad host range plasmids that have been introduced in this way in *Azospirillum* are pRK290 (pLAFR1), pRK252, pRK404 (IncP-type) and pKT230, pKT231, pKT240 (IncQ-type). Plasmid pRK252 (10 kbp) was selected for transformation studies in *A. brasilense* Sp7. Although transformants could be obtained using the $CaCl_2$ method, the efficiency obtained sofar is very low (10^{-9}). Recently Fani *et al* (16) demonstrated transformation of *A. brasilense* Sp6 with plasmid pRK290 (20 kbp).

3. BACTERIAL DETERMINANTS OF *AZOSPIRILLUM*-PLANT INTERACTION

Biochemical and physiological studies on *Azospirillum* in laboratory conditions have demonstrated that azospirilla have many characteristics in common with other phytobacteria. The common properties listed below have been shown to be essential for other phytobacteria in their interaction with plants. All azospirilla produce phytohormones: auxins and cytokinins (17). In test tubes some *A. brasilense* strains produce up to 400 ng of indole-3-acetic acid (IAA) per ml culture medium, without exogenously added tryptophane (S. Horemans, personal communication). These levels of IAA are comparable to those found in pathogenic strains of *Pseudomonas savastanoi* (18). We have attempted to identify *Azospirillum* genes involved in auxin and cytokinin production using DNA hybridizations on restricted total DNAs of various *Azospirillum* strains with *Agrobacterium* DNA probes containing genes encoding enzymes for auxin and cytokinin biosynthesis. No homology with any of the probes tested was found. Biochemical studies indicated that the pathway for IAA biosynthesis in *Azospirillum* is likely to be different from he pathway found in *P. savastanoi* since no indole-3-acetamide could be found in their supernatant. Most *Azospirillum* strains produce exopolysaccharides that stain with Calcofluor white in liquid (19) and on solid media. *Exo* mutants of *R. meliloti* cannot produce extracellular β-linked polysaccharides and interact infrequently with root hairs (20). Homology of *Azospirillum* DNA with *R. meliloti* probes containing *exo* genes was found and will facilitate the cloning of the corresponding *exo* genes of *Azospirillum* (Michiels *et al.*, in preparation).

Some *Azospirillum* strains show pectinolytic activity. Tien *et al* (21) tested six strains for pectinolytic activity in an *in vitro* assay with pectin and polygalacturonic acid as substrates. We have tested twenty three strains of *Azospirillum* for maceration of potato tuber tissue by the well method of Keen *et al.* (22). Some strains macerated tuber slices within two days, while other strains did not attack potato tuber slices at all, even after five days of incubation. Interestingly a correlation was found between the presence of pectinolytic activity and the intercellular location of a particular *Azospirillum* strain in wheat roots.

In another approach, similarities between *Azospirillum* and other phytobacteria, in particular with members of the *Rhizobiaceae*, regarding plant recognition and/or attachement could be demonstrated. Recently, Fogher *et al.* (23) found homology between *Azospirillum* DNA and DNA probes containing *nod* and *hsn* genes of *R. meliloti*. Clones containing the *Azospirillum hsn* homologues were isolated. Preliminary observations suggest that the *hsn* homologous regions might be plasmid borne in *Azospirillum* (Elmerich *et al.*, in preparation).
A similar approach was made by our laboratory, using a DNA probe containing the chromosomal virulence genes of *A. tumefaciens*, identified and cloned by Douglas *et al* (24). All azospirilla tested show significant homology with a 9 kbp DNA probe containing the *chv*B region of *Agrobacterium* (Figure 1). Surprisingly, in most strains tested the *chv*B homologous regions in *Azospirillum* amount up to about 50 kbp of DNA.

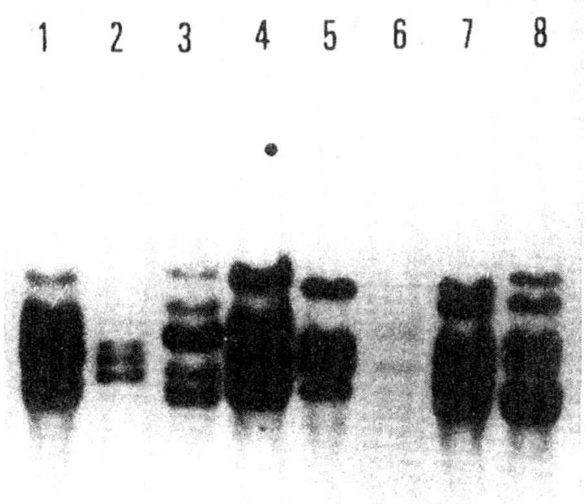

Figure 1.
Hybridization of *Sal*I digested total DNAs of various *Azospirillum* strains with a 9 kbp DNA probe of *A. tumefaciens* containing the *chv*B locus. 1: Sp7; 2: SpBr17; 3: RO7; 4: Sp13; 5: SpBr14; 6: SpS28; 7: Sp245; 8: Sp107.

4. CONCLUSIONS

Biochemical, physiological and molecular biological studies on *Azospirillum* seem to indicate fundamental similarities with other phytobacteria (pathogens and symbionts), regarding plant attachment and/or recognition phenomena, exopolysaccharide synthesis, hormone production and pectinolytic activity. Although it is too early to draw conclusions about the significance of the homology to *nod*, *hsn* and *chv* in *Azospirillum spp*, these observations suggest that some of the early steps of the recognition between bacteria and plants proceed from common mechanisms.
The techniques of at random transposon mutagenesis (this paper, 12) and site directed mutagenesis (7, 8) will allow us to start functional analysis of these homologous sequences and to identify *Azospirillum* genes involved in plant hormones production, exopolysaccharide synthesis and pectinolytic activity.

5. ACKNOWLEDGEMENTS

This research was supported by grants from the "Fonds voor Kollectief Fundamenteel Onderzoek, F.K.F.O. 2.0013.85" and the CEC "Contract no TSD-A-255-B (RS)". We are grateful to C. Elmerich for helpful suggestions during the course of this work, to S. Horemans for communication of results prior to publication. We greatly acknowledge the skillful assistance of A. Vermassen in preparing the manuscript. MV is recipient of a fellowship from the Belgian Ministry of Scientific Affairs (PREST Program). KM is recipient of a fellowship from the N.F.W.O.

6. REFERENCES

1. Tarrand, J., Krieg, N. and Döbereiner, J. (1978) Can. J. Microbiol. 24, 967-980.
2. Döbereiner, J. and Day, J.M. (1978) in Proceedings of the First International Symposium on Nitrogen Fixation (Newton, W.E. and Nyman, C.J. ed.), pp. 518-538, Washington State University Press, Pullman.
3. Patriquin, D.G., Döbereiner, J. and Kain, D.K. (1983) Can. J. Microbiol. 29, 900.
4. Okon, Y. (1985) Trends in Biotechnology 3, 223-228.
5. Elmerich, C. (1984) Bio/Technology 2, 967-978.
6. Quiviger, B., Franche, C., Lutfalla, G., Rice, D. et al. (1982) Biochimie 64, 495-502.
7. Perroud, B., Bandhavi, S.K. and Elmerich, C. (1985) in Azospirillum III: Genetics, Physiology, Ecology (Klingmüller, W., ed.), pp. 10-19,
8. Singh, M. and Klingmüller, W. (1983) Exp. suppl. 48, 47-55.
9. Selveraj, G. and Iyer, V.N. (1983) J. Bacteriol. 158, 1580-1589.
10. Simon, R., Priefer, U. and Pühler, A. (1983) Biotechnology 1, 784-790.
11. Vanstockem, M., Michiels, K., Vanderleyden, J. and Van Gool, A. (1985) in Azospirillum III: Genetics, Physiology, Ecology (Klingmüller, W., ed), pp. 74-84, Springer Verlag, Berlin.
12. Singh, H. and Klingmüller, W. (1986) Mol. Gen. Genet. 202, 136-142.
13. Putnoky, P., Kiss, G.B., Ott, I. and Kondorosi, A. (1983) Mol. Gen. Genet. 191, 288-294.
14. Franche, C. and Elmerich, C. (1981) Ann. de Microbiol. 132, 3-17.
15. Simon, R. (1984) Mol. Gen. Genet. 196, 413-420.
16. Fani, R., Bazzicalupo, M., Coianiz, P. and Polsinelli, M. (1986) FEMS Microbiology Letters 35, 23-27.
17. Tien, T., Gaskins, M. and Hubbell, D. (1979) Appl. Envir. Micr. 37, 1016-1024.
18. Smidt, M. and Kosuge, T. (1978) Phys. Plant. Path. 13, 203-214.
19. Sadasivan, L. and Neyra C.A. (1985) J. Bacteriol.163, 716-723.
20. Leigh, J.A., Signer, E.R. and Walker, G.C. (1985) P.N.A.S. (U.S.A.) 82, 6231-6235.
21. Tien, T.M., Diem, H.G., Gaskins, M.H. and Hubbell, D.H. (1981) Can. J. Microbiol. 27, 426-431.
22. Keen, N.T., Dahlbeck, D., Staskawicz, B. and Belser, W. (1984) J. Bacteriol. 159, 825-831.
23. Fogher, C., Dusha, I., Barbot, P. and Elmerich, C. (1985) FEMS Microbiol. Letters 30, 245-249.
24. Douglas, C.J., Staneloni, R.J., Rubin, R.A. and Nester, E.W. (1985) J. Bacteriol 161, 850-860.

Section VI

SUPPLEMENT TO SECTIONS I AND II

ROLE OF VIR GENES IN THE EXCISION OF T-DNA FROM THE TI-PLASMID

K. Veluthambi, R. K. Jayaswal, & S. B. Gelvin

The ability of Agrobacterium tumefaciens to transfer the T-DNA portion of the Ti-plasmid to the nuclear genome of plant cells has been extensively used to engineer desirable genes into plants. We are interested in studying the mechanism by which this DNA transfer occurs and also in understanding the nature of the interaction between A. tumefaciens and the plant cell that eventually leads to the DNA transfer. The T-DNA is flanked by 25-bp imperfect direct repeat sequences called 'borders' which appear to define the portion of the DNA destined for transfer (1, 2, 3). In addition to the T-DNA region, a second region called the virulence (vir) region is also important for the virulence of the bacterium. The genes in this region, which are organized in 6 complementation groups (vir A, B, C, D, E and G), are expressed in the bacteria and have been proposed to play an important role in the excision, transfer and perhaps integration of the T-DNA(4).

It has earlier been shown that upon co-cultivation of Agrobacterium with regenerating plant cells, circular copies of the T-DNA generated from the Ti-plasmid can be recovered in E. coli(5). Model systems have recently been developed to study the formation of T-region circles in E. coli to facilitate the study of the role of vir genes in this process(6).

Our objective is to study directly the early events of T-DNA transfer within Agrobacterium itself. We have addressed the question whether T-DNA excision occurs at or around the borders under the conditions favorable for gene transfer. In order to induce conditions favorable for gene transfer, Agrobacterium was co-cultivated with regenerating tobacco leaf protoplasts. We routinely co-cultivated Agrobacterium strains containing Tn-3-lacZ fusions in the vir genes(4) and measured β-galactosidase activity to assess the levels of vir gene induction.

Figure 1 shows the results of such experiments using lacZ fusions in the vir genes B and E, as well as the closely associated gene pinF(4). Within 8-12 hours, a large increase in β-galactosidase activity is seen using either regenerating tobacco leaf protoplasts (Fig 1, left panel) or 10 μM acetosyringone (AS; Fig 1, right panel). A strain harboring a fusion in vir G, however, did not show an induction of β-galactosidase activity: the lacZ gene was fully induced in K_3 medium without protoplasts or acetosyringone (data not shown).

DNA was extracted from the octopine-type A. tumefaciens strain A348 grown in AB minimal medium pH 7.0 (B_o), or incubated for 24 hours in K_3 medium pH 5.6 without (K_3) or including (P) regenerating protoplasts. The DNA was digested with various restriction endonucleases and subjected to Southern blot analysis. DNA sequences consisting of the T-DNA borders

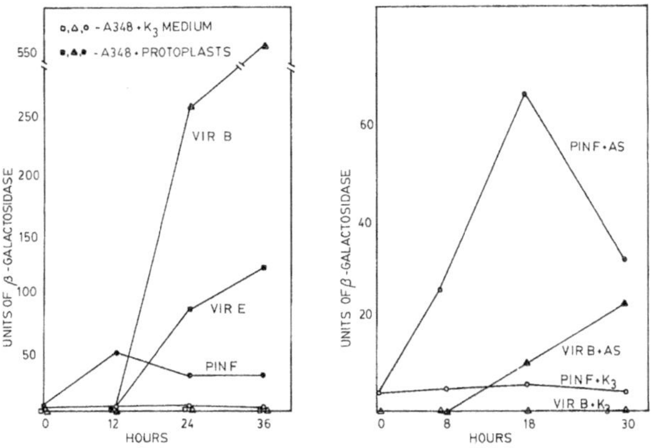

Fig.1 Induction of various *vir* genes by co-cultivation with regenerating tobacco leaf protoplasts (left panel) or acetosyringone (right panel)

(see Fig 2, upper panel) were used as hybridization probes to study the changes in restriction endonuclease patterns resulting from the cleavage of the T-DNA at the borders. New plant inducible fragments were detected using borders A, B, and C as probes. No detectable new fragments homologous to Border D were induced (Fig 2, lower panel).

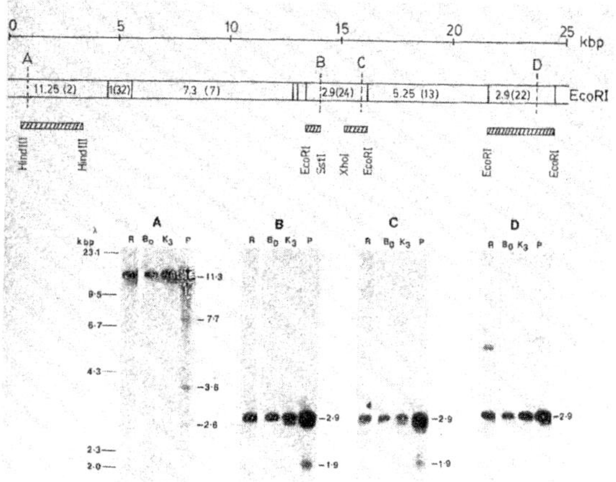

Fig.2 Cutting at the T-DNA borders in *Agrobacterium*. Upper panel shows a restriction endonuclease map of the T-DNA, including the four borders (A, B, C, D) and the hybridization probes (hatched bars). Lower panel shows bacterial DNA digested with EcoRI and probed with fragments covering each border. Bacteria have been co-cultivated with tobacco protoplasts for 24 hours. R=reconstruction; B_o=bacteria grown in AB minimal medium, pH7.0; K_3=bacteria transferred to K_3 medium, pH5.6; P=bacteria transferred to K_3 medium plus protoplasts. Note the subfragments (1-10% of the total signal) indicating cutting at borders A, B, and C.

If a cleavage has occurred at border A, the resulting DNA, after digestion with EcoR1, should show two fragments homologous to the border A probe. One will be 7.7 kbp in size and derived from sequences to the left of the border (this fragment has limited homology to the probe and is therefore difficult to detect) and the other will be 3.6 kbp in size and derived from sequences to the right of the border. These fragments are indeed seen, as well as fragments of sizes 2.6 and (in some experiments) 2.0 kbp (Fig 2, 4, and 5). These additional fragments may represent cleavage of the T-DNA at pseudoborders. The cleavage of the T-DNA at border A has been confirmed using the enzymes PstI and HindIII (data not shown). Cleavage at border A occurs in 1-10% of the T-DNA molecules, and can be enhanced if extra copies of the vir genes are present in the A. tumefaciens cells (data not shown).

Cleavage at borders B and C is complex. Bacteria incubated in K_3 medium without protoplasts exhibit a new T-DNA-derived fragment approximately 2.75 kbp in size. The amount of this fragment is increased in the presence of protoplasts (Fig 2). Recent experiments have shown that this fragment results from an internal deletion of 140-145 bp of T-DNA, probably to the right of border C (data not shown). This deletion may involve very short directly repeated sequences in this region and reflect a stimulation of generalized recombination when A. tumefaciens cells are co-cultivated with plant cells. In addition to this deletion, incubation of bacterial cells with protoplasts results in the generation of a 1.9 kbp fragment homologous to borders B and C (Fig 2). Although the size of this fragment (using EcoR1 to cleave the DNA) is consistent with cleavage at both borders B and C, the results of other experiments, using HindIII and XhoI, are not compatible with this interpretation (data not shown). The origin of this 1.9 kbp fragment is still under investigation.

If the cutting of the T-DNA upon incubation with protoplasts were specific for border (or perhaps pseudoborder) sequences, one should not see cleavage of internal fragments of the T-DNA. That this is so is shown in Fig 3, Panels 1 and 2. Neither the T_L internal EcoR1 fragment 7 (7.3 kbp) nor the T_R internal EcoR1 fragment 13 (5.3 kbp) are cleaved under plant induction conditions. In addition, DNA from plant induced A.

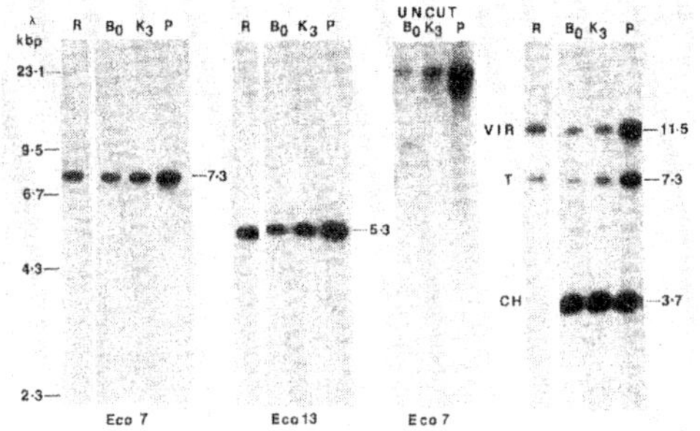

Fig.3 Control Experiments. Panel 1 shows lack of cutting of the T_L internal fragment EcoRI 7. Panel 2 shows lack of cutting of the T_R internal fragment EcoRI 13. Panel 3 shows (using EcoRI fragment 7 as a probe) lack of a detectable free T-DNA intermediate in Agrobacterium DNA not digested with a restriction endonuclease. Panel 4 shows total Agrobacterium DNA digested with EcoRI and hybridized simultaneously with probes representing the T-DNA, the vir region, and the Agrobacterium chromosome.

tumefaciens cells that has not been cleaved by a restriction endonuclease does not show the presence of free low molecular weight sequences homologous to a T_L probe (Fig 3, Panel 3). The DNA from plant-induced bacteria in Figures 2 and 3 shows an increased signal when hybridized with the various T-DNA probes. Equal amounts (0.5 μg) of DNA were loaded onto each lane for Southern blot analysis. In order to determine whether this increased signal was a result of amplification of the entire Ti-plasmid or just the T-DNA relative to the *A. tumefaciens* chromosome, three different probes representing the *A. tumefaciens* chromosome (CH), the *vir* region (VIR) and the T-DNA (T) were hybridized simultaneously to the bacterial DNA. Figure 3, Panel 4 shows that both the T-DNA and *vir* region are amplified 5-10 fold relative to the bacterial chromosome in this induction experiment. These data suggest that, under favorable induction conditions, the entire Ti-plasmid may over-replicate relative to the chromosomal DNA.

In order to determine which *vir* genes are necessary for T-DNA cleavage, various *vir* gene mutants were analyzed individually for the ability to cleave the T-DNA at border A following plant induction. Figure 4 indicates that, whereas cleavage can occur normally in the wild type strain or in strains harboring *vir* B, C, or E mutations, cleavage could not occur in strains harboring mutations in *vir* A, G, or D. Cleavage at border A can be restored to strains harboring *vir* G or D mutations (Fig 5) or a *vir* A mutation (data not shown) by complementation *in trans* by cosmids containing wild-type copies of the mutated genes. Because it has been proposed that *vir* A and G are regulatory and encode functions essential for the expression of the other *vir* genes(4), the *vir* D product(s) is most likely responsible for the T-DNA cleavage.

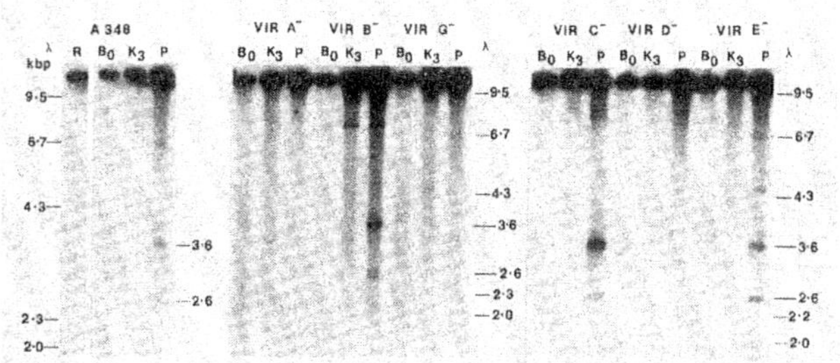

Fig.4 Cutting at Border A in various *vir* region mutants. Bacteria were grown in AB minimal medium (B_O), or transferred to $^oK_3$ medium pH5.6 (K_3) or K_3 medium plus tobacco protoplasts (P) for 24 hours. The DNA was digested with EcoRI and hybridized with HindIII fragment Y.

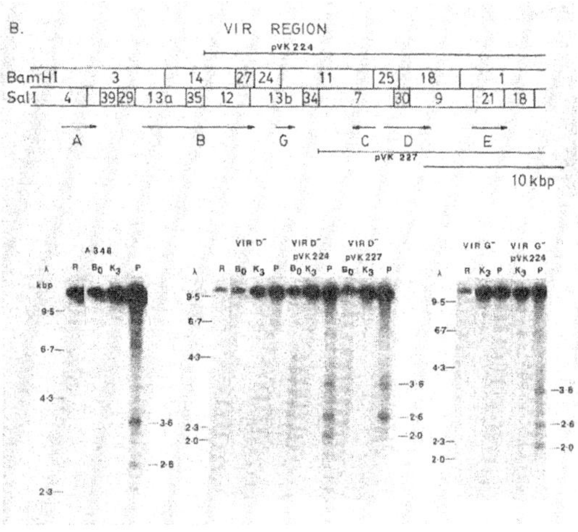

Fig.5 Restoration of cutting at border A in vir D or vir G mutants complemented in trans by the cosmids pVK224 or pVK227. Upper panel shows a restriction endonuclease map of the vir region and the respective cosmids. Lower panel shows cutting at border A in the wild-type strain A348, or in vir D or G mutants lacking or harboring the complementing cosmids.

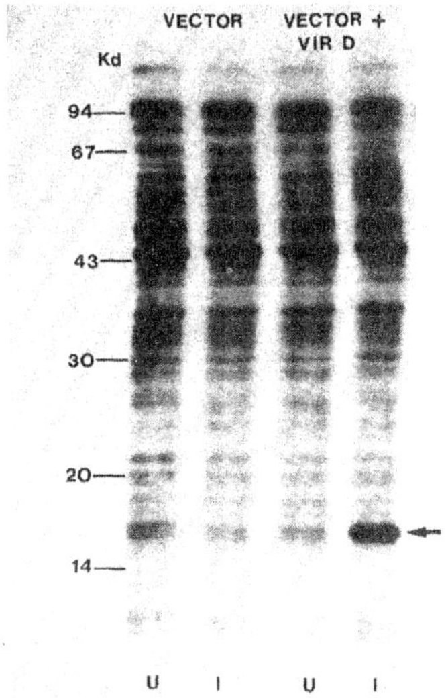

Fig.6 Expression of vir D gene products (SmaI-HpaI fragment) in E. coli. Fluorogram of SDS-polyacrylamide gel electrophoresis analysis of total proteins labelled with ^{35}S-methionine from bacterial strains containing the vector-pKK223-3 or the vector + vir D (SmaI-HpaI fragment). U-uninduced, I-induced for 2 hours with 2.5 mM IPTG. The arrow notes a new vir D protein of approximately 16.5 kd which is induced by IPTG.

As a first step in identifying and characterizing the vir D product(s), we have cloned the vir D region (a 5.6 kbp SmaI-HpaI fragment) into the E. coli expression vector pKK223-3 under TAC promoter control. After transformation into E. coli strain JM105, the bacteria containing the vector alone or the vector plus the vir D fragment were labelled for 5 min with ^{35}S-methionine either under uninduced or IPTG-induced conditions. Total bacterial proteins were extracted and subjected to electrophoresis on a 15% SDS-polyacrylamide gel (Fig 6). A 16.5 kd protein was specifically induced from the strain harboring the vir D construction. The size of this protein is consistent with the size of the first open reading frame of vir D as determined by DNA sequence analysis (J. Slightom, personal communication). We are continuing to investigate the expression and function of this and other vir D encoded proteins, both in E. coli and in A. tumefaciens.

Acknowledgements-This work was supported by a NSF Presidential Young Investigator Award to S. B. G., as well as by matching funds from Agrigenetics Research Associates.

REFERENCES

1. Wang K, Herrera-Estrella L, Van Montagu M, Zambryski P (1984) Cell 38:455-462.
2. Peralta EG, Ream LW (1985) Proc. Natl. Acad. Sci. USA 82:5112-5116.
3. Jen GC, Chilton M-D (1986) J. Bacteriol. 166:491-499.
4. Stachel SE, Nester EW (1986) EMBO J. 5:1445-1454.
5. Koukolikova-Nicola Z, Shillito RD, Hohn B, Wang K, Van Montagu M, Zambryski P (1985) Nature (London) 313:191-196.
6. Alt-Moerbe J, Rak B, Schroder J (1986) EMBO J. 5:1129-1135.

CLONING VECTORS FOR CORYNEFORM BACTERIA

Maya KOZLOWSKI, Marta SRULOVICZ and R. Wayne DAVIES

Allelix Inc., Mississauga, Ontario, L4V 1P1 Canada

Coryneform bacteria constitute a broad morphological group of Gram-positive bacteria which is widely distributed in nature. It is a taxonomically ill-defined group which also includes members of such closely related genera as : Arthrobacter, Brevibacterium, Corynebacterium and Rhodococcus.

Many species of Coryneform bacteria inhabit the soil and a number of them cause disease in plants varying from wilts to abnormal growths. Phyto-pathogenic Corynebacteria have been isolated from economically important plants such as alfalfa, bean, corn, tomato, potato, wheat and sugar cane.

The genetics of plant pathogenic Corynebacteria is largely unexplored. The development of recombinant DNA techniques would therefore provide powerful tools to study these organisms especially at the molecular level.

In this report we describe the construction of novel cloning vehicles for Coryneforms, as well as the adaptation to Coryneform bacteria of various molecular biology techniques such as : a PEG-mediated transformation, purification of plasmid molecules and a colony hybridization.

1. Identification of plasmids in Rhodococcus.

In order to establish a gene cloning system for Coryneform bacteria, it was necessary to isolate a plasmid replicon which could be useful in the development of a vector. A strain of Rhodococcus was screened for the presence of plasmid molecules. Two plasmid species (5 and 19 kb) were isolated (Photo 1). The protocol which we used to purify large amounts of plasmid DNA is outlined in Fig. 1.

1. Grow bacterial cultures in 1L of 2YT at 30° for about 48 hours.
2. Spin the cells down and resuspend in 45ml of 10mM Tris pH 8.0.
3. Spin the cells again at 5000 rpm for 15 min.
4. Remove supernatant and freeze the pellet for 1 hour at -70°C.
5. Thaw the cells at room temperature and resuspend in total volume of 50ml ST buffer [ST = 0.5M sucrose, 100 mH Tris pH 8.0]
6. Add 10 ml of lysozyme [10 mg/ml] made in ST buffer.
7. Incubate the mixture for 1 hour at 37°C.
8. To 60 ml suspension add: 12 ml 5M NaCl
3 ml 0.5 M EDTA
60 ml 2% SDS-0.78M NaCl
Leave on ice overnight.
9. Clear the cell debris and the chromosomal DNA by spinning at 5000 rpm for 1 hour.
10. Transfer the supernatant to a new tube add PEG to 10% and leave overnight at 4°C.
11. Spin down at 5000 rpm for 15 min, save the pellet.
12. Resuspend the pellet in 7.0 ml of TE buffer. Add 8.0 gms of CsCL and 1 ml of Et/Br (10 mg/ml).
13. Place the sample in an ultra-centrifuge and run overnight at 60,000 rpm, 15°C.

Photo 1

Fig.1 Method for plasmid isolation from Corynebacterium.

2. Transformation of Corynebacterium.

Plasmid molecules isolated from Rhodococcus were used to transform a plasmid-free strain of Corynebacterium. The transformation protocol used was a modification of the procedure developed by Yoshihama et al. (J. Bact. 162, 591; 1985) and is outlined in Fig. 2. The method resulted in 90% of protoplast formation, 20% protoplast regeneration, and 100-1000 transformants per µg DNA. Plasmid species used for transformation were cryptic. In order to determine that the transformation process had taken place, the total cell suspension was plated on sorbitol-media to allow the cells to form colonies. They were then pooled and the plasmid DNA purified. In this manner we were able to show that the 19 kb plasmid (pMS1) can stably replicate in Corynebacterium, while the 5 kb plasmid species is lost. In (Photo. 2) parallel the colonies containing "putative" transformants were subjected to colony hybridization using pMS1 plasmid as a radioactive probe. Fig. 3 outlines a protocol for colony hybridization.

Procedure for Transformation of *Corynebacteria* with Plasmid DNA.

Modification of the procedure of Yoshihama et al., 1985, J. Bact. 162, 591-597)

1. Inoculate 20 ml of 2YT, 2% glucose, 2% glycine with bacteria. Grow the cells with vigorous shaking for 16-18 hrs at 37°C.
2. Centrifuge the cell suspension at 7000 rpm for 10 minutes.
3. Wash the cells with 5 ml of 1M sorbitol, and spin again down the cell suspension at 7000 rpm for 10 minutes.
4. Resuspend the cells in 1 ml of sorbitol - lysozyme mixture (lysozyme 2.5 mg/ml).
5. Incubate the cells without shaking for 90 minutes at 37°C.
6. Spin the cells suspension at 7000 rpm for 10 minutes. Resuspend protoplasts in 1 ml of 0.5M sorbitol and aliguot on L-agar and sorbitol agar plates.
7. Transfer 0.3 ml of protoplasts into 10 ml test tube, add plasmid DNA. Incubate for 10 minutes at RT. Add 10µl of 10% glucose and 0.7 ml of 50% PEG - (prepared in 0.5M sorbitol).
8. Dilute immediately the mixture 2x by adding 2 ml of sorbitol medium (LB medium containing 0.5 M sorbitol, 20 mM $MgCl_2$ and 20 mM $CaCl_2$).
9. Incubate the mixture without shaking at 30°C for 2-1/2-3 hrs.
10. Transformants are plated on SB media supplemented with antibiotics and overlayed with soft agar containing antibiotics.
11. Incubate the plates at 30°C for 2-5 days.

Fig.2 Transformation procedure for Corynebacterium.

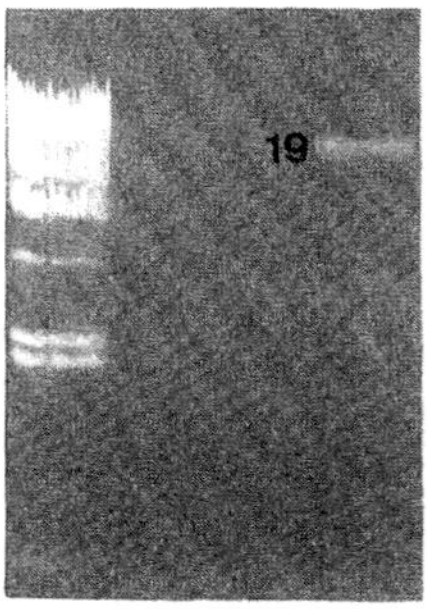

Photo 2

Colony Hybridization

(Modification of the procedure of Grunstein and Hogness, for use in *Coryneform* bacteria).

1. Store a plate containing the colonies at 4° prior to use.
2. Place a colony screen disc onto the agar plate (NEN-colony hybridization "Colony screen" filters).
3. Allow the disc to sit on agar plate to adhere for 10 min. (Mark the orientation of the disc).
4. Lift the filter and place it in a Petri-dish containing 1 ml of TE-10mg/ml lysozyme solution.
5. Soak the filter for 1 hr at 37°C.
6. Dry the filter on Whatman paper.
7. Denature the DNA in IM NaOH 1% SDS for 20 min RT.
8. Wash in 5M NaCl 1M Tris-HCl pH7.5.
9. Dry the filter at RT and use it for hybridization.

Fig.3 Colony hybridization.

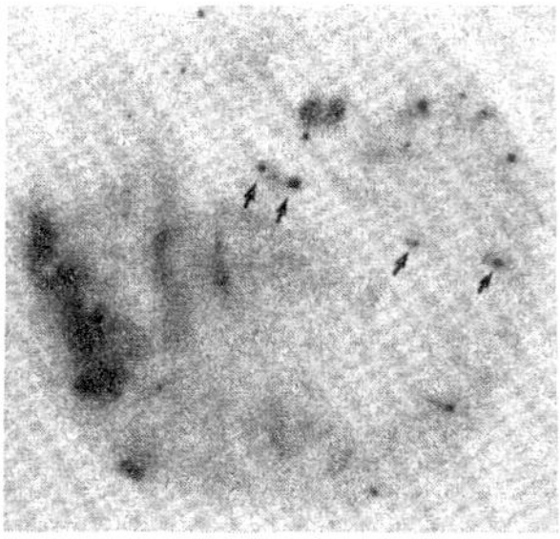

3. Restriction and functional map of pMS2.

The DNA of pMS2 was digested to completion with each of several restriction enzymes and in various pair-wise combinations of enzymes. The analysis of DNA fragments generated allowed us to construct a cleavage map for BamH1, EcoR1, Hind3, Kpn1, Sal1 and Sph1 (Fig. 4). The plasmid has two unique cleavage sites for BamH1 and EcoR1.

In an effort to reduce the size of pMS2 as well as map the location of Km^r, several deletions of the plasmid were generated. Fig.5 illustrates the extent of deletions in various derivatives of pMS2. Deletions extending from 18.2 to 20.7 pMS2 map co-ordinates resulted in the loss of the Km^r phenotype. Km^r is therefore included in a 2.5 kb DNA fragment located between 18.2 and 20.7 position on the pMS2 map.

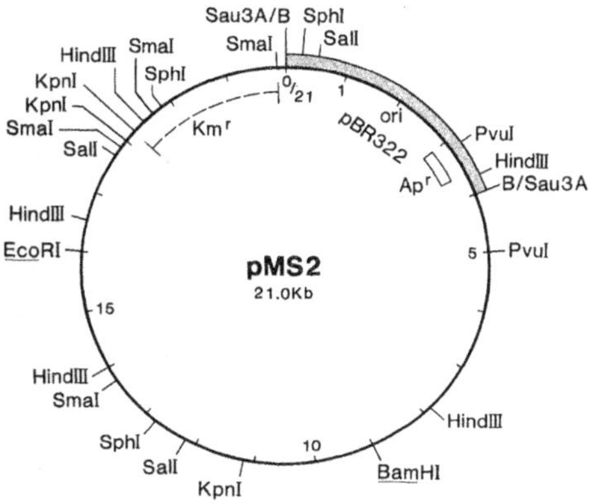

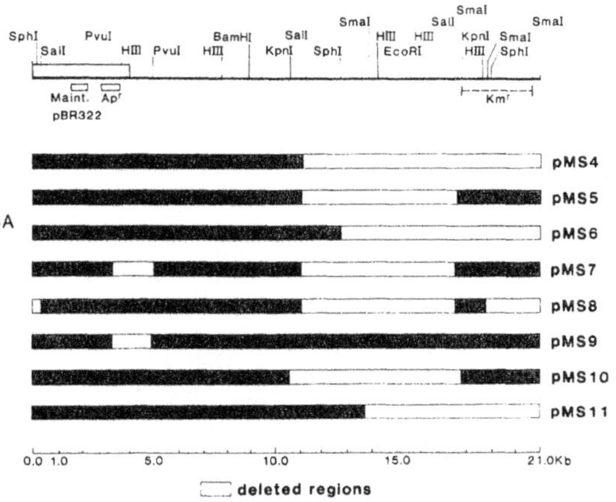

Fig.4 Restriction and functional map of pMS2.

Fig.5 Deletion derivatives of pMS2.

In an effort to extend the use of pMS2, an EcoR1 fragment derived from a B. subtilis plasmid pTV53 (youngman et al., 1985, Science 19, 285-291) was cloned into an EcoR1 site of pMS2. The DNA fragment derived from B. subtilis carries Tc^r and a transposable element Tn917. Fig. 6 illustrates the functional map of pMS21. pMS21 can function as a promoter-probe vector.

Insertion of DNA fragments derived from Corynebacterium in front of a lac Z gene in the Tn917 portion of the plasmid should result in the production of B-galactosidase whose transcription is initiated by a promoter located within a cloned DNA fragment. Experiments are in progress to test pMS21 as a promoter-probe vector and as a transposon delivery vehicle.

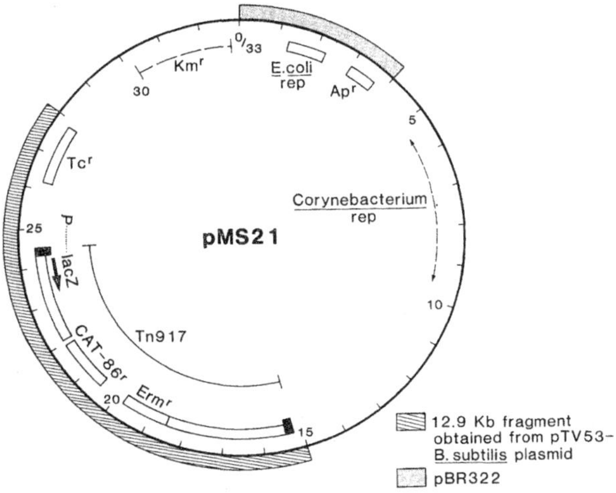

Fig.6 Functional map of pMS21.

CLONING OF SERRATIA LIQUEFACIENS CHITINASE GENE(S)

Sadhna JOSHI and Maya KOZLOWSKI
Allelix Inc., 6850 Goreway Drive, Mississauga Ontario,
L4V 1P1 Canada

The production of plant chitinase(s) is induced upon attack by several plant pathogens (i.e. fungi, nematodes, viruses, etc.). This process can take up to several days and in the meanwhile the infection can become well established. Because cell walls of many of the plant pathogens are made up of chitin but chitin is absent in plants, such enzymes could possibly be involved in plant defense mechanisms. We have cloned chitinase gene(s) from *Serratia liquefaciens* in order to provide the plant roots and/or the entire plant with the ability to affect plant pathogens *in situ*. The ability of *S. liquefaciens* chitinase(s) to inhibit fungal growth is shown in Fig. 1.

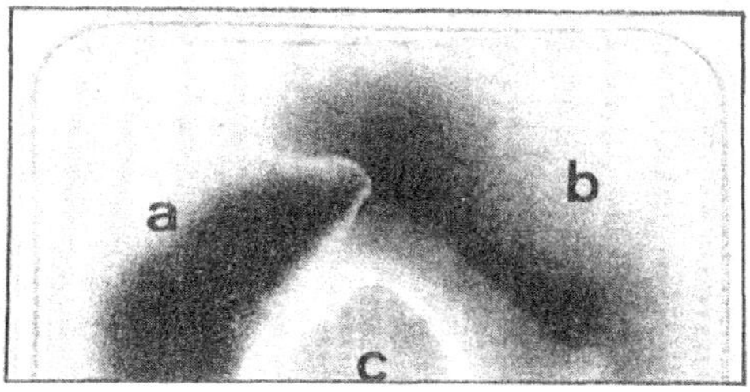

Fig. 1 Inhibition of fungal growth by *S. liquefaciens* chitinase(s); a) *Aspergillus niger*; b) *A. nidulans* and c) *S. liquefaciens*.

The overall strategy used for cloning and characterization of chitinase gene(s) is shown in Fig. 2.

Tn5 MUTAGENESIS OF S. LIQUEFACIENS GENOME
↓
SCREENING FOR MUTANTS WITH ALTERED CHITINASE ACTIVITY
↓
CHITINASE MINUS (SJ1) MUTANT
↓
CLONING OF CHITINASE SEQUENCES FLANKING THE Tn5 INSERTION IN SJ1;
SELECT FOR Km+ CLONES OF E. COLI (pSJ1)
↓
→ USE pSJ1 AS A PROBE FOR COLONY HYBRIDIZATION → PSJ12

CONSTRUCTION OF S. LIQUEFACIENS
GENOMIC DNA LIBRARY IN E. COLI

→ SCREENING FOR CHITINASE ACTIVITY → pSJ12 and pSJ21

Fig. 2 Strategy used for cloning of chitinase gene(s).

The S. liquefaciens genome was mutagenized with suicide vectors (pGS9, pMKK10, pMKK50) containing the Tn5 element. This element is known to insert randomly, thus mutagenizing the genome. It confers Km resistance which makes it easier to select for mutants containing Tn5 as well as allowing an easy cloning of the DNA sequences containing the Tn5 element. About 4000 transconjugants were screened for an altered chitinase activity : both chitinase "-" (SJ1) and chitinase overproducing (SJ101) mutants of S. liquefaciens were obtained.

In the mutant SJ1 the chitinase gene is presumably inactivated due to Tn5 insertion. From this mutant the chitinase sequences flanking the Tn5 element were cloned into E. coli. Clones were selected for Km resistance contained on the Tn5 element. The presence of the Tn5 element was further confirmed by restriction analysis. This plasmid pSJ1 contains 2 inserts of 3 and 6 kb on either side of the Tn5 element (Fig. 3a); the 6 kb fragment was used as a probe for colony hybridization.

In parallel to the above experiments, a wild type genomic DNA library of S. liquefaciens was constructed as follows : partial EcoR1 digestion of the S. liquefaciens genome, size fractionation of the different fragments on a sucrose gradient, ligation of 10-20 kb fragments into pBR329, and transformation of E. coli RR1 cells.

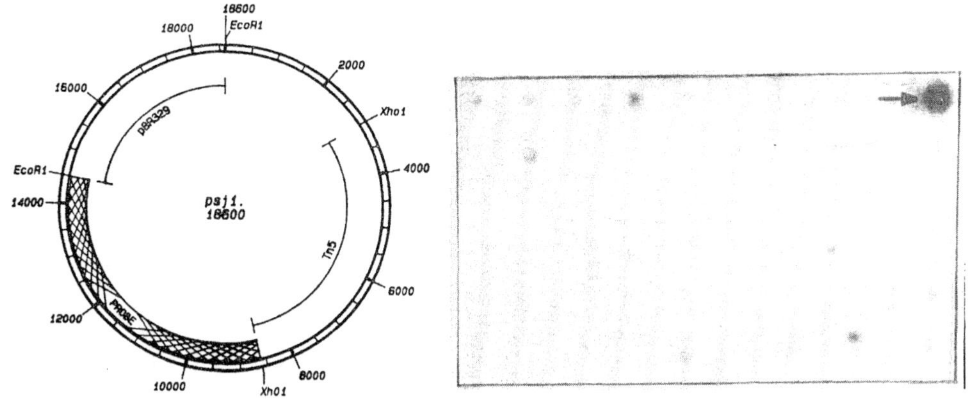

Fig. 3a. Probe used for colony hybridization (6kb Xho1-EcoR1 fragment from pSJ1). 3b. Colony hybridization to the wild type genomic DNA library of S. liquefaciens.

The E. coli clones containing the wild type copy of the chitinase gene(s) were identified by : a) colony hybridization to the 6kb probe derived from pSJ1 and, b) screening for the chitinase activity on minimal chitin-agar plates (containing leu, pro, thiamin and mannose).

Out of 300 colonies screened, 2 chitinase "+" clones were identified. One clone containing pSJ12 was selected by both colony hybridization (Fig. 3b) and by the presence of a zone of clearance on minimal chitin-agar plates (20 mm^2). Another clone containing pSJ21 was negative in colony hybridization but showed a larger zone of clearance (154 mm^2) on chitin-agar plates.

The ability of the cloned chitinase sequences to complement the Tn5 induced chitinase "-" mutation of S. liquefaciens (SJ1) was also confirmed. In this respect the cloned chitinase insert was ligated to a broad host range plasmid (pRK290) and the resulting construct pSJ13 transferred to S. liquefaciens SJ1 by triparental mating in the presence of pRK2013. The chitinase activity of SJ1 harbouring pSJ13 is shown in Fig. 4.

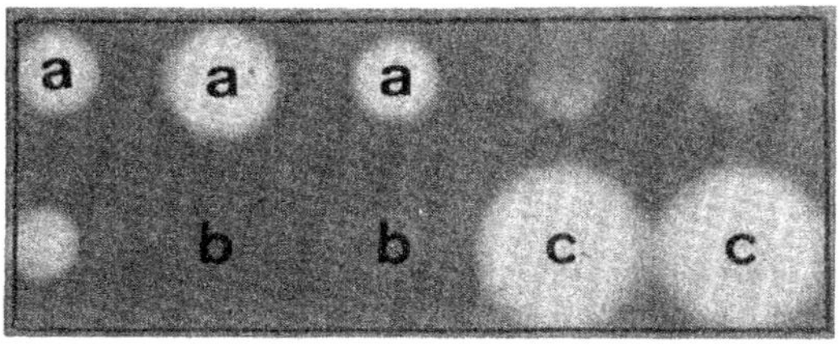

Fig. 4 Complementation of the Tn5 mutation in SJ1 by cloned chitinase sequences; a) SJ1/pSJ13; b) SJ1; c) wild type S. liquefaciens.

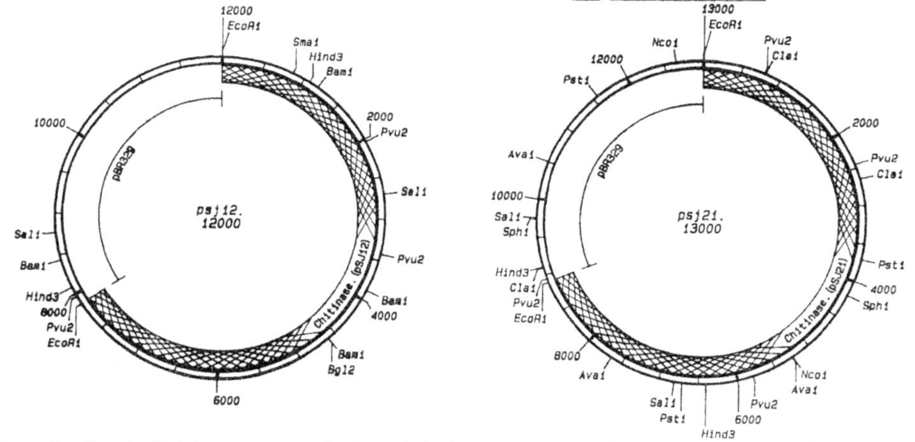

Fig. 5 Restriction maps of the chitinase clones (pSJ12 and pSJ21).

A comparison of the restriction maps of both plasmids containing cloned chitinase sequences (pSJ12 and pSJ21; see Fig. 5) reveals that we are probably dealing with 2 distinct chitinase gene(s). Tn5 mutagenesis of the plasmid pSJ12 reveals the presence of a regulatory gene whose inactivation leads to overproduction of chitinase(s).

Experiments are underway to better characterize and to sequence these gene(s).

AUTHOR INDEX

	Page
Alvarez-Morales, A.	182
Arnold, W.	243
Atherly, A.G.	202
Aubry, C.	4
Ausubel, F.M.	211
Bachem, C.	205
Bakker, P.	49
Baldwin, T.O.	270
Banfalvi, Z.	188,205,280
Barran, L.R.	148
Bassam, B.J.	179
Batley, M.	164
Batut, J.	246
Bender, G.L.	179
Bennetzen, J.L.	63
Berger, W.H.	18
Beyou, A.	4
Bhagwat, A.	113
Birkenhead, K.	283
Bisseling, T.	84,89
Boesten, B.	283
Boistard, P.	246
Borthakur, D.	160
Boulanger, F.	4
Bradley, D.J.	134
Bressan, R.A.	63
Brewin, N.J.	134
Brink, B.A.	158
Brisson, N.	138
Brom, S.	142
Bromfield, E.S.P.	148
Broughton, W.J.	220
Butcher, G.W.	134
Campos, F.	108
Cannon, F.	238
Carroll, B.J.	78,87
Cava, J.R.	158
Cevallos, M.A.	142,204a
Charles, D.J.	63
Chen, H.C.	164
Chiu, J.	63
Chua, K.Y.	226
Close, T.J.	12
Clover, R.	173
Colonna-Romano, S.	243
Condon, C.	283
Crawford, M.S.	39
Dart, P.J.	289
Davalos, A	204a
Daveran, M.L.	246
David, M.	246
Davila, G.	142
Davis, R.W.	309
Dazzo, F.B.	162
Day, D.A.	78
Debelle, F.	220
De Bruijn, F.	254
Defez, R.	243
de Lajudie, P.	123
De'Narie, J.	217
Delves, A.C.	78
Denny, T.P.	57
Deshmane, N.	188
Djordjevic, M.A.	164,217
Djordjevic, S.P.	164
Domergue, O.	246
Donald, R.G.K.	260
Donaldson, P.A.	33
Dow, J.M.	50
Dowling, D.N.	220
Downie, J.A.	213
Dreyfus, B.	199
Ebeling, S.	182
Enriquez, C.	108
Estramareix, C.	4
Ettinger, W.F.	39
Evans, I.J.	213
Feenstra, W.J.	84
Fees, H.	267
Filser, M.	243
Finan, T.M.	173
Firmin, J.L.	213
Fischer, H.-M.	182
Fleury-Guerout, A.-M.	4
Flores, M.	142
Folkerts, O.	193
Fortin, M.G.	95,116
Fox, D.	295
Franche, C.	291
Franssen, H.	89
Frenk, S.	142
Fuggi, A.	243
Galfre, G.	134
Garciarrubio, A.	142
Garnerone, A.M.	246
Gelvin, S.B.	303
Gerhold, D.	188
Ghai, J.	246
Giroux, H.	138
Gloudemans, T.	89
Goethals, K.	199
Goldthwaite, J.	105
Golinska, B.	120
Gonzalez, V.	142
Göttfert, M.	182,205
Gottlob-McHugh, S.G.	111

Page

Govers, F. 89
Gray, J.X. 164
Gresshoff, P.M. 78,87,226
Gröner, P. 264
Gubler, M. 182
Guida, M. 243
Gunning, B.E.G. 291
Gyorgypal, Z. 205
Hahn, M. 182
Handa, A.K. 63
Hanks, J.F. 105
Hellmiss, R. 18
Hennecke, H. 182
Hertig, C. 246
Heycke, N. 280
Higgisson, B. 283
Hilgert, U. 254
Hirsch, A.M. 105,170
Hollingsworth, R.I. 162
Holsters, M. 199
Honma, M.A. 211
Hontelez, J. 229
Hooykaas, P.J.J. 153
Horvath, B. 205
Hua, S.-S.T. 131
Huguet, T. 123
Iaccarino, M. 243
Infante, D. 246
Innes, R. 217
Iyer, V.N. 33
Jacobs, F. 116
Jacobsen, E. 84,229
Jansma, J.-D. 229
Jayaswal, R.K. 63,303
Ji, J.M. 18
Jochimsen, B. 126
Johansen, E. 150
John, M. 205
Johnson, D.A. 111
Johnston, A.W.B. 160,213
Joshi, S. 312
Kado, C.I. 12,53
Kahn, D. 246
Kamp, R. 120
Kane, H. 226
Kanvinde, L. 295
Kerr, A. 1
Kieber, J. 173
Kiely, B. 283
Kiss, G.B. 101
Kitts, C.L. 286
Klein, S. 170
Kneen, B.E. 72
Kolattukudy, P.E. 39

Page

Kondorosi, A. 205,280
Kondorosi, E. 205
Kozlowski, M. 309,312
Krotzky, A. 78
Kuempel, P. 217
Kulpaca, B. 158
Kuykendall, D. 276
Laetsch, W.M. 131
Laliberte, J.-F. 30
Lamb, J.W. 160,182
Lamberti, A. 243
Lankhorst, R.K. 229
Lara, M. 135
Larsen, K. 126
Larue, T.A. 72
Lee, L. 63
Lefebvre, D.D. 30
Legocki, A.B. 120
Legocki, M. 270
Legocki, R.P. 270
Leigh, J.A. 156
Lewin, A. 220
Lewis, D.M. 148
Lopez, M.F. 176
Loroch, A.I. 260
Lotz, W. 267
Ludwig, R.A. 260,286
Lugtenberg, B.J.J. 232,235
Macol, L.A. 105
Madrzak, C.J. 120
Manen, J.-F. 220
Manian, S. 283
Manian, S.S. 264
Marineau, C. 138
Maris, M. 299
Martinez, E. 142
Martinez, J 204a
Marugg, J. 47
Mathews, A. 78,87
Matthysse, A.G. 9
McLaughlin, L. 162
McLoughlin, T.J. 150
McQueen, D.A.R. 69
Merlo, A.O. 150
Michiels, K. 299
Mignotte, C. 4
Miller, K.L. 131
Moerman, M. 89
Morris, P. 162
Murphy, P.J. 280
Nap, J.-P. 89
Nayudu, M. 179
Nees, D.W. 260
Nelson, L.M. 267

Page

Nieuwkoop, A.J. 188
Noel, K.D.. 158
Noti, J.D.. 193
Nowell, L.. 295
Nuti, M.P.. 153
O'Gara, F.. 283
Okker, R.H.J. 232,235
O'Regan, M. 283
Olsson, J.E.. 78
Ophel, K. 1
Ortega, J.L.. 135
Pachori, P. 158
Padilla, J. 108
Palacios, R.. 142
Pawlowski, K. 254
Pees, E.. 232,235
Peralta, E.G. 18
Peralta, Y. 204a
Philip, S.. 162
Pinero, D.. 142
Plazinski, J. 291
Popin, T.T. 286
Postma, J.G.. 84
Price, G.D. 78,226
Priefer, U.B. 243,264
Pühler, A.. 243,264
Putnoky, P. 205
Quinto, C.. 142
Ramakrishnan, N.. 202
Ramseier, T.. 182
Ratet, P. 254
Ream, W.. 18
Redmond, J.W. 164
Reed, J.. 156
Regensburger, B.. 182
Renalier, M.H.. 246
Reyes, O. 4
Riccio, A.. 243
Richaud, F. 4
Ripke, H.M. 267
Roberts, D.P. 59
Rodriguez-Quinones, F.. 205
Rogowsky, P.. 12
Rolfe, B.G. . . . 164,179,217,291
Romero, D.. 142
Rosenberg, C. 220
Rossen, L.. 213
Saad, M.. 226
Salzwedel, J. 162
Sanchez, F. 108
Sastry, G.R.K.. 293,295
Schell, J.. 205,254,280
Schell, M.A.. 57
Schell, M.G.. 188

Page

Schild, C.. 267
Schippers, B. 47
Schmidt, J. 205
Schottel, J.L.. 69
Schuller, K.A.. 78
Scofield, G.. 50
Scott, K.F. 226,289
Sebastian, J. 39
Segovia, L. 142
Shah, K.S.. 276
Shaw, J.J.. 53
Shearman, C.A.. 213
Shen, B.-F. 252
Shen, S.C.. 252
Signer, E.R.. 170,173,176
Sikorski, M.. 120
Simpson, R.B. 25
Sinclair, M.. 179
Smith, C.A. 170
Smith, K.B. 162
So, J.-S. 188
Soliman, M.H. 293,295
Spaink, H.P.. 232,235
Spielmann, A. 25
Squartini, A. 153
Srulovicz, M. 309
Stacey, G.. 188
Stanley, J. 220
Strozycki, P. 120
Studer, D.. 182
Surin, B.P. 213
Szalay, A.A.. 193,270
Szeto, W. 238
Szybiak-Strozycka, U. 120
Tempé, J. 280
Thomas, J.. 113
Thöny, B. 182
Tichy, H.V. 267
Torok, I. 193
Tucker, W.T.. 289
Upadhyaya, N.M. 289
Valderrama, B.. 135
Vam Vikites, D.V. 72
van Brussel, A.A.N. 232
van den Bos, R.C. 229
Vandenbosch, K.A. 158
van den Eede, G.. 199
van der Hofstad, G. 49
Vanderleyden, J.. 299
van Gool, A.P.. 299
van Kammen, A.. 89,229
van Montagu, M. 199
Vanstockem, M.. 299
Vazquez, M. 108

Page

Vegh, Z.. 101
Veluthambi, K. 303
Verma, D.P.S. 95,116
Vincze, E.. 101
Vreeland, V.J.. 131
Walker, G.C.. 156
Wang, S.-P. 252
Wardhan, H. 295
Weinman, J.J. 164

Page

Weisbeek, P.. 47
Welsch, M.A.. 162
Wijffelman, C.A.. 232,235
Wittmann-Liebold, B.. 120
Wood, E.A.. 134
Yun, A.C. 193
Zaat, B.A.J.. 232
Zhang, M. 116

SUBJECT INDEX

A

Acetosyringone, 13
Aeschymonene indica, 221
Aeschymonene scabra, 270
Agrobacterium tumefaciens, 4,9, 18,25,93,143,203,293,295,300
Agrobacterium radiobacter, 293
Agrobacterium rhizogenes, 4,19
Alder, 111
Allantoic acid, 126
Allantoin, 126
Allantoinase, 126
Almonds, 1
Anabaena, 291
Arachidonic acid, 138
Arachis hypogaea, 1,221
Aspartate aminotransferase, 128
Attachment, 9,28
Auxins, 9,10,300
Avirulence, 57
Azolla, 291
Azorhizobium sesbaniae, 260,286
Azospirillum, 299

B

Bacteroids, 84,85,134,135,171,173, 224,226,243,229,248
Bioluminescence, 270
Black rot, 50
Bradyrhizobium japonicum, 79, 174,188,189,207,239,276
Brassica campestris, 30

C

Cadmium, 30
Cajanus, 221
Calopogonium caeruleum, 222
cAMP, 283
CaMV promoters, 31
Capsular polysaccharides, 113
Carboxymethyl cellulose, 59
cDNA
  metallothionein, 30
  nodulins, 89,111,116
Cellubiose, 59
Colletotrichum capsici, 41
Corynebacterium, 44
Cosmid, 58,156,202,220,228,264, 267,280,290,297
Crotalaria pumila, 142
Crown gall tumor, 9,18,25
Cuticle, 39
Cutin, 69
Cutinases, 39
Cyanobacteria, 291
Cytokinins, 300

D

Dalea leporina, 142
Daucus carota, 4,9
Desmodium intortum, 164,179
Desmodium uncinatum, 164
Direct repeats, 18

E

Eicosapentaenoic acid, 138
Elicitors, 114,138
Endoglucanase, 57
Endopolygalacturonase, 43
Enhancer, 18
Erwinia carotovoras, 63
Esterase, 69
Exopolygalacturonase, 43
Exopolysaccharide
  involvement of nodulation genes, 162
  modifications, 156
  mutations, 156
  succinylation, 157

F

Fix^-, 199,254
Flavanones, 232
Frankia, 111
Fusarium solani, 40

G

Glutamine synthetase, 109,128,135,243
Glutamate synthetase mutants, 257
Glycine max, 95,102,123,126, 131,179,182,188
Grapevine, 1

H

Hairy-roots, 4
B-Hydroxybutyrate dehydrogenase, 196
Hypersensitive response, 138,164

I

Indole-acetic acid (IAA), 9,10,300
Iron, 47

K

Kalanchoe diagremontiana, 4,35
Klebsiella pneumoniae, 182,197, 229,243,252,254,264,297

L

Lablab purpureus, 164,179,220
Late blight, 138
Leaf curling, 289
Lectins, 113
Leghemoglobin, 89,105,109,111,116, 120,123,125
Leucaena esculenta, 142
Leucaena leucocephala, 164,179, 220
Lipopolysaccharide, 10,134,158,173
Luciferase, 13
Lupinus luteus, 102,120
Luteolin, 232
Lux cassette, 13,53
*Lux*AB, 270

M

Macroptilium atropurpureum, 164,179,188,190,208,220,226
Macroptilium gibbosifolium, 142
Medicago, 207
Medicago sativa, 101,105,113, 148,156,176,211,220,252,280
Melilotus alba, 211
Metallothioneins, 30
Mimosa pudica, 222
moc genes, 280
Moghania congesta, 222
mos genes, 280
Mutants
- fix^- plants, 85
- nod_3, 84
- Nodulation-defective, 84
- Non-nodulating soybeans, 87
- nts382, 87,79
- ri-plasmid, 7
- rj_1 gene, 87
- rj_6 gene, 87
- *sym* genes, 72

N

Narigenin, 232
Nicotiana tabacum, 35
Nicotinic acid oxidation, 286
Nif and Fix genes, 182
Nif^-, 254
Nitrogen fixation, 39,72,148,199
Nitrogen fixation (nif) genes (see also *Rhizobium)*, 199
Nitrogenase, 86,174,229
Nod^-, 199
Nodule
- development, 89,158,189,212,256
- morphology, 80
- number, 73
- stem, 123,199,254

Nodulins, 84,89,95,101,105,108, 111,116,120,123
Nostoc cycas, 291
5'-nucleotides, 126

O

Opines, 280
Overdrive, 18

P

Parasponia, 179,184,222,226
Patatins, 138
Pathogenicity, 8,39,50,57,63,69
Pectate lyase, 43,63
Pectate synthesis, 132
Pectinases, 39
Peribacteroid membrane, 95,117, 134,135,171,
Phaseolus coccineus, 142
Phaseolus vulgaris, 35,102,108, 135,160
Phosphinothricin, 245
Phytoalexin, 138
Phytophthora infestans, 138
Pigeonpea, 289
Pisum sativum, 40,72,84,89,102, 229,267
Plant-inducing factors, 114,209, 212,213
Plasmids
- pGM1515, 217
- pGS9, 276
- pIJ1089, 189
- pIJ3020, 189

pRiA4, 8,19
pRK2, 28
pRL1J1, 153
pRL6J1, 264
pRme, 156
pRP2J1, 160
pRt032, 217
pSA, 9
pSym, 93,143,153,165,179,202, 208,220,229,244,246,249,280, 289
pTi, 18,22
pTiA6, 19
pTiA6NC, 19
pTiC58, 12
pTiK27, 1
pTiK309, 1
pTiK377, 1
pTiT37, 19
pVW5J1, 264
Ri, 4
Polygalacturonases, 63
Polygalacturonate lyase, 50
Potato tuber, 47,63,69,138,300
Pseudomonas fluorescens, 47
Pseudomonas putida, 44,47
Pseudomonas savastanoi, 300
Pseudomonas solanacearum, 57
Psophocarpus tetragonolobus, 221
Psychotria nairobiensis, 293
Purine nucleotidase, 126

R

Repeated sequences, 184
Rhizobium
Adenyl cyclase gene, 284
competitiveness, 150
curl⁻, 289
dct genes, 284
EPS, 156,160,164,166,170,177
Exo⁻, 106,170,171
Fix genes, 229,246,264
Fix⁻ strains, 106,174,179, 229,284,
flavonoids, 214
Hac⁻, 189
host range, 142,181,217,220,228
Hsn genes, 179,190,205,217, 220,224
Hup⁺, 267
infection thread, 106,131,134, 158,164,170,176,226
LPS, 158,164
Muc⁻, 164
Nif genes, 142,170,179,182,190, 205,224,229,239,260,264,291, 293,297
nifA, 248,264
nitrogenase, 182,295
Nod box, 188,189,205,226
nod genes, 153,165,170,179,181, 184,188,202,205,211,214,226, 235,291
Nod⁺, 170,171,267,284
Nod⁻, 170,171
nodA promoter, 153
nodD DNA sequence, 154
nodF, 215
nodI, 214
nodM, 215
nodulation, 202,218
nodule development, 158,189, 212,226
nodule invasion, 156
ntr loci, 238,245,255
Pea, 213
psi, 160
psr,160
pss,160
root-hair curling, 189,213
strain 5740, 125
strain ANU265, 165,179
strain ANU289, 226
strain B151, 34
strain B316, 34
strain IC3342, 289
strain LPR5035, 34
strain MPIK3030, 208
strain NGR234, 164,179,220
strain ORS571, 123,199,254,296
strain PD101, 34
strain pD102, 34
strain RCR3622, 34
strain TOM, 72
symbiotic genes, 285
transposon mutagenesis, 156, 160,165,179,189,202,208,224, 267,280,256,289
Rhizobium fredii, 144,189,202, 207
Rhizobium leguminosarum, 72,84, 153,162,189,203,207,213,229, 243,264,267,281
Rhizobium loti, 282
Rhizobium meliloti, 106,148, 156,170,176,189,190,200,202, 205,211,217,224,229,239,246, 264,280,283

Rhizobium phaseoli, 108,142,
158,160,203,207,
Rhizobium trifolii, 162,167,
181,203,207,215,217,226,
228,292
Rifampin, 148
RNA polymerase, 148
rolA locus, 7
rolB locus, 7
rolD locus, 8
Root-exudate, 88,79
Root-hair curling,
78,87,113,131,220
Root hairs, 162
Ros mutant, 14

S

Sesbania punctata, 221
Sesbania rostrata, 123,199,
220,254,286
Siderophores, 47
Signal peptide, 116
Soft rot disease, 63
Soybean, 188,202
Streptomyces, 69
Streptomyces coelicolor, 69
Streptomyces lividans, 69
Streptomyces scabis, 69
Stylosanthes hamata,, 222
Suberin, 69
Sucrose synthetase, 116
Symbiotic genes organization, 181
Synergistic nitrogen fixation, 287
Systemic infection, 30

T

T-DNA, 18,4
borders, 18,19,21
copy number, 25
integration, 25,27
transmission, 14,18
Tephrosia candida, 222
Ti-plasmid, 1,18,25
tms loci, 7
Transcriptional fusions, 160,181
Transformation
CaMV, 30
Ti-plasmid, 25
Transgenic plants, 25
Translational fusions, 195
Trifolium repens, 217
Trigonella foenum-graecum, 113
Tsr factor, 232

U

UDP-glucose pyrophosphorylase, 64
Ureides, 126
Uricase, 109,116,126

V

Vector
binary, 25
CaMV, 30
unitary, 25
Vibrio fischeri, 13,53,270
Vicia faba, 102,268
Vicia hirsuta, 215
Vicia sativa, 90
Vigna radiata, 113,220
Vigna unguiculata, 113,179,220,
Vir region, Ti-plasmid, 12,22,25,
291
Virulence, 9,12,18,19,57,63
Viral DNA, 30

X

Xanthine dehydrogenase, 126
Xanthomonas campestris, 50,55,
160

GPSR Compliance

The European Union's (EU) General Product Safety Regulation (GPSR) is a set of rules that requires consumer products to be safe and our obligations to ensure this.

If you have any concerns about our products, you can contact us on ProductSafety@springernature.com

In case Publisher is established outside the EU, the EU authorized representative is:

Springer Nature Customer Service Center GmbH
Europaplatz 3
69115 Heidelberg, Germany

Batch number: 10761599

Printed by Printforce, the Netherlands